Transport Phenomena

Transport Phenomena

Second Edition

**W. J. BEEK
K. M. K. MUTTZALL
J. W. VAN HEUVEN**

JOHN WILEY & SONS, LTD
Chichester • New York • Weinheim • Brisbane • Singapore • Toronto

Copyright © 1975, 1999 John Wiley & Sons Ltd,
Baffins Lane, Chichester,
West Sussex PO19 1UD, England

National 01243 779777
International (+44) 1243 779777
e-mail (for orders and customer service enquiries): cs-books@wiley.co.uk
Visit our Home Page on http://www.wiley.co.uk
or http://www.wiley.com

Reprinted November 2000

Other Wiley Editorial Offices

John Wiley & Sons, Inc., 605 Third Avenue,
New York, NY 10158-0012, USA

WILEY-VCH Verlag GmbH, Pappelallee 3,
D-69469 Weinheim, Germany

Jacaranda Wiley Ltd, 33 Park Road, Milton,
Queensland 4064, Australia

John Wiley & Sons (Asia) Pte Ltd, 2 Clementi Loop #02-01,
Jin Xing Distripark, Singapore 129809

John Wiley & Sons (Canada) Ltd, 22 Worcester Road,
Rexdale, Ontario M9W 1L1, Canada

Library of Congress Cataloging-in-Publication Data

Beek, W. J.
    Transport phenomena. — 2nd ed/W. J. Beek, K. M. K. Muttzall, J. W.
van Heuven.
        p. cm.
    Includes index.
    ISBN 0–471–99977–6 (cloth) : alk. paper). — ISBN 0–471–99990–3
(pbk. : alk. paper)
    1. Transport theory. 2. Heat — Transmission. 3. Mass transfer.
I. Muttzall, K. M. K. (Klaus Max Karl) II. Heuven, J. W. van (Jan
Willem) III. Title.
QC175.2.B4 1999
530.13′8 — dc21   99–13161
                                                                        CIP

British Library Cataloguing in Publication Data

A catalogue record for this book is available from the British Library

ISBN 0 471 99977 6 (cloth)
ISBN 0 471 99990 3 (paper)

Contents

Preface to the First Edition

Momentum, heat and mass transport phenomena can be found everywhere in nature. Even early morning activities such as boiling an egg[†] or making tea are governed by laws which will be treated here. A solid understanding of the principles of these transport processes is essential for those who apply this science in practice, e.g. chemical and process engineers.

The history of teaching transport phenomena went from a practical but less fundamental approach via a short period of a practical and academic approach, to the present sophisticated approach. Our experience in education and in industry is that today's abstraction does not appeal to all students and engineers, who may feel themselves easily lost in vast literature and difficult mathematics. Hence, our objective in writing this book was to digest the enormous amount of new knowledge and present it in a form useful for those who work as professional engineers or who study engineering.

The present book incorporates much fundamental knowledge, but we have also always tried to illustrate the practical application of the theory. On the other hand, we have included practical information and have not shied away from giving one or two useful empirical correlations, where theory would have been too difficult. The book is based on the text for a course in transport phenomena given by W. J. Beek at Delft University from 1962 to 1968. Parts of the last draft have been used, together with most problems encountered in three postgraduate courses.

Each chapter ends with a number of problems which form an integral part of the book. We would like to ask the student to try to solve as many of these problems as possible — this is the best way to absorb and digest the theory. We have given the answers to all of the problems so that the reader can check his results. Where we expected difficulties to arise, we have explained some problems in greater detail. Furthermore, we have invented 'John', our scientific sleuth, and we hope that the reader likes his way of solving problems. The reason for his ability to do so is evident: he has read the present book and has worked through the problems!

[†] For identical temperature equalization in the egg, the Fourier number must be $F_0 = at/d^2 = $ constant, i.e. for all eggs:

$$\text{boiling time} \times (\text{mass})^{-2/3} = \text{constant}$$

is valid. Knowing the optimal boiling time for one species of egg, we can thus predict the boiling times for other eggs, even ostrich eggs!

Preface to the Second Edition

When the authors initially discussed the idea of a second edition of this book with the editor, they were reluctant at first to consider the effort that would be involved in such a project. In most universities, the text *Transport Phenomena* by R. B. Bird, W. E. Stewart and E. N. Lightfoot (John Wiley & Sons, 1960), had become a standard text, and rightly so. This particular book, as well as the present text, found its roots in the lecture notes of H. Kramers, edited in Delft between 1956 and 1961. However, we have received a number of requests for a new edition from many teachers all over the world, especially from those motivated by a vocational curriculum, because they felt that their students, particularly those not opting for an academic career, needed such a text as a first introduction to this subject.

Thus resulted this new and updated version which differs from the first edition in the following aspects:

(a) The introduction of the law of energy conservation for flow systems has been elaborated upon by linking it with the first law of thermodynamics for closed systems (Section I.1);
(b) Sections I.4 and I.5, concerning units, dimensions and dimensional analysis, have both been completely revised in order to bring them into line with current practice;
(c) Section II.4 on traditional flow meters and Section II.8 on residence time distribution have both been shortened, but without loosing their essential features;
(d) Section III.9, describing heat transfer by radiation, has been revised in an attempt to better elucidate Poljak's scheme for the 'bookkeeping' of radiation fluxes;
(e) Chapter IV on mass transport, which originally stressed its analogy with heat transport, has been significantly expanded with the aim of clarifying the fundamental differences between the two phenomena;
(f) Section IV.5, concerning mass transfer with chemical reaction, has been slightly extended by incorporating biochemical reactions which obey Michaelis–Menten kinatics

In addition, the text has been improved and a number of errors have been corrected as a result of the contributions from various users of the 1st Edition, with special thanks being due here to K. Ch. A. M. Luyben, R. G. J. M. van der Lans, J. J. M. Potters and W. F. J. M. Engelhard. Finally, some new problem exercises have been added, and a comprehensive list of the symbols used, plus their appropriate units, has been provided.

Delft, January 1999
W. J. B
K. M. K. M
J. W. V. H

CHAPTER I

Introduction to Physical Transport Phenomena

During the designing of industrial process plant, both qualitative and quantitative considerations play an important role.

On the basis of qualitative (sometimes semi-quantitative) considerations, a preselection of feasible concepts of processes suitable for carrying out the desired production in an economical way is made. The type of operation, e.g. distillation against extraction or the choice of a solvent, will also be fixed by this type of reasoning, in which experience and a sound economic feeling play an important role.

As soon as one or two rough concepts of a production unit are selected, the different process steps will be analyzed in more detail. This asks for a quantitative approach with the aid of a mathematical model of the unit operation. The experience that mass, energy and momentum cannot be lost, provides the three conservation laws, on which the quantitative analysis of physical and chemical processes wholly relies and on which the process design of a plant is based. This kind of design, which aims at fixing the main dimensions of a reactor or an apparatus for the exchange of mass, momentum and energy or heat, is the purpose of the disciplines known as 'chemical engineering' and 'chemical reaction engineering'. The basic ideas behind these disciplines are found under the headings 'transport phenomena' and 'chemical (reaction) engineering science', which rely on deductive science and, hence, have the advantage of analytical thought but which, because of this, lack the benefit of induction based on experience when aiming at a synthesis.

Qualitative and quantitative reasoning cannot be separated when setting up a plant, or to put it in another way:

no apparatus, however good its process design might be, can compete with a well-designed apparatus of a better conception,

which can be the device for a process designer, or:

no research, however brilliant in conception it might be, can result in a competitive production plant without having a quantitative basis,

which could well be a suitable motto for a research fellow.

Examples of questions, in which feeling and reasoning have to match well before science is used to some profit, are to be found in the following areas: the potential possibilities of raw materials, intermediate and end products, the choice of materials and especially materials of construction, the influence of side reactions on the performance of subsequent process steps and the considerations on quality and end-use properties of a product. This type of question, although of importance for the integral approach of a design engineer, will not be dealt with in this present book, which finds its limitations at this point. This book treats the practical consequences of the conservation laws for the chemical engineer in an analytical way, trying not to exaggerate scientific 'nicety' where so many other important questions have to be raised and answered, but by also pretending that a solid understanding of the heart of the matter at least solves a part of all questions satisfactorily.

The laws of conservation of mass, energy and momentum are introduced in Section I.1. They are extended to phenomena on a molecular scale in Sections I.2 and I.3. Section I.4 is concerned with the dimensions of physical quantities, especially SI units, whereas Section I.5 discusses the technique of dimensional analysis. We will end this chapter (and most paragraphs of the following chapters) with some proposals for exercising and with comments on the solution of some of the problems given.

After this, three main chapters follow, each of which concentrates on one of the conserved physical quantities: hydrodynamics (mainly momentum transfer) in Chapter II, energy transfer (mainly heat transfer) in Chapter III and mass transfer in Chapter IV. These chapters elaborate the ideas and concepts which are the subject of the following introduction.

I.1 Conservation laws

John looked at the still smoking ashes of what once had been the glue and gelatine factory. The fire had started with an explosion in the building where bones were defatted by extraction with hexane. John remembered that the extraction building had a volume of 6000 m³ and that the temperature in the building was always 30°C higher than outside. He knew that per 24 h, 70 ton steam were lost as well as 9 ton hexane. He made a quick calculation and concluded that the steady-state hexane vapour concentration in the plant was well below the explosion limit of 1.2 vol% and that some accident must have happened which subsequently led to the explosion.

Physical technology is based on three empirical laws, namely matter, energy and momentum cannot be lost.

The law of conservation of matter is based, among other things, on the work of Lavoisier, who proved that during chemical reactions no matter, i.e. no mass

(mass being the most important property of matter), is lost. The law of conservation of matter is not always valid: in nuclear technology, matter is transformed into energy but for chemical or physical technology this exception is of no importance. It is, of course, possible that matter is transferred from a desired form into an undesired one (e.g. the degradation of a polymer, which finally leads to only CO_2, H_2O, etc.).

The law of conservation of energy is based, among other things, on the work of Joule, who proved that mechanical energy and heat energy are equivalent. His work finally led to the first law of thermodynamics, which, when formulated for a flowing system, is the law of energy conservation that we are looking for. It is historically remarkable that it took more than two centuries before this law, formulated initially for a closed system, was translated into a form in which it could be applied to flowing systems.

The law of conservation of momentum was finally formulated in its simplest form for a solid body by Newton: if the sum of the forces acting on a body is different from zero, this difference is (in size and direction) equal to the acceleration of that body. Together with his second law, i.e. action equals reaction, this formed the basis for dynamics and hydrodynamics. This time it did not take much more than one century to transpose the concept, originally formulated for a rigid body, to the more general case of flow in fluids.

These conservation laws play the same role in daily life as the experience that a pound cannot be spent twice and that a difference exists between the pound you owe somebody and the one somebody owes you. The economic rules and the conservation rules of our study are used in the same manner: balance sheets are set up which account for inflow and outflow and for the accumulation of the quantity under consideration.

Let us denote by X a certain amount of money, mass, energy or momentum. Then the general law of conservation, on which all phenomenological descriptions of change in the physical world are based, reads as follows:

$$\frac{\text{accumulation of } X \text{ in system}}{\text{unit time}}$$

$$= \frac{\text{flow of } X \text{ into system}}{\text{unit time}} - \frac{\text{flow of } X \text{ out of system}}{\text{unit time}}$$

$$+ \frac{\text{production of } X \text{ in system}}{\text{unit time}} \tag{I.1}$$

The system may be a country, a concern, a factory, an apparatus, a part of an apparatus (e.g. a tray), a pipe, or an infinitely small element of volume, etc. This sounds very general and easy, but daily practice proves that we have to develop the qualitative judgement for defining the system such that the analysis stays as easy as possible. To this end, in order to develop a feeling for the qualitative aspects of analytical science, we have to go through many a quantitative exercise. Introductions into a discipline, such as this one, may easily confuse the reader if these points are not made clear in the beginning; the subject of our study is the three laws of conservation, which will appear in many forms because we will study many different systems, and not because the basic rules are many.

If we are only interested in macroscopic properties, such as mean concentration in the chosen control volume or the rate of change of the mean temperature in this volume, we can choose a macroscopic control volume, e.g. a complete reactor, a complete catalyst particle, etc., for setting up a balance of the desired physical property. If, on the other hand, we are interested in temperature or concentration distributions, we then have to start by setting up a microbalance, i.e. a balance over an infinitesimally small volume element, and to integrate the differential equations obtained over the total (macroscopic) volume. Both balances will be treated in detail in this and in the following section.

Let us now try to formulate the law of conservation which we have just defined in words (equation (I.1)) in a more precise way. In order to do this, we need some symbols:

V for the volume of the system (space volume, number of inhabitants, etc.), $\phi_{v,\text{in}}$ and $\phi_{v,\text{out}}$ for the ingoing and outgoing volumetric flow rates (see Figure I.1),

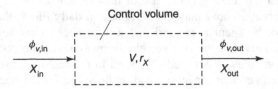

Control volume

Figure I.1 Macrobalance

r_X the volumetric production of X per unit of time and X for the volumetric concentration of the physical quantity under study.

The accumulation per units of volume and time can now be denoted by dX/dt; hence, the accumulation in the system per unit of time is given by $V\,dX/dt$. A flow at a rate ϕ_v, containing a volumetric concentration X of the considered quantity, represents a flow rate $\phi_v X$ of this quantity. Hence, our formulation of the conservation law reads as follows:

$$V\frac{dX}{dt} = \phi_{v,\text{in}}X_{\text{in}} - \phi_{v,\text{out}}X_{\text{out}} + Vr_X^{\dagger} \tag{I.2}$$

[†] In order to be completely exact we have to write this equation as:

$$V\frac{dX^v}{dt} = \phi_{v,\text{in}}X_{\text{in}}^f - \phi_{v,\text{out}}X_{\text{out}}^f + Vr_X$$

where v indicates a volume average and f a flow average.

We will now apply this law to the following quantities: money, mass, energy and momentum.

The money balance

Let us start with the most common daily practice, the conservation law for pocket money (the pocket being defined as the system, although some operate it very unsystematically). The number of pence in the system is given by $X(V = 1)$, and equation (I.2) can be read as follows: the accumulation of pence in my pocket per week (which may prove to be negative!) equals the difference between the number of pence I have taken in during this week and the number I have spent in this time, increased by the number of pence I have produced in the meantime. The last contribution sounds somewhat cryptic or even illegal, but an honest production of pence would be to change other coins into pence. Similar laws can be expressed for the other coins, as well as for the overall contents of the pocket (moneywise). From all these statements it follows that the sum of all the production terms r must equal zero (expressed as an intrinsic value and not as numbers) or, to put it in another way, changing in its own currency never results in a positive gain. To find out how simple these statements might be, try and see what happens to your thinking when X no longer stands for money, but for mass, energy or momentum!

The mass balance

The mass balance still looks familiar and comes close to the money balance. Here $X = \rho_A$, the volumetric concentration of component A in a mixture, with the units in which ρ_A is measured being kg/m^3.[†]

Hence, in a rayon factory, for instance, where NaOH (in the form of viscose) and HCl (in the form of the spinning bath) are used, the conservation law for NaOH (A) reads as follows: the accumulation of NaOH on the site in a month (or any other chosen time unit) is equal to the delivery of NaOH to the factory in that time, subtracting the amount of NaOH distributed from the site in that unit of time as NaOH and adding the amount of NaOH produced in the factory during that time (a production which is negative where NaOH is used as a reactant).

A similar mass balance can be set up for the other reactant, HCl (B), as well as for the products NaCl (C) and H_2O (D). Again, the sum of all individual production rates must be zero: $r_A + r_B + r_C + r_D = 0$ (Lavoisier). More often than not, the accountants of a factory are more aware of the implications of the conservation laws and the dynamic consequences of varying inflows and outflows of a factory than the engineers in charge of production.

[†] From the onset, we will accept kg, m, s, °C or K, and mol as the basic units for comparing physical phenomena (Section I.4). The symbol c will be used for concentration in (k)mol/m^3.

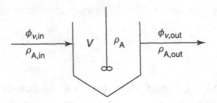

Figure I.2   Mass balance of a stirred vessel

The well-stirred continuous flow tank reactor of Figure I.2 gives an example of the applications of the mass balance. By 'well-stirred' we mean that the concentration ρ_A of a certain compound (e.g. salt) is the same at all places in the vessel so that the salt concentration in the effluent stream equals the concentration in the vessel. Let us assume that the salt concentration in the vessel at time $t = 0$ is ρ_{A_0} and that, from $t = 0$ onwards, a continuous stream of pure water (salt concentration $\rho_A = 0$) is passed through the vessel. The question is then how does the salt concentration in the vessel change with time?

We can now set up two mass balances, one for the water and one for the salt. The first balance says that, if the liquid volume in the reactor is constant, then the mass flow rate of liquid out of the reactor must equal the mass flow rate into the reactor, $\phi_{v,\text{out}} = \phi_{v,\text{in}}$. The second balance says that the decrease in the amount of salt in the vessel ($V\rho_A$) per unit of time must equal the mass flow rate of salt from the vessel ($\phi_v\rho_A$). Thus:

$$\frac{\mathrm{d}(V\rho_A)}{\mathrm{d}t} = 0 - \phi_v\rho_A$$

or:

$$\frac{\mathrm{d}\rho_A}{\mathrm{d}t} = -\frac{\phi_v\rho_A}{V} = -\frac{\rho_A}{\tau}$$

where $\tau(= V/\phi_v)$ is the mean residence time of the fluid in the vessel. Integration between $t = 0$, c_0 and t, c yields:

$$\frac{\rho_A}{\rho_{A_0}} = \exp\left(-\frac{t}{\tau}\right)$$

This function is shown in Figure I.3, which shows that at the time $t = \tau$ 37% of the initial salt is still present in the reactor. For $t = \frac{1}{2}\tau$ and $\frac{3}{2}\tau$, these values are 65 and 22%, respectively. Apparently, the liquid in the reactor has a large distribution of residence time (see also Section II.6.3).

The energy balance

Before we can attempt to interpret equation (I.2) in terms of an energy balance, we must first define X as a concentration of energy. This quantity will comprise

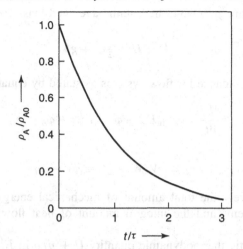

Figure I.3 Change of salt concentration in stirred vessel

the energy of state available for work, the kinetic energy and the potential energy of a unit of mass. In order to do this the familiar principles of thermodynamics have to be extended from a closed system (a non-flowing mass losing available energy by doing work, W, on and exchanging heat, Q, with its environment) to an open (flow) system, the exhaust from which still carries energy available for work.

The thermodynamics for a closed system starts with the definition of internal energy per unit mass, U:

$$dU = d(Q - W) \tag{I.3a}$$

defined for an equilibrium (point) change. If the (constant) mass is accellerated and moving in a potential field the total energy is expressed by:

$$dU + \tfrac{1}{2}d(v^2) + g\,dz = d(Q - W) \tag{I.3b}$$

for cases in which the velocity of the mass (v) changes as well as its position (z) in the gravitational or an other potential field (g). This expression can be integrated as long as the pathway of the mass considered is defined and reversibility is therefore assumed.

In flow systems this is not the case. However, if we accept that the only difference between closed and open systems is the work added to a unit of mass when introducing it at the systems inlet and the work extracted when removing it at the outlet, given by $[(p/\rho)_{in} - (p/\rho)_{out}]$, this infers that the flow-averaged mass concentration of energy E_t^f/ρ:

$$\frac{E_t^f}{\rho} = U + \frac{p}{\rho} + \tfrac{1}{2}v^2 + gh \tag{I.4a}$$

differs by an amount p/ρ from the volume-averaged mass concentration E_t^v/ρ:

$$\frac{E_t^v}{\rho} = U + \tfrac{1}{2}v^2 + gh \tag{I.4b}$$

Hence, the energy balance for flow systems as stated by equation (I.2) has to be read as:

$$V\frac{dE_t^v}{dt} = \phi_{v,in}E_{in}^f - \phi_{v,out}E_{out}^f + \phi_A - \phi_H \tag{I.5a}$$

with

$$Vr_X = \rho V(W - Q) \equiv \phi_A - \phi_H \tag{I.5b}$$

denoting respectively the total amount of mechanical energy added (e.g. by a pump in the system) and the integral amount of heat flow from the systems control volume.

The characteristic thermodynamic quantity $U + p/\rho$ in E_t^f for flow systems, as compared to U in E_t^v for closed systems, is termed the enthalpy H:

$$H \equiv U + \frac{p}{\rho} \tag{I.6}$$

In the foregoing derivation we have implicitly assumed that the mass passing the system does not change its entropy. Isentropic flow is a fair approximation in many cases (see Section II.3). We refrain from elaborating on non-isentropic cases, because Bernoulli has shown how in general the amount of mechanical energy converted into heat by friction may be accounted for (see Section II.2.3).

If heat is transferred between the control volume and its surroundings it is accepted practice to put the heat transfer rate per unit of boundary surface area proportional to the temperature difference between the surroundings and the system. The proportionality factor thus defined has been named the heat transfer coefficient and is given the symbol h; the practical significance of this coefficient will be treated in more detail in Chapter III. In this present chapter we shall indicate how the heat transfer coefficient can be calculated from first principles in a number of heat transfer situations. In complicated cases, however, this coefficient is nothing more than a factor which has to be obtained from formerly reported, consistent experience. It is one of the goals of the discipline to keep scientifically well-organized records of this experience.

The momentum balance

The amount of momentum per unit mass is $mv/m = v$, whereas the amount of momentum per unit of volume is $mv/V = \rho v$. Thus the momentum flux per unit volume is ρvv. Because momentum has direction as well as size, we must use the three components of momentum in the x-, y- and z-directions, namely ρv_x, ρv_y and ρv_z. For each direction, the law of conservation of momentum is valid and thus we obtain three separate equations. The momentum-producing terms must be interpreted as forces (Newton). The forces which can occur in a flow system are pressure forces, friction forces (caused by shear stresses) and potential forces (weight forces due to gravitational acceleration). Thus the momentum balance in the x-direction is found by replacing $X = \rho v$ in equation (I.2):

$$V\frac{d\rho v_x}{dt} = \phi_{v,in}\rho v_{x,in} - \phi_{v,out}\rho v_{x,out} + \sum F_x \qquad (I.7)$$

where $\sum F_x$ indicates the sum of all forces acting in the x-direction. The balances for the y- and z-directions can be written analogously. The dimension of all terms in the above balance is the Newton (N).

Application of balances to a pipe corner

We will now, at the end of this section, illustrate the application of the conservation laws with a practical example. Let us consider a horizontal pipe corner of cross-section A which is well insulated against heat loss. The mass balance then reads for an incompressible liquid flowing through the pipe (equation I.2 with $X = \rho$, the total concentration, and $r_X = 0$) under steady-state conditions:

$$V\frac{d\rho}{dt} = 0 = \phi_{v,in}\rho_{in} - \phi_{v,out}\rho_{out}$$

and, with $\phi_{v,1} = v_1 A_1$ and $\phi_{v,2} = v_2 A_2$ and constant density:

$$v_1 A_1 = v_2 A_2$$

indicating that at a constant cross-section of the pipe the entrance and exit velocities of the liquid must be equal.

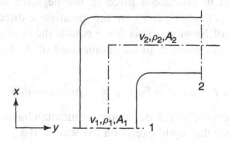

Figure I.4 Pipe corner

The energy balance reads in the stationary state:

$$V\frac{dE_t}{dt} = 0 = \phi_{v,in}E_{t,in} - \phi_{v,out}E_{t,out} + \phi_A - \phi_H \qquad (I.8)$$

and since $\phi_A = \phi_H = 0$ and $\phi_{v,\text{in}} = \phi_{v,\text{out}}$:

$$E_{t,\text{in}} = E_{t,\text{out}}$$

or, with the definition of E_t, if $A_1 = A_2$ and thus $v_1 = v_2$ and if there is no phase transition ($h_1 = h_2$, because the pipe is horizontal):

$$p_1 - p_2 = U_2 - U_1 = \rho c_p (T_2 - T_1)^\ddagger \tag{I.9}$$

Equation I.9 indicates that the pressure drop can only be predicted if the amount of frictional heat liberated per unit of volume is known. This is a principal problem of practical hydrodynamics that can only be solved theoretically in a few cases. In all other cases we have to rely on empirical correlations for estimating the pressure drop in flow systems (see Sections II.2.2 and II.2.3).

The momentum balance over the pipe corner in the x-direction yields for the stationary state:

$$V \frac{d\rho v_x}{dt} = 0 = \rho v_1 A_1 v_1 - \rho_0 A_2 v_2 + \sum F_x \tag{I.10}$$

The second term on the right-hand side of this equation expresses the fact that no momentum in the x-direction is taken away with the fluid. The forces acting in the x-direction on the liquid are the pressure force $p_1 A_1$ and the reaction force that the wall exerts on the liquid, $F_{x,w-f}$, thus:

$$\sum F_x = p_1 A_1 + F_{x,w-f} \tag{I.11}$$

and combining equations (I.10) and (I.11) we find for the momentum balance in the x-direction:

$$V \frac{d\rho v_x}{dt} = 0 = \rho v_1^2 A_1 + p_1 A_1 + F_{x,w-f}$$

Thus the wall has to produce a force in the negative x-direction and withstand a force $F_{x,f-w}(=-F_{x,w-f})$ in the positive x-direction (action equals reaction — first law of Newton). This force equals the sum of the pressure force and the force which is caused if all the momentum of the flowing liquid is taken up by the wall[†]:

$$F_{x,w-f} = -F_{x,f-w} = -\rho v_1^2 A_1 - p_1 A_1 \tag{I.12a}$$

Analogously to the above, we can develop a momentum balance for the y-direction and find that the force the liquid exerts on the wall in the y-direction is given by:

$$-F_{y,f-w} = F_{y,w-f} = +\rho v_2^2 A_2 + p_2 A_2 \tag{I.12b}$$

[†] Practical technical flow velocities in pipes are 1 m/s for liquids and 15 m/s for gases. The reader may check that the momentum force on a pipe corner is mostly negligible when compared to pressure forces.

[‡] See note on page 149

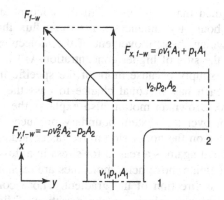

Figure I.5 Forces acting on a pipe corner

With this information, the resulting force of the fluid on the wall can be constructed as shown in Figure I.5. The pipe corner must be fixed in such a way that these forces can be taken up by the holding structure.

In the above example we have assumed that the velocity distributions of the liquid entering and leaving the pipe are uniform and that there are no velocity differences over the cross-section. In this case, the momentum of the inflowing liquid is indeed $\rho v_1^2 A_1$. If the above assumption of uniform flow velocities is not allowed, we have to write $\rho \langle v_1^2 \rangle A_1$, indicating that v_1^2 must be averaged over the cross-section. Since $\rho A_1 \langle v_1^2 \rangle = \phi_m \langle v_1^2 \rangle / \langle v_1 \rangle$ we say that the momentum per unit mass is given by $\langle v_1^2 \rangle / \langle v_1 \rangle$. For the same reason we write $\langle v^3 \rangle / 2 \langle v \rangle$ for the kinetic energy per unit mass. Such expressions for physical quantities averaged over the flow will be found regularly in our discipline.

I.2 Rate of molecular transport processes

> *In the previous section we have defined the various terms in the balances based on the three laws of conservation. In so doing we have concentrated on transport with the flowing fluid, the so-called convective transport caused by movement of the molecules, and have neglected the statistical transport. We have mentioned briefly the conduction of heat through the boundaries of a system as a means of heat transport, viscous friction as a means of momentum transport and viscous dissipation as a possible contribution to heat transport.*

The opportunity of introducing diffusion of matter through the system boundaries as a means of transporting a component has been missed until now. Together with heat transport by conduction and momentum transport due to viscosity, diffusion of matter is the third molecular process which is of interest to us.

It might be imagined that a difference in the concentration of a component on both sides of a boundary enhances a net mass flux through that boundary because of the Brownian (heat) movement of the molecules of this component: the bruto flux from the side of higher concentration will override the bruto flux from the other side, simply because more of the specific molecules are present on that side while each has an equal chance to pass the boundary at a given uniform temperature. Brownian motion also explains the conduction of heat: a temperature gradient over the system boundary produces a net flux of energy, because the molecules on the hotter side move faster and carry a higher energy.

Molecular movement again is a reason for stress in a flow field with a velocity gradient. Due to Brownian movement, molecules are exchanged in all directions and hence also in the direction of the gradient. Those coming from the faster moving side have a higher velocity in the direction of flow, and thus carry a bigger momentum than those coming from the slower moving side. This results in a net flux of momentum from the faster flowing area to the slower moving parts or, in other words, slow neighbours try to decelerate fast neighbours and vice versa. So, a net transversal transport of momentum-in-the-direction-of-flow can be interpreted as stress. It is good to notice again the vectorial character of momentum flow: both the direction of transport (transversal) and the direction of the quantity transferred (longitudinal) are important. This means that in an arbitrary plane, stresses in all three directions can be found when velocity gradients in all these directions are present.

Transport of heat, mass or momentum by molecular processes is sometimes called statistical transport because it is caused by the random Brownian movement of the molecules. [†] Statistical transport is usually much smaller than convective transport, unless very low flow velocities occur. For this reason statistical transport in the direction of flow can mostly be neglected.

Our elaboration of the conservation concept has to account for mass, energy and momentum fluxes due to processes on a molecular scale. Generally, these fluxes are found to be proportional to the gradient dX/dn, where n denotes the coordinate in the direction perpendicular to the considered plane. The proportionality constant between the flux and the driving force dX/dn is known as:

the mass diffusivity, $\mathbb{D}$, in the case of mass transfer (Fick's law), the thermal diffusivity, $a = \lambda/\rho c_p$, with λ being the heat conductivity, in the case of heat transfer (Fourier's law)[‡] and the kinematic viscosity, $\nu = \eta/\rho$, with η being the dynamic viscosity (Newton's law).

The units for each of these constants are m^2/s.

[†] Besides statistical molecular transport there can also occur statistical transport by eddies (Sections II.2, III.3.2 and IV.3.3). We speak of laminar flow if there is only statistical molecular transport.

[‡] Here, heat transfer has been used deliberately as a description which is not completely similar to energy transfer. Heat transfer accounts only for the transport of perceptible and latent heat, and not for the transport of other forms of energy.

When ϕ'' denotes the flux (the flow rate per m^2) of the quantity X due to molecular movement in a gradient of X, we thus write:[†]

$$\phi'' = -(\mathbb{D} \text{ or } a \text{ or } \nu)\frac{dX}{dn} \qquad (\text{I.13})$$

The minus sign denotes that the net flux is positive in the direction of diminishing concentrations of X. Thus the mass, heat and momentum fluxes in the positive direction of n are given by:

mass flux: $\qquad \phi''_m = -\mathbb{D}\dfrac{d\rho_A{}^{\ddagger}}{dn}$ (kg/m^{2}s) (Fick)

heat flux: $\qquad \phi''_H = -a\dfrac{d(\rho c_p T)}{dn}$ (W/m^2) (Fourier)

momentum flux: $\quad F'' = \tau = -\nu\dfrac{d(\rho v)}{dn}$ (N/m^2) (Newton)

Figure I.6 illustrates the above relationships for mass and heat transport caused by a concentration or temperature gradient.

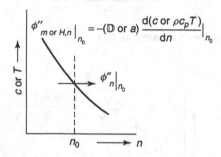

Figure I.6 Diffusion and heat conduction

In the case of momentum transport there is the extra complication that we have not only to specify the plane through which the transport occurs (e.g. $x = x_0$) but also the direction of the momentum considered (e.g. y). The corresponding momentum flux is then written as F''_{xy}, indicating that there is transport of y-momentum in the x-direction at the plane $x = x_0$ caused by a velocity gradient

---

[†] The relationship between momentum flux and velocity gradient discussed here is restricted to Newtonian liquids. For non-Newtonian liquids more complicated relationships exist which will be discussed in the next chapter.

[‡] If we wish to find the molar flux instead of the mass flux we write: $\phi''_{mol} = -\mathbb{D}\dfrac{dc}{dn}$ (kmol/m^{2}s). For the implications of this, see the introduction of Chapter IV.

dv_y/dx. In other words, viscous shear is transversal transport of longitudinal momentum, as indicated in Figure I.7. This momentum transport represents itself as shear which, per unit of surface, is called the shear force τ. Thus the momentum flux F''_{xy} corresponds with a shear force τ_{xy} in the y-direction at $x = $ constant. This is illustrated in Figure I.8. The exact definition of the shear force needs, therefore, a statement of the plane and direction being considered, e.g.

$$\tau_{xy} = -v\frac{d(\rho v_y)}{dx}$$

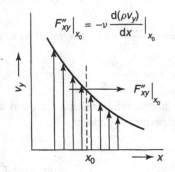

Figure I.7   Shear in terms of momentum flux

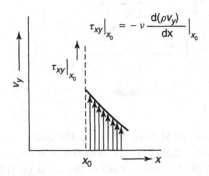

Figure I.8   Shear in terms of shear forces

The shear force is defined in such a way that the force the left-hand fluid exerts on the right-hand fluid is positive if the gradient (from left to right) is negative.

We have seen in the foregoing discussion that the statistical molecular transport through a plane can be described quantitatively by equation (I.13) if the proportionality constant ($\mathbb{D}$, a or v) and the gradient at that plane is known. In practice, we are interested in transport through a plane forming the boundary of a system (e.g. a wall, an interface, etc.), but, alas, often the concentration gradient at the boundary is not known beforehand and a practical approach has to be followed when qualifying the flux. In these cases, empirical phenomenological transfer coefficients are used, which are defined as follows:

$$\phi_m'' = -\mathbb{D}\frac{d\rho_A}{dn}\bigg|_w = k(c_{A\infty} - c_{Aw})^\dagger \tag{I.14a}$$

$$\phi_H'' = -a\frac{d(\rho c_p T)}{dn}\bigg|_w = h(T_\infty - T_w) \tag{I.14b}$$

$$\tau_w = -\nu\frac{d(\rho v)}{dn}\bigg|_w = \left(\frac{f}{2}\rho v_\infty\right)(v_\infty - v_w)^\ddagger \tag{I.14c}$$

in which w stands for wall, ∞ for system (indicating a characteristic position in the system, such as, for instance, the axis), k for mass transfer coefficient, h for heat transfer coefficient and $f\rho v_\infty/2$ for momentum transfer coefficient (which looks awkward for historical reasons); f itself is better known as the Fanning friction factor.

These empirical relationships, to which much background can be added, as will be seen in the following chapters, are of practical use for establishing transport rates through boundaries only if we can resort to a physically completely similar situation for which quantitative data are available. This has already been brought out when we discussed equation (I.2) in terms of energy as heat, introducing the concept of the heat transfer coefficient at that stage. Physical similarity, however, does not mean being identical, as will become clear when the concepts are worked out in subsequent chapters.

Since order-of-magnitude estimations are of great importance in our discipline it is useful to realize the order of magnitude of $\mathbb{D}$, a and ν (or $\eta = \nu\rho$) for gases, liquids and solids. According to the kinetic theory for gases:

$$\mathbb{D} \approx a \approx \nu \approx \tfrac{1}{3}\bar{v}\bar{l}.$$

where $\bar{v}$ is the mean velocity of the molecules ($\approx \sqrt{2RT/M}$) and $\bar{l}$ the mean free path of the molecules ($\bar{l} \sim M/\rho = RT/\rho$). At normal pressure and temperature for gases, $\mathbb{D}$, a and ν are of the order $0.5 - 2 \times 10^{-5}$ m^2/s. The kinetic theory further predicts that η and λ(the thermal conductivity $= a\rho c_p$) for ideal gases are

---

† The definition for the molar flux reads:

$$\phi_{mol}'' = -\mathbb{D}\frac{dc}{dn}\bigg|_w = k(c_\infty - c_w)$$

---

‡ The velocity at a solid wall is zero, which is one of the hypotheses on which hydrodynamics rests; hence v_w nearly always equals zero.

independent of pressure and that they are approximately dependent on the square root of the absolute temperature.

For the diffusion coefficients of liquids at room temperature we find values of the order $\mathbb{D} = 10^{-8} - 10^{-9}$ m^2/s; for other temperatures, diffusivities can be estimated by means of the Einstein-Nernst-Eyring relationship, i.e. $\mathbb{D}\eta/T = $ constant, where η is the viscosity of the solvent. The viscosity η of liquids varies widely (e.g. at 20 °C, for water $\eta = 10^{-3}$ Ns/m^2, for glycerol $\eta = 1.5$ Ns/m^2) and is strongly dependent on temperature; in the first approximation, $\ln \eta \sim 1/T$. The thermal diffusivity (a) of liquids is about $\approx 10^{-7}$ m^2/s with the exception of molten salts and liquid metals, which show considerably higher thermal diffusivities.

The diffusion coefficients $(\mathbb{D})$ of solid materials at room temperature are of the order of $10^{-11} - 10^{-13}$ m^2/s. The thermal diffusivity of non-metallic solids (a) is roughly 10^{-7} m^2/s, while metals belong (because of the effective heat transport by the free electrons) to the best heat conductors $(a \approx 5 - 100 \times 10^{-6}$ m^2/s). In addition, solid materials have a very high viscosity. The ratio of the viscosity and the elasticity modulus (having the dimension of time) gives an impression of the time necessary before an initially elastic deformation can be observed as flow. On the other hand, many liquids (e.g. molten polymers) show elastic behaviour. For these liquids the time necessary before flow is observed is generally much greater than 0.1 s and it is therefore not possible to describe the flow of these materials by means of a simple shear stress-velocity gradient relationship, because the previous history of each particle has to be included in the description. These problems belong to the area of rheology and we will not go into more detail here.

I.3 Microbalances

In some cases the concentration, temperature and flow velocity distributions in a system can be calculated by starting with the principle of conservation and applying this principle to every small volume element of the system. This leads to the so-called microbalances, as compared to the macrobalances of the foregoing section (I.1.) which give no insight into the distribution of the quantity over the system. These microbalances can be formulated in a general way, as will be shown below. However, the partial differential equations which will emerge from this microconcept can be solved analytically only for relatively simple situations. Therefore, dependence on equations (I.14) is daily practice, but the idea behind a practical correlation for the transfer coefficients is often based on a fundamental analysis of a somewhat simplified and idealized situation which can still be treated analytically. However, numerical solutions of these equations for complicated problems are possible by making use of fast computers. Let us, therefore, see how we arrive at a microbalance.

Let us consider a Cartesian element (Figure I.9) of volume $dx\,dy\,dz$, where x, y and z are the three coordinates in space. The accumulation of the quantity X considered per time interval dt in the system is now given by:

$$\frac{\partial X}{\partial t} dx\,dy\,dz$$

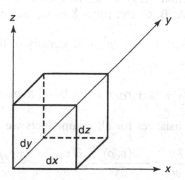

Figure I.9 Volume element in rectangular coordinates

The net inflow of X in the x-direction is:

$$\phi_x''|_x\,dy\,dz - \phi_x''|_{x+dx}\,dy\,dz = -\frac{\partial \phi_x''}{\partial x}\,dx\,dy\,dz \tag{I.15}$$

where ϕ_x'' is the flux of the quantity X in the x-direction. Similar expressions hold for the net flow into the system in the y- and z-directions. We can therefore write equation (I.1) for an infinitesimal volume element as:

$$\frac{\partial X}{\partial t}dx\,dy\,dz = -\frac{\partial \phi_x''}{\partial x}dx\,dy\,dz - \frac{\partial \phi_y''}{\partial y}dx\,dy\,dz - \frac{\partial \phi_z''}{\partial z}dx\,dy\,dz - r_X\,dx\,dy\,dz$$

$$\tag{I.16}$$

or

$$\frac{\partial X}{\partial t} = -\frac{\partial \phi_x''}{\partial x} - \frac{\partial \phi_y''}{\partial y} - \frac{\partial \phi_z''}{\partial z} + r_X \tag{I.17}$$

Equation (I.17) is the basic microbalance which can be further evaluated. When doing so, we have to realize that the flux of X on this microscale consists of a convective transport term and a statistical transport term. These fluxes in the n-direction are therefore given by:[†]

$$\phi_n'' = -(\mathbb{D} \text{ or } a \text{ or } \nu)\frac{\partial X}{\partial n} + v_n X \tag{I.18}$$

[†] Interpreted as the momentum balance, equation (I.18) is only valid for Newtonian liquids. We will discuss this problem in more detail in the following.

Thus, if we take $\mathbb{D}$ as a representative of $\mathbb{D}$, a or v, we obtain from equation (I.17) with equation (I.18):

$$\frac{\partial X}{\partial t} = -\frac{\partial(v_x X)}{\partial x} - \frac{\partial(v_y X)}{\partial y} - \frac{\partial(v_z X)}{\partial z} + \mathbb{D}\frac{\partial^2 X}{\partial x^2} + \mathbb{D}\frac{\partial^2 X}{\partial y^2} + \mathbb{D}\frac{\partial^2 X}{\partial z^2} + r_X \quad (I.19)$$

If we interpret equation (I.19) as a mass balance for a component A in a mixture of substances A, B, C, etc. (thus $X = \rho_A$, ρ_B, etc.), we know that:

$$\rho_A + \rho_B + \rho_C + \cdots = \rho \quad \text{(the density of that mixture)}$$

and

$$r_A + r_B + r_C + \cdots = 0 \quad \text{(Lavoisier)}$$

Summing up the mass balances for all components we find:

$$\frac{\partial \rho}{\partial t} = -\frac{\partial(v_x \rho)}{\partial x} - \frac{\partial(v_y \rho)}{\partial y} - \frac{\partial(v_z \rho)}{\partial z} \quad (I.20)$$

which is known as the equation of continuity. For a stationary situation, $d\rho/dt$ equals zero. When ρ is constant over the flow field, we find furthermore:

$$\frac{\partial v_x}{\partial x} + \frac{\partial v_y}{\partial y} + \frac{\partial v_z}{\partial z} = 0 \quad (I.21)$$

which is a very familiar expression for the conservation of total mass.

If equation (I.19) is to be used as a microbalance for the conservation of heat (or of energy if the transport of heat exceeds the transport of other forms of energy, as given in equation (I.4)), X stands for $\rho c_p T$. This microbalance will be further evaluated in Section III.1.4, whereas the micromass balance for one component will be discussed in more detail in Section IV.1.3.

We will now concentrate on the further development of equation (I.17) as a micromomentum balance. In order to obtain a relationship which is valid for liquids of different rheological behaviour (thus Newtonian and non-Newtonian fluids) we will, instead of equation (I.18), use the more general expression for the momentum flux in the n-direction:

$$\phi_n'' = \tau_n + v_n X \quad (I.22)$$

which is valid for all rheologies. Here X stands for ρv. The momentum production term in the n-direction is the sum of the pressure and gravity forces acting in that direction on the control volume:

$$r_n = -\frac{\partial p}{\partial n} + \rho g_n \quad (I.23)$$

With the above expressions we obtain from equation (I.17) for the micromomentum balance in the x-direction:

$$\frac{\partial(\rho v_x)}{\partial t} = -\frac{\partial(\rho v_x v_x)}{\partial x} - \frac{\partial(\rho v_x v_y)}{\partial y} - \frac{\partial(\rho v_x v_z)}{\partial z} - \frac{\partial \tau_{xx}}{\partial x} - \frac{\partial \tau_{yx}}{\partial y} - \frac{\partial \tau_{zx}}{\partial z} - \frac{\partial p}{\partial x} + \rho g_x$$
(I.24)

With the aid of the continuity equation (I.20) this expression can be simplified to:

$$\rho\frac{\partial v_x}{\partial t} = -\rho v_x\frac{\partial v_x}{\partial x} - \rho v_y\frac{\partial v_x}{\partial y} - \rho v_z\frac{\partial v_x}{\partial z} - \frac{\partial \tau_{xx}}{\partial x} - \frac{\partial \tau_{yx}}{\partial y} - \frac{\partial \tau_{zx}}{\partial z} - \frac{\partial p}{\partial x} + \rho g_x$$
(I.25)

Analogous expressions can be written for the micromomentum balance in the y- and z-directions as follows:

$$\rho\frac{\partial v_y}{\partial t} = -\rho v_x\frac{\partial v_y}{\partial x} - \rho v_y\frac{\partial v_y}{\partial y} - \rho v_z\frac{\partial v_y}{\partial z} - \frac{\partial \tau_{xy}}{\partial x} - \frac{\partial \tau_{yy}}{\partial y} - \frac{\partial \tau_{zy}}{\partial z} - \frac{\partial p}{\partial y} + \rho g_y$$
(I.26)

and

$$\rho\frac{\partial v_z}{\partial t} = -\rho v_x\frac{\partial v_z}{\partial x} - \rho v_y\frac{\partial v_z}{\partial y} - \rho v_z\frac{\partial v_z}{\partial z} - \frac{\partial \tau_{xz}}{\partial x} - \frac{\partial \tau_{yz}}{\partial y} - \frac{\partial \tau_{zz}}{\partial z} - \frac{\partial p}{\partial z} + \rho g_z$$
(I.27)

If, instead of a Cartesian volume element in rectangular coordinates, we had considered a volume element in cylindrical coordinates, as indicated in Figure I.10, we would have obtained the following micromomentum balances in the r-direction

$$\rho\frac{\partial v_r}{\partial t} = -\rho\left(v_r\frac{\partial v_r}{\partial r} + \frac{v_\theta}{r}\frac{\partial v_r}{\partial \theta} - \frac{v_\theta^2}{r} + v_z\frac{\partial v_r}{\partial z}\right) - \frac{1}{r}\frac{\partial(r\tau_{rr})}{\partial r}$$

$$- \frac{1}{r}\frac{\partial \tau_{r\theta}}{\partial \theta} + \frac{\tau_{\theta\theta}}{r} - \frac{\partial \tau_{rz}}{\partial z} - \frac{\partial p}{\partial r} + \rho g_r$$
(I.28)

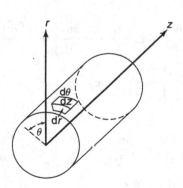

Figure I.10 Volume element in cylindrical coordinates

in the θ-direction:

$$\rho \frac{\partial v_\theta}{\partial t} = -\rho \left(v_r \frac{\partial v_\theta}{\partial r} + \frac{v_\theta}{r} \frac{\partial v_\theta}{\partial \theta} + \frac{v_r v_\theta}{r} + v_z \frac{\partial v_\theta}{\partial z} \right) - \frac{1}{r^2} \frac{\partial (r^2 \tau_{r\theta})}{\partial r}$$

$$- \frac{1}{r} \frac{\partial \tau_{\theta\theta}}{\partial \theta} - \frac{\partial \tau_{\theta z}}{\partial z} - \frac{1}{r} \frac{\partial p}{\partial \theta} + \rho g_\theta \tag{I.29}$$

and in the z-direction:

$$\rho \frac{\partial v_z}{\partial t} = -\rho \left(v_r \frac{\partial v_z}{\partial r} + \frac{v_\theta}{r} \frac{\partial v_z}{\partial \theta} + v_z \frac{\partial v_z}{\partial z} \right) - \frac{1}{r} \frac{\partial (r \tau_{rz})}{\partial r}$$

$$+ \frac{1}{r} \frac{\partial \tau_{\theta z}}{\partial \theta} + \frac{\partial \tau_{zz}}{\partial z} - \frac{\partial p}{\partial z} + \rho g_z \tag{I.30}$$

These micromomentum balances can naturally also be formulated in spherical coordinates. The relationships obtained can be found in many handbooks.

The above momentum balances in terms of shear stresses are valid for all fluids, because the shear stresses are independent of the rheological behaviour of the liquid. In the special case of Newtonian liquids τ is proportional to the velocity gradient (equation (I.13)), e.g.

$$\tau_{xy} = -\nu \frac{\partial (\rho v_y)}{\partial x}$$

Introducing this expression into the above microbalances we obtain, for rectangular coordinates, in the x-direction:

$$\rho \frac{\partial v_x}{\partial t} = -\rho \left[v_x \frac{\partial v_x}{\partial x} + v_y \frac{\partial v_x}{\partial y} + v_z \frac{\partial v_x}{\partial z} \right] + \nu \left[\frac{\partial^2 (\rho v_x)}{\partial x^2} + \frac{\partial^2 (\rho v_x)}{\partial y^2} \right.$$

$$\left. + \frac{\partial^2 (\rho v_x)}{\partial z^2} \right] - \frac{\partial p}{\partial x} + \rho g_x \tag{I.31}$$

in the y-direction:

$$\rho \frac{\partial v_y}{\partial t} = -\rho \left[v_x \frac{\partial v_y}{\partial x} + v_y \frac{\partial v_y}{\partial y} + v_z \frac{\partial v_y}{\partial z} \right] + \nu \left[\frac{\partial^2 (\rho v_y)}{\partial x^2} + \frac{\partial^2 (\rho v_y)}{\partial y^2} \right.$$

$$\left. + \frac{\partial^2 (\rho v_y)}{\partial z^2} \right] - \frac{\partial p}{\partial y} + \rho g_y \tag{I.32}$$

and in the z-direction:

$$\rho \frac{\partial v_z}{\partial t} = -\rho \left[v_x \frac{\partial v_z}{\partial x} + v_y \frac{\partial v_z}{\partial y} + v_z \frac{\partial v_z}{\partial z} \right] + \nu \left[\frac{\partial^2 (\rho v_z)}{\partial x^2} + \frac{\partial^2 (\rho v_z)}{\partial y^2} \right.$$

$$\left. + \frac{\partial^2 (\rho v_z)}{\partial z^2} \right] - \frac{\partial p}{\partial z} + \rho g_z \tag{I.33}$$

In cylindrical coordinates, Newton's law (equation (I.13)) reads in the z- and r-directions:

$$\tau_{\theta z} = -v \frac{\partial(\rho v_\theta)}{\partial z} \quad \text{and} \quad \tau_{zr} = -v \frac{\partial(\rho v_z)}{\partial r}, \quad \text{respectively,}$$

and in the θ-direction:

$$\tau_{r\theta} = -vr \frac{\partial \left(\frac{1}{r} \rho v_\theta \right)}{\partial r} \tag{I.34}$$

Introducing this into the general microbalance in terms of shear stresses we obtain in the r-direction:

$$\rho \frac{\partial v_r}{\partial t} = -\rho \left\{ v_r \frac{\partial v_r}{\partial r} + \frac{v_\theta}{r} \frac{\partial v_r}{\partial \theta} - \frac{v_\theta^2}{r} + v_z \frac{\partial v_r}{\partial z} \right\} + v \left\{ \frac{\partial}{\partial r} \left[\frac{1}{r} \frac{\partial(\rho r v_r)}{\partial r} \right] \right.$$
$$\left. + \frac{1}{r^2} \frac{\partial^2(\rho v_r)}{\partial \theta^2} - \frac{2}{r^2} \frac{\partial(\rho v_\theta)}{\partial \theta} + \frac{\partial^2(\rho v_r)}{\partial z^2} \right\} - \frac{\partial p}{\partial r} + \rho g_r \tag{I.35}$$

in the θ-direction:

$$\rho \frac{\partial v_\theta}{\partial t} = -\rho \left\{ v_r \frac{\partial v_\theta}{\partial r} + \frac{v_\theta}{r} \frac{\partial v_\theta}{\partial \theta} + \frac{v_r v_\theta}{r} + v_z \frac{\partial v_\theta}{\partial z} \right\} + v \left\{ \frac{\partial}{\partial r} \left[\frac{1}{r} \frac{\partial(\rho r v_\theta)}{\partial r} \right] \right.$$
$$\left. + \frac{1}{r^2} \frac{\partial^2(\rho v_\theta)}{\partial \theta^2} + \frac{2}{r^2} \frac{\partial(\rho v_r)}{\partial \theta} + \frac{\partial^2(\rho v_\theta)}{\partial z^2} \right\} - \frac{\partial p}{\partial \theta} + \rho g_\theta \tag{I.36}$$

and in the z-direction:

$$\rho \frac{\partial v_z}{\partial t} = -\rho \left\{ v_r \frac{\partial v_z}{\partial r} + \frac{v_\theta}{r} \frac{\partial v_z}{\partial \theta} + v_z \frac{\partial v_z}{\partial z} \right\} + v \left\{ \frac{1}{r} \frac{\partial}{\partial r} \left[r \frac{\partial(\rho v_z)}{\partial r} \right] \right.$$
$$\left. + \frac{1}{r^2} \frac{\partial^2(\rho v_z)}{\partial \theta^2} + \frac{\partial^2(\rho v_z)}{\partial z^2} \right\} - \frac{\partial p}{\partial z} + \rho g_z \tag{I.37}$$

The micromomentum balances given above form the basis for the calculation of velocity distributions and flow rate-pressure drop relationships during laminar flow; we will apply them many times in the next chapter.

I.4 Dimensions and magnitude

> *Quantitative physics uses concepts labelled "physical quantities" and equations relating such quantities. We are accustomed to operating algebraically with these quantities,*

e.g. a car speeding at 100 km per hour travels 50 km in half an hour. Is this worth a discussion? Yes, not because 1/2 × 100 = 50, but because it also implies the operation "distance divided by time is speed". If we choose independent yardsticks to measure all three quantities separately, their relationship would become 'distance divided by time' is proportional to speed, the proportionality factor of which would be arbitrary and without any physical significance. This happened, for example in the 1970s, when mass was inferred by measuring its weight in the (variable) field of gravitation and energy from heat was assessed in a different way to energy from work. As a consequence of this, the gravitational 'constant' became oddly involved in relationships in which "mass", and not its 'weight', was fundamental and the 'mechanical equivalent of heat' in thermodynamic reasoning.

Physical theory has to deal with a consistent set of physical quantities, although this set can still be chosen arbitrarily. To inquire into the "real" dimension of a physical quantity has no more measuring than to inquire into the real name of a rose. A rose is a rose, whatever it is called. This entails that other concepts are involved in defining the consistency of a set of fundamental physical quantities, e.g. they have to be measurable both reproducibly and accurately (the metrical standard), and the set has to be so encompassing that the physical situations being studied can be expressed in functional relationships between physical quantities measurable with the fundamental set.

This form of thinking acquired a solid basis when abstract sets or group theory were applied to evaluate some of the different systems of units that were still in use in various countries up until the 1970s. We will not elaborate further on this here, but suggest that the reader keeps this in mind: a physical quantity is represented adequately only by a number *and* a unit or a dimension.

Obtaining unanimity on the fundamental set of physical quantities took much time. It was only in 1964 that such a set was selected by the scientific community, and subsequently by about 1980 by many legal authorities, e.g. the relevant Food and Drug administrations. Apart from the units used for electrical current (A, ampere), which have no role in this present text, and those used for luminosity (cd, candela), which have a small role here (e.g. see Section III.9), the basic (and some derived) units of this SI (système international or Unite's) set are given in Table I.1.

However, many physicists will associate more readily their experiences with heat with the (k)cal and not with the joule, and pressure with the bar or mmHg (in clinical physics) and not with the pascal (Nm^{-2}). For this reason, conversion factors are given in Table I.2, in order to facilitate reporting in SI units.

To conclude this section, Table I.3 provides information on some common properties of gases, liquids and solids, which should prove useful for approximate calculations, while Table I.4 lists the adopted prefixes used for SI units.

Table I.1 Basic and derived SI units

Quantity	SI unit	Unit symbol
Length	metre	m
Mass	kilogram	kg
Amount of substance	mole	mol
Time	second	s
Temperature	Kelvin[a]	K
Force	Newton	$N = kgm/s^2$
Work, energy, quantity of heat	Joule	$J = Nm$
Power	Watt	$W = J/s$
Pressure	Pascal	$Pa = N/m^2$
Dynamic viscosity		Ns/m^2
Kinematic viscosity		m^2/s
Surface energy or tension		J/m^2 or N/m
Enthalpy		J/kg
Heat capacity		$J/kg\ K$
Heat transfer coefficient		$W/m^2\ K$
Mass transfer coefficient		m/s
Thermal conductivity		$W/m\ K$

[a]Temperature difference is commonly expressed in degrees Celsius instead of Kelvin.

Table I.2 Conversion factors

Magnitude	Expressed in	Multiply by / Divide by	In SI units
Length	inch (in)	0.0254	m
	foot (ft)	0.305	
	yard (yd)	0.914	
	mile	1609	
	Angstrom (Å)	10^{-10}	
Area	in^2	6.45×10^{-4}	m^2
	ft^2	0.0929	
	yd^2	0.836	
	acre	4047	
	$mile^2$	2.59×10^6	
Volume	in^3	1.64×10^{-5}	m^3
	ft^3	0.0283	
	yd^3	0.765	
	UK gallon	4.55×10^{-3}	
	US gallon	3.785×10^{-3}	
Time	minute (min)	60	s
	hour (h)	3600	
	day	8.64×10^4	
	year	3.16×10^7	
Mass	grain	6.48×10^{-5}	kg
	ounce(oz)	2.84×10^{-2}	

(continued overleaf)

Table I.2 (*continued*)

Magnitude	Expressed in	Multiply by / Divide by	In SI units
	pound (lb)	0.454	
	hundredweight (cwt)	50.8	
	ton	1016	
Force	poundal (pdl)	0.138	N
	pound force (lbf)	4.45	
	dyne (dyn)	10^{-5}	
	kg force (kgf)	9.81	
Volumetric flow	ft^3/min	4.72×10^{-4}	m^3/s
	UK gal/min	7.58×10^{-5}	
	US gal/min	6.31×10^{-5}	
Mass flow	lb/min	2.10×10^{-6}	kg/s
	ton/h	0.282	
Density	lb/in^3	2.77×10^4	kg/m^3
	lb/ft^3	16.0	
Pressure	lbf/in^2	6.89×10^3	$N\,m^2$
	lbf/ft^2	47.9	
	dyn/cm^2	0.1	
	kgf/cm^2 ($=$ at)	9.81×10^4	
	atm (standard)	1.013×10^5	
	bar	10^5	
	in water	2.49×10^2	
	ft water	2.99×10^3	
	in Hg	3.39×10^3	
	mm Hg (torr)	1.33×10^2	
Dynamic viscosity	lb/ft h	4.13×10^{-4}	Ns/m^2
	lb/ft s	1.49	
	poise (P $=$ g/cm s)	0.1	
	centipoise (cP)	10^{-3}	
Kinematic viscosity	ft^2/h	2.58×10^{-5}	m^2/s
	stokes (St $=$ cm^2/s)	10^{-4}	
	centistokes (cSt)	10^{-6}	
Surface tension	dyn/cm ($=$ erg/cm^2)	10^{-3}	N/m
Temperature difference	degree F (or R)	5/9	°C (K)
Energy (work, heat)	ft pde	0.0421	$J = Ws = Nm$
	ft lbf	1.36	
	BTU	1.06×10^3	
	CHU	1899	
	hph	2.68×10^6	
	erg	10^{-7}	
	kgf m	9.81	
	kcal	4.19×10^3	
	kWh	3.60×10^6	

Table I.2   (*continued*)

Magnitude	Expressed in	Multiply by ⇄ Divide by	In SI units
Power (energy flow)	BTU/h	0.293	W
	CHU/h	0.528	
	ft lbf/s	1.36	
	hp (British)	746	
	hp (metric)	736	
	erg/s	10^{-7}	
	kcal/h	1.163	
	cal/s	4.19	
Heat flux	BTU/ft^{2}h	3.15	W/m^2
	cal/cm^{2}s	4.19×10^4	
	kcal/m^{2}h	1.163	
Specific heat	BTU/lb °F	4.19×10^3	J/kg °C
	kcal/kg °C	4.19×10^3	
Latent heat	BTU/lb	2.33×10^3	J/kg
	kcal/kg	4.19×10^3	
Heat conductivity	BTU/ft h °F	1.73	W/m °C
	cal/cm s °C	4.19×10^2	
	kcal/m h °C	1.163	
Heat transfer coefficient	BTU/ft^2 h °F	5.68	W/m^2 °C
	cal/cm^2 s °C	4.19×10^4	
	kcal/m^2 h °C	1.163	

Table I.3 Physical properties of some materials

	Air 20 °C	Water 20 °C	Bean oil 100 °C	Stainless steel 20 °C
ρ (kg/m^3)	1.20	998	870	7750
η (Ns/m^2)	17×10^{-6}	10^{-3}	7×10^{-3}	—
$\nu = (\eta/\rho)$ (m^2/s)	14.2×10^{-6}	10^{-6}	8×10^{-6}	—
c_p (J/kg °C)	1.03×10^3	4.19×10^3	2.12×10^3	0.45×10^3
λ (W/m °C)	0.025	0.6	0.156	26
σ (N/m)	—	7×10^{-2}	2.9×10^{-2}	—
$\mathbb{D}_{H_2}$ (m^2/s)	$\sim 2 \times 10^{-5}$	5×10^{-9}	2.4×10^{-9}	—
$a = \lambda/\rho c_p$ (m^2/s)	20×10^{-6}	0.143×10^{-6}	0.084×10^{-6}	7.45×10
Pr $= \nu/a = c_p\eta/\lambda$	0.71	7.0	95	—
Sc $= \nu/\mathbb{D}$	0.71	200	300	—
ΔH_v (J/kg)	—	2.45×10^6	—	—
ΔH_m (J/kg)	—	3.35×10^5	—	—

Gas law constant $R = 8314$ J/kmol K
Avogadro constant $N = 6.023 \times 10^{26}$ kmol^{-1}
Gravitational acceleration $g = 9.81$ m/s^2
Stefan-Boltzmann constant $\sigma = 5.67 \times 10^{-8}$ W/m^{2}K^4
Volume of ideal gas at STP = 22.41 m^3/kmol
Standard temperature and pressure (STP) : 1.013×10^5 N/m^2 and 273.15 K

Table I.4   Prefixes used for SI units[a]

Prefix	Symbol	Multiple/fraction
teva	T	10^{12}
giga	G	10^{9}
mega	M	10^{6}
kilo	k	10^{3}
milli	m	10^{-3}
micro	μ	10^{-6}
nano	n	10^{-9}
pico	p	10^{-12}
femto	f	10^{-15}

[a]e.g. 1 GJ = 10^{9} J

I.5 Dimensional analysis

> *The application of set theory leads to the conclusion that a physical problem determined by n physical quantities, which are related to a set of r basic units, is reducable to a problem determined by $n - r = p$ (dimensionless) physical quantities.*

We will illustrate this theoretically obtained result by discussing examples of the somewhat unsophisticated ways in which it is used in daily practice, both for 'inductive' and 'straightforwardly simple' cases. The reader should be aware, however, that for much more complicated problems than those presented here, the solid logical background of such reasoning is best studied first in order to avoid the sometimes serious errors that may present themselves in dimensional analysis.

We will illustrate the use of dimensionless numbers by a few examples. Figure I.11 shows the end of a pipe through which gas flows at a rate v. Another gas A present at a concentration c_∞ outside the pipe will diffuse into the pipe.

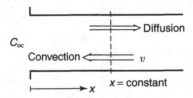

Figure I.11   Transport by diffusion and convection in a pipe

Under stationary conditions the rate of diffusion into the pipe equals the rate of discharge of A with the flow. This situation occurs, for example, in pneumatic control equipment where a small gas leak is applied (bleeding) in order to prevent dust or water vapour entering the apparatus. The concentration distribution c of the gas A in the pipe as a function of the distance x from the opening is easily calculated if we realize that in the stationary state the amount diffusing through each cross-section x = constant into the pipe equals the amount of A removed with the flow, thus:

$$\phi''_{mol} = 0 = -\mathbb{D}\frac{dc}{dx} - vc \qquad (I.38)$$

with $c = c_\infty$ at $x = 0$. The solution is:

$$\frac{c}{c_\infty} = \exp\left(-\frac{vx}{\mathbb{D}}\right) \qquad (I.39)$$

which is a dimensionally homogeneous equation. The relative change in concentration appears to be a function of the dimensionless number $vx/\mathbb{D}$ (the so-called Péclet number, Pe), which is a measure of the ratio of mass transport by convection and mass transport by diffusion. If this ratio is large, the relative concentration of A is small. Each dimensionless number can be interpreted as the ratio of the effects of two physical mechanisms.

If we had not been able to produce a complete analysis of the above problem we could have written formally on the basis of physical intuition:

$$c = f(x, c_\infty, v, \mathbb{D}) \qquad (I.40)$$

indicating that the concentration of A must be dependent on the distance from the opening, x, the outside concentration of A, c_∞, the rate of gas flow v and the diffusivity of A, $\mathbb{D}$. Since the quantity 'amount of substance' occurs only in c and c_∞, dimensional homogeneity is only possible if the ratio c/c_∞ of these two variables occurs in the solution. Analogously, the dimension 'time' occurs only in v and $\mathbb{D}$, indicating that these two variables must occur as the ratio $v/\mathbb{D}$ in the solution. This ratio has the dimension of 'length' and this dimension can be eliminated by introducing distance into the ratio. Thus the solution we find from dimensional considerations is:

$$\frac{c}{c_\infty} = f\left(\frac{vx}{\mathbb{D}}\right) \qquad (I.41)$$

This is in agreement with the result obtained earlier. That the function f is an exponential function cannot be found from dimensional considerations; this result can only be obtained from a complete analysis. Dimensional analysis has resulted in reducing the problem with initially five variables to a problem with only two parameters. Furthermore, comparisons between analogous situations are possible without the complete solution. If we consider, for example, two situations where the gas velocities differ by a factor of 2, we find the same relative concentrations at distances which differ by a factor of $\frac{1}{2}$ (i.e. the term vx stays constant).

As a second illustration of the procedure followed during dimensional analysis, we will consider the force F acting on a sphere which is placed into a flowing fluid with velocity v_r relative to the sphere (see also Section II.5.2). Based on physical considerations we can state that the force on the sphere will be given by:

$$F = f(D, v_r, \rho, \eta) \qquad (I.42)$$

where D is the diameter of the sphere, ρ the density and η the viscosity of the fluid. This relationship must be dimensionally homogeneous, i.e. the dimensions occurring on the left-hand side must equal the dimensions on the right-hand side of the equation. This must also be the case if we write:

$$F = D^a v_r^b \rho^c \eta^d \text{ constant}$$

The dimensions of these quantities can be expressed in length (L), time (T) and mass (M) and we obtain:

$$LT^{-2}M = L^a(LT^{-1})^b(L^{-3}M)^c(L^{-1}T^{-1}M)^d \text{ constant}$$

The condition of homogenity leads thus to the following three expressions (one for each dimension) between the exponents a, b, c and d:

$$L = L^a L^b L^{-3c} L^{-d} \text{ or } 1 = a + b - 3c - d$$
$$T^{-2} = T^0 T^{-b} T^0 T^{-d} \text{ or } -2 = -b - d$$
$$M = M^0 M^0 M^c M^d \text{ or } 1 = c + d$$

Solving these equations for a, b and c we find:

$$a = 2 - d$$
$$b = 2 - d$$
$$c = 1 - d$$

The following expression for the force F is thus dimensionally homogeneous

$$F = D^{(2-d)} v_r^{(2-d)} \rho^{(1-d)} \eta^d \text{ constant}$$

or, also:

$$\frac{F}{\rho v_r^2 D^2} = \left(\frac{\eta}{\rho v_r D}\right)^d \text{ constant} \qquad (I.43)$$

Since both $F/\rho v_r^2 D^2$ and $\eta/\rho v_r D$ are dimensionless combinations of the variables, any relationship between these two groups is also dimensionally homogeneous. We can therefore conclude that:

$$\frac{F}{\rho v_r^2 D^2} = f\left(\frac{\eta}{\rho v_r D}\right) = f\left(\frac{1}{Re}\right) \qquad (I.44)$$

Here $Re(= \rho v_r D/\eta)$ is the Reynolds number, which can be interpreted as the ratio of momentum transport by convection ($\sim \rho v_r v_r$) and by internal friction ($\sim \eta v_r/D$).

Again, the number of variables in this problem of initially five has been reduced to two. In most cases, the number of dimensionless groups obtained by dimensional analysis equals the number of variables occurring in the problem minus the number of independent dimensional equations which can be set up. In flow problems the number of dimensional equations is three (L, T and M) and in heat/mass transport problems four (L, T, M and temperature/amount of substance).

The function f in the result obtained is again unknown and can only be found from a complete analysis of the problem. Sometimes the result obtained from dimensional analysis can be simplified further by making use of former physical experience. In the case of the sphere, for example, experience tells us that at very low flow velocities the force F on the sphere is independent of the density of the fluid. Thus the relationship between the two dimensionless groups must be:

$$\frac{F}{\rho v_r^2 D^2} = \frac{\eta}{\rho v_r D} \text{ constant}$$

or

$$\frac{F}{\eta v_r D} = \text{constant} \tag{I.45}$$

This equation represents Stokes' law. The fact that the numerical value of the constant is 3π can only be found by solving the micromomentum balances for this problem.

As a further example we will discuss the droplets formed, for instance, during condensation of water vapour at the underside of a horizontal surface. The hanging droplets grow until they reach a critical volume V_{cr} and fall down. A dimensionally homogeneous relationship for V_{cr} follows from the consideration that V_{cr} depends on the density of the liquid ρ, the gravitational acceleration g, the surface tension σ of the liquid and the degree to which the surface is wetted by the liquid. For the great number of cases where the surface is well wetted by the liquid we can write:

$$V_{cr} = f(\rho, g, \sigma) \tag{I.46}$$

The reader may check that the result is given by:

$$V_{cr}\left(\frac{\rho g}{\sigma}\right)^{3/2} = \text{constant} \tag{I.47}$$

The numerical value of the constant follows only from an exact analysis or from experimental results. Analysis yields here a constant of 6.66. The problem with initially four variables has been reduced to one dimensionless group, i.e. the Laplace number which represents the ratio of the weight of the droplet ($\sim \rho g V_{cr}$) and the capillary force ($\sim \sigma V_{cr}^{1/3}$).

As a final example, we will show how dimensional analysis may be used to introduce, with some physical insight and ingenuity, new concepts to deal with some chemical engineering problems. One has a notion that at large Re-numbers turbulent eddies will exist. The question than is to determine at which (small) scale the transfer processes to small particles and droplets are no longer influenced by these eddies. This means that we are looking for the size λ_K of the smallest eddies in which the remainder of the turbulent energy is dissipated. Physical intuition tells us that the energy dissipation per unit of mass, ε (W/kg) and the kinematic viscosity ν (m^2/s) are involved as the principal determinants. Using only this purely kinematic hypothesis we may write:

$$\lambda_K \cong \varepsilon^a \nu^b, \text{ or } (m) = (m^2/s^3)^a (m^2/s)^b$$

or

$$\lambda_K \cong \varepsilon^{-1/4} \nu^{3/4} \tag{I.48}$$

This is known as the Kolmogoroff-scale of turbulence, which divides problems into two regions for particles or droplets smaller or bigger than this scale. Each regime has its own laws for describing transfer processes (see also Sections II.4 and II.6).

If dimensional analysis is applied, all relevant variables have to be included in the considerations. If we had, for example, forgotten the gravitational acceleration in the analysis of the hanging droplet, we would not have found a solution. If, on the other hand, we include too many variables, the result becomes unnecessarily complicated. Most profit is obtained if, besides dimensional analysis, a fragmentary piece of physical experience is applied, as illustrated in the above examples.

Sometimes, one finds variations in this approach, which also prove to be fundamentally correct. The first of these is to use a derived unit to replace a fundamental unit. The second is to treat the dimension 'length' as a vector. Vectorial dimensional analysis may be of use in situations in which fundamental phenomena can be distinguished with different directions of preference. Both variations can be illustrated by the example of heating up a fluid flowing with a mean velocity v in a channel between hot, wide horizontal plates (length l_x, height H_y). The physical quantities involved are λ, ρc_P, V, H, l and h, the wall heat flux per unit area and per unit of temperature difference (heat-transfer coefficient), between wall and fluid. We choose to work with the fundamental dimensions, length (L), time (T) and temperature (Θ), plus a derived unit, namely 'heat' energy $(Q, L^2T^{-2}M)$, to replace mass (M) in our set. The latter is acceptable as long as mechanical energy considerations do not play a role in the problem of finding a dimensionless relationship for the heat transfer coefficient h. By following the rules of scalar dimensional analysis, we can now construct the following table of exponents:

	h	λ	ρc_P	v	H	l
L	−2	−1	−3	1	1	1
T	−1	−1	0	−1	0	0
Q	1	1	1	0	0	0
Θ	−1	−1	−1	0	0	0

The last two rows in this table are interdependent, so consequently $6 − 3(= 3)$ exponents can be chosen freely. The reader should now easily derive either, by inspection or by analysis, that these exponents refer to the following dimensionless numbers or combinations thereof:

$$hH/\lambda = \text{Nu (Nusselt)}, \quad vH/a = \text{Pe (Péclet) and } H/l.$$

If we now take into account the fact that the heat flow is perpendicular to the material flow, a more refined table of exponents can be obtained:

	h	λ	ρc_p	v	H	l
L_x	−1	−1	−1	1	0	1
L_y	0	1	−1	0	1	0
L_z	−1	−1	−1	0	0	0
T	−1	−1	0	−1	0	0
Q	1	1	1	0	0	0
Θ	−1	−1	−1	0	0	0

A nice puzzle to leave to the reader is to find the rationale behind this reasoning. (Hint: the dimension of ρC_P is $Q(L_X L_Y L_Z \Theta)^{-1}$, etc. and three of the six rows are interdependent). So, now $6 − 4(= 2)$ exponents can be chosen freely, thus leading to the dimensionless groups:

$$\frac{hH}{\lambda} \quad \text{and} \quad \frac{vH^2}{al}$$

which is a more restricted result than that found by using scalar dimensional analysis.

Scalar and vectorial dimensional analysis can be used advantageously when planning experiments, including those involving model fluids, and for presenting results in a universal way.

1.6 Problems

In the following the reader will find a number of exercises. These problems form an important part of the book and the reader is asked to solve as many of them as possible. They provide a check on the understanding of the principles discussed, they illustrate the practical application of these principles and sometimes they extend the matter presented. In the three chapters which follow each section will end with a collection of problems. The answers to these problems are given in order to enable the reader to check his own results. With a few problems (marked with an asterisk) we have also included some comments in order to illustrate an important point or to show that the solution is really not very difficult if the basic principles are applied in the right way.

When solving a problem, the most important thing is to establish the physical mechanism that governs the situation studied (e.g. is convection, conduction and radiation of heat important or can one or two of these transport mechanisms be neglected in this special case?). Secondly, we have to decide over which control volume the balance is made (e.g. over a microscopic volume if information about the temperature distribution is required; over a pipe or heat exchange or over a total reactor with overall transfer coefficients if only information about the mean temperature is required). Next, we have to determine whether the situation is in steady state or not (sometimes a situation can be made steady state by assuming the observer to move with the system, e.g. the flowing fluid). Having come so far we are now in a position to solve the problem with the help of the basic information provided in this book. This all sounds very difficult — and it is — but the best way to solve any difficulties is to try very hard.

1. A closed vessel, provided with a piston allowing expansion, rests on a carrier running downhill. Write down the law of energy conservation for 1 kg of the vessel's content, assuming near-equilibrium expansion and zero heat exchange with the environment.

   Answer: $0 = dU + p\,d\left(\dfrac{1}{\rho}\right) + v\,dv + g\,dh$

2. A student compares the specific heat content of his central heating (1.5 atm, 85 °C, 85 kcal/l) with that of steam (10 atm, 184 °C, 3.68 kcal/l) and concludes that a thermodynamic machine based on a water cycle would be preferable to a steam engine. What is the error in his reasoning and which engineering objectives does he confuse?

   Answer: Hot water is the more efficient for the transport of heat, but not for the conversion of heat into mechanical energy. The latter needs an almost reversible cyclic operation between two distinct thermodynamic states.

*3. 1 l/s water is forced through a horizontal pipe. The pressure difference over the length of the pipe is 2.10^5 N/m^2.

(a) What is the amount of power ϕ_A necessary and what is the temperature increase of the water ΔT if there is no heat exchange with the surroundings?

(b) What are ϕ_A and ΔT if the water flow is doubled?

Answer: (a) $\phi_A = 200$ W, $\Delta T = 0.048\,°C$
     (b) $\phi_A = 1600$ W, $\Delta T = 0.19\,°C$

4. Two spheres of equal weight fall through air. What is the ratio of their stationary rates of free fall if the ratio of their diameters, D_1/D_2, equals 3?

Answer: $v_1/v_2 = \frac{1}{3}$

5. Show that the power ϕ_A necessary to rotate a stirrer (diameter d) with a speed $n(\mathrm{s}^{-1})$ in a fluid of density ρ and viscosity η is given by:

$$\frac{\phi_A}{\rho n^3 d^5} = f\left(\frac{\rho n d^2}{\eta}, \frac{n^2 d}{g}\right)$$

*6. Write down the conservation laws for a kettle of water on the fire:
(a) during heating up;
(b) during boiling.

*7. At the end of a glass capillary (outer diameter 2 mm) single water droplets are formed very slowly. What is the diameter of the droplets which fall from the capillary?

Answer: $d = 4.4$ mm

8. Through a well-stirred 15 m^3 tank flows 0.01 m^3/s of coconut oil. After a certain time palm kernel oil is passed through the vessel at the same speed. After what time does the effluent oil contain less than 1% coconut oil?

Answer: $t = 6900$ s

9. Vessels are filled with a liquid while standing on a scale. The filling time is 20s, and the velocity of the inflow is 10 m/s.

(a) Which percentage of overweight has to be accounted for in order to achieve a correct weight?

(b) How does this percentage change if the filling time is reduced by 40%?

Answer: (a) 5%
     (b) 14%

10. A pulp containing 40 wt% moisture is fed into a countercurrent drier at a rate of 1000 kg/h. The pulp enters at a temperature of 20 °C and leaves at 70 °C, with the moisture content at the exit being 8%. Air of humidity of 0.007 kg water per kg dry air enters the drier at 120 °C while the temperature

of the air leaving the drier is 80 °C. What is the rate of air flow through the drier and the humidity of the air leaving the drier?

Data: heat losses from drier $= 5 \times 10^6$ J/h, specific heat of water vapour $c_p = 2 \times 10^3$ J/kg °C, specific heat of dry pulp $c_p = 10^3$ J/kg °C, ΔH_v at 80 °C $= 2.3 \times 10^6$ J/kg

Answer: $\phi_{mg} = 2.3 \times 10^4$ kg/h, water content $= 0.022$ kg water per kg dry air

*11. A 100 kmol batch of aqueous methanol containing 45 mol% CH_3OH is to be subjected to simple atmospheric distillation until the residue in the kettle contains only 5 mol% CH_3OH. What is the bulk composition of the distillate?
Vapour-liquid equilibria at 1 atm (X and Y = mole fraction CH_3OH in liquid and vapour phases, respectively)

X	Y	X	Y
0.02	0.13	0.40	0.73
0.05	0.27	0.50	0.78
0.10	0.42	0.60	0.83
0.20	0.58	0.80	0.91
0.30	0.67	0.90	0.96

Answer: 63 mol%

12. Modern dutch windmills are reported to generate $1.57\, J\rho D^2 v^3$ kWh/year (D is the diameter of the cross-section for flow in m, and v the air velocity at the blades in m/s).
    (a) Prove that this reporting is dimensionally consistent and furthermore that the mean efficiency of these windmills is 44%.
    (b) What are the main reasons for this (not really) disappointing result?

13. Prove that a complete set of dimensionless numbers describing relationships for subsonic, incompressible, Newtonian flow in, or from geometrically similar, straight channels is given by:
    vL/v (Reynolds), v^2/Lg (Fronde), $\rho g L^2/\sigma$ (Bond/or Weber), and $f/2 = \tau_w/\rho v^2$ (Fanning).
    Compare equations (I.44), (I.47) and (II.35) (see Section II.2.2) with these results. Define each result as the ratio of two physical processes, using a similar argument to that adopted when discussing equation (I.44).

14. Table III.5 (see Section III.3.1) gives dimensionless relationships for various heat transfer situations. Why is the relationship for the scraped surface heat exchanger not written as $Nu = 1.13\Psi(Re\ Pr\ n)^{1/2}$?

15. The literature gives the following for the diameter, d, of single air bubbles emerging from a nozzle with diameter D submersed in water:

$$d = 5.55 \times 10^{-3}(vD^{2.5})^{0.289}.$$

Translate this into a dimensionless homogeneous relationship.

Answer: $\dfrac{d}{D} = 1.7 \times 10^{-2}\mathrm{Fr}^{0.145}\mathrm{We}^{-0.065}$

16. Electrohydrodynamic atomization is a method used to produce very fine droplets from a liquid by using an electric field (electric current, I (A)). Depending on the properties of the liquid (density ρ, surface tension σ), and its flow rate (ϕ_v), different modes of droplet formation will occur. The droplet size (d) changes if

$$I \geq 27\sqrt{\frac{\varepsilon_0}{\rho}}\sigma$$

from $\quad d = 2\left(\dfrac{\rho\varepsilon_0\phi_v^4}{I^2}\right)^{1/6} \quad$ to $\quad d = 2\left(\dfrac{\varepsilon_0\sigma\phi_v^2}{I^2}\right)^{1/3}$

in which ε_0 is the permittivity of vacuum. $(8.85 \times 10^{-12}\mathrm{C}^2/\mathrm{Nm}^2)$, where 1C (coulomb) = 1As).

Prove that these relationships are dimensionally homogeneous.

*17. Show that John reached the right conclusion in the case of the burnt-down factory discussed at the beginning of this chapter.

Comments on problems

Problem 3

A macroscopic energy balance over the entire pipe yields for the stationary state and without phase transition, $v_1 = v_2$ (constant cross-section of pipe), $h_1 = h_2$ (horizontal pipe), $\phi_{v,\mathrm{in}} = \phi_{v,\mathrm{out}}$ and $\phi_H = 0$ (no heat losses):

$$0 = \phi_v(\Delta p + \rho c_p \Delta T) + \phi_A$$

Now, the power necessary to bring the liquid to the initial pressure of 2×10^5 N/m^2 is (no temperature change, $\Delta T = 0$):

$$\phi_A = -\phi_v \Delta p = -\frac{\phi_m}{\rho}\Delta p = 200 \text{ W}$$

This power is converted into heat in the pipe (where $\phi_A = 0$), thus:

$$\Delta T = -\frac{\Delta p}{\rho c_p} = 0.048\,^{\circ}\mathrm{C}$$

If the water flow is doubled, the pressure drop which is proportional to the momentum ($\Delta p \sim \phi_v v \sim v^2$) is increased by a factor of 4. Thus $\phi_A(\sim \phi_v \Delta p \sim v^3$) is increased by a factor of 8 and $\Delta T(\sim \Delta p \sim v^2$) by a factor of 4.

Problem 6

The drawing shows the situation for the kettle on the fire. We will set up our balances over the entire kettle. During heating up, no mass leaves the kettle, and thus the mass and momentum balances reduce to:

$$V\frac{d\rho_L}{dt} = 0 \text{ and } V\frac{d(\rho v)}{dt} = 0, \text{ respectively}$$

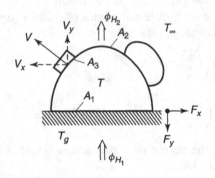

whereas the energy balance becomes:

$$V\frac{d(\rho_L c_p T)}{dt} = \phi_{H_1} - \phi_{H_2} = h_1(T_g - T)A_1 - h_2(T - T_\infty)A_2$$

During boiling, the balances become:

$$\text{mass} : V\frac{d\rho_L}{dt} = -\rho_g v A_3 = -\phi_m$$

$$\text{momentum} : V\frac{d(\rho v)}{dt} = 0 = \sum F$$

where the force in the y-direction is given by $F_y = Mg + \phi_m v_y$, and the force in the x-direction by $F_x = \phi_m v_x$ and:

$$\text{energy} : V\frac{d(\rho_L c_p T)}{dt} = \phi_{H_1} - \phi_{H_2} - \phi_m \Delta H_v - \phi_m c_p T \approx 0$$

(ΔH_v being the heat of evaporation) if we neglect all other forms of energy because they are so small.

Problem 7

A macromomentum balance over the entire droplet will provide the answer that we are looking for. Because the flow velocities are very low we can neglect the momentum of the fluid and our momentum balance becomes a force balance:

$$\sum F = 0$$

Now, the forces acting are the weight force of the droplet and the surface tension of the liquid in contact with the capillary, as shown in the drawing. Thus the force balance becomes:

$$\sum F = 0 = \frac{\pi}{6} d^3 \rho g - 2\pi r \sigma$$

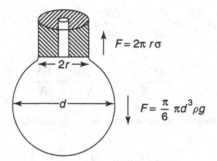

and thus:

$$d = \sqrt[3]{\frac{12 r \sigma}{\rho g}} = 4.4 \text{ mm}$$

Problem 11

This problem is easily solved by application of the law of conservation of matter. The complication here is that the composition (Y) of the vapour phase changes as the distillation proceeds, because of the changing composition of the residue. Apparently, the right thing to do therefore is to set up an unsteady state material balance for a short time interval dt and to integrate the result found over the total process time t_e. The mass balance (equation I.1) for this case becomes (no flow into the system, no production of methanol but variable contents of the evaporator):

$$\frac{d(NX)}{dt} = -\phi_{mol} Y = \frac{dN}{dt} Y$$

where the symbols have the meanings given in the figure. After partial differentiation we can write:

$$\frac{d(NX)}{dt} = N \frac{dX}{dt} + X \frac{dN}{dt}$$

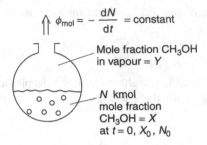

and combining these two equations we obtain:

$$N\frac{dX}{dt} = \frac{dN}{dt}Y - \frac{dN}{dt}X = -\phi_{mol}(Y - X)$$

or:

$$\frac{dX}{Y - X} = -\frac{\phi_{mol}dt}{N} = \frac{dN}{N}$$

which yields after integration between $t = 0$, x_0 and t, x:

$$N = N_0\exp\left(\int_{X_0}^{X}\frac{dX}{Y - X}\right)$$

This equation gives the relationship between the total number of moles in the evaporator and the methanol content of this residue. We are, however, interested in the average composition of the distillate $\langle Y \rangle$ which is given by:

$$\langle Y \rangle = \frac{\int_0^{t_e}\phi_{mol}Y\,dt}{\phi_{mol}t_e} = \frac{-\int_0^{t_e}d(NX)}{N_0 - N} = \frac{-(NX)|_{0,X_0}^{t_e,X_e}}{N_0 - N}$$

$$= \frac{-N_0}{N_0 - N}\left\{X\exp\left(\int_{X_0}^{X}\frac{dX}{Y - X}\right)\right\}\Big|_{X_0}^{X_e} = \frac{X_0 - X_e\exp(\int_{X_0}^{X_e}dX/Y - X)}{1 - \exp\left(\int_{X_0}^{X_e}dX/Y - X\right)}$$

This is the desired result; all that remains is for us to evaluate graphically the integral, as shown in the figure. We find for the area under the curve:

$$-\int_{X_0}^{X_e}\frac{dX}{Y - X} = 1.17$$

and thus:

$$\langle Y \rangle = \frac{0.45 - 0.05\exp(-1.17)}{1 - \exp(-1.17)} = \frac{0.435}{0.69} = 63.1\%$$

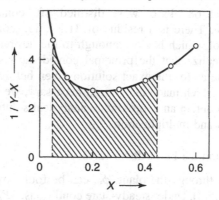

The amount of distillate is found as:

$$N_0 - N = N_0 \left[1 - \exp \left(\int_{X_0}^{X_e} \frac{dX}{Y - X} \right) \right] = 69 \text{ kmol}$$

We see here that sometimes the solution of a simple mass balance can be quite involved. There is, however, a much simpler way to solve this problem approximately by carrying out the distillation stepwise. We start with 100 kmol of a mixture containing a mole fraction of $X_0 = 0.45$ methanol. Let us distil off so much distillate that X drops to $X_1 = 0.35$. Apparently, in this range the average mole fraction of methanol in the vapour phase will be $Y_0 = 0.73$ and thus we can find the total number of moles distilled off from a simple mass balance as follows:

$$X_1 = \frac{N_0 X_0 - (N_0 - N) Y_0}{N_0 - (N_0 - N)} = 0.35 = \frac{100 \times 0.45 - 0.73(N_0 - N)}{100 - (N_0 - N)}$$

We find $N_0 - N = 26.3$ kmol, containing 19.2 kmol of methanol. Now we repeat this exercise as illustrated in the following table until we reach the desired methanol fraction of $X = 0.05$ in the liquid phase.

Total number of kmol in pot N	Mole fraction of methanol in pot X	Mole fraction of methanol in vapour Y	Number of kmol distilled off $N_0 - N$	Number of kmol of methanol distilled off
100	0.45 ⟶ 0.35	0.73	26.3	19.2
73.7	0.35 ⟶ 0.25	0.67	17.6	11.8
56.1	0.25 ⟶ 0.15	0.58	13.0	7.6
43.1	0.15 ⟶ 0.05	0.42	11.6	4.9
31.5				
Total			68.5	43.5

Thus we find that 68.5 kmol were distilled off, containing 43.5 kmol of methanol, i.e. 63.5%. There is a residue of 31.5 kmol, containing $45 - 43.5 = 1.5$ kmol of methanol, which is close enough to the desired 5 mol%.

The reader will realize that the principal considerations in this approximate solution are the same as for the exact solution given before. Instead of infinitesimal steps dX or dt which made integration necessary, we just used larger steps which enabled us to obtain an approximate analytical solution after a reasonable amount of additions and multiplications.

Problem 17   John and the burnt-down factory

The flow rate of air through the plant, ϕ_v, can be found by means of an energy balance (equation (I.5a)). Under steady-state conditions, $dE_t/dt = 0$ and realizing that the change in heat content of the air, i.e. $\rho c_p \Delta T$, is much bigger than the changes in the other forms of energy, we can write:

$$V\frac{dE_t}{dt} = \phi_v \rho c_p \Delta T + \phi_A - \phi_H = 0$$

Now, no mechanical energy is added ($\phi_A = 0$) and the amount of heat supplied equals $\phi_H = \phi_m \Delta H_v$, where ϕ_m is the rate of steam loss. Thus:

$$\phi_v = \frac{\phi_m \Delta H_v}{\rho c_p \Delta T} = \frac{(70 \times 10^3)(2.3 \times 10^6)}{(24)(3600)(1.2 \times 10^3)(30)} = 52 \text{ m}^3/\text{s}$$

If we assume that the air in the building is well mixed, the hexane concentration in the building will equal the concentration in the exit stream. A material balance then yields with $c_{in} = 0$ for the steady state:

$$V\frac{dc}{dt} = -\phi_v c + r = 0$$

where r equals the rate of hexane loss. Thus we find:

$$c = \frac{r}{\phi_v} = \frac{9000}{(24)(3600)(52)} = 0.002 \text{ kg/m}^3$$

and, regarding the hexane vapour as an ideal gas, we find:

$$\rho = \frac{86}{22.4} = 3.83 \text{ kg/m}^3$$

and thus $c = 0.052 \text{vol}\%$, which is indeed well below the explosion limit.

CHAPTER II
Flow Phenomena

In this chapter the transport of momentum is studied and we will discuss the prediction of flow resistance in pipe systems and of the resistance due to obstacles placed in a flow field.

The analysis of pipe flow is divided into two sections, laminar and turbulent flow, the difference being that in the first case shear stresses in the fluid are predictable from velocity gradients (deformation rates) but in the latter they are not because of flow fluctuations. This means that the 'theory' of turbulent pipe flow relies strongly on empirical knowledge; however, this knowledge can be made generally applicable by scientific reasoning.

In stationary flow through pipes of a uniform cross-section, the momentum balance degenerates to a balance between shear forces and pressure forces, but this is no longer the case in flow systems in which the cross-section available for flow changes. Hence, in this case a more general and practical approach will be needed.

A class of flow problems, in which the energy dissipation due to shearing is small in comparison to the amount of mechanical energy which is transformed, e.g. from kinetic energy into potential energy, will be dealt with separately.

Flow around obstacles will not only be extended to flow through beds of particles but also to stirrers. This brings us to the problem of mixing and of the non-uniform distribution of residence time of the fluid elements which pass through a continuous flow system.

In all hydrodynamic theory we will make use of the three fundamental conservation laws which were introduced in Chapter I. The idea of conservation of mass, momentum and energy provides us with five relationships for a three-dimensional flow field, from which the velocity distribution (v_x, v_y and v_z) and the temperature and pressure distributions can be obtained in principle, as soon as:

these quantities are defined at the boundaries of the system and relationships between shear stresses and deformation rates are known.

The whole problem in using the concepts of conservation of mass, energy and momentum in predicting flow fields and flow resistances is to define the control volume in such a way that it facilitates the analysis, and to make clever guesses about the order of magnitude of the several terms in the balances in order to make them amendable for an algebraic treatment. Hence, although the whole theory on transport phenomena does not go beyond the concepts which have already been introduced in Chapter I, the professional engineer needs considerable experience before he is able to turn these concepts into practical use. Therefore, this chapter has been written. It treats a number of problems in an elementary way in order to illustrate the practical application of the basic principles and it provides a number of elementary solutions which can be the starting point to the solution of more complex problems.

II.1 Laminar flow

John looked at the spoon in his hand. Syrup dripped from it into his porridge at a decreasing rate. He thought about writing a criminal story in which someone was imprisoned until the very last drop of syrup would leave the spoon. John had a feeling that this would last very long — months, perhaps, or even years. Being a scientific story writer and working efficiently, he read the following introduction and found out that the thickness of the syrup layer on his spoon decreased about 100 times in 25 minutes. Consequently, he thought, this layer would diminish to molecular dimensions in about 3×10^5 years — a long sentence indeed.

During laminar flow, fluid elements move along parallel streamlines. Consider, for example, the flow of a liquid film (e.g. syrup) on a vertical plane surface. Here the weight of the fluid constitutes the driving force, resulting (if the viscosity of the fluid is not too low) in laminar flow. At the wall, flow velocity will be zero and there will be a velocity gradient over the thickness of the film. Thus, adjacent fluid elements will have different flow velocities, causing shear forces between the fluid elements.

A great number of laminar flow problems can be solved with the aid of the balances discussed in the foregoing chapter if, for the fluids used, the relationship between shear forces and the velocity gradient is known. This relationship reads for unidirectional flow of Newtonian fluids:

$$\tau_{yx} = -v\frac{d(\rho v_x)}{dy} \tag{II.1a}$$

and if the fluids are non-compressible:

$$\tau_{yx} = -\eta\frac{dv_x}{dy} \tag{II.1b}$$

τ_{yx} is the shear force in the x-direction in a plane xz, normal to the y-axis. This relationship is shown, together with two other possible relationships between τ_{yx} and the velocity gradient, in Figure II.1.

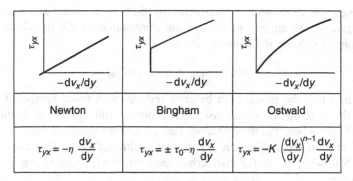

Figure II.1 Relationship between shear force and velocity gradient

To solve a flow problem we start with a momentum balance which, for channels with constant cross-section, takes the form of a force balance, from which we find a relationship between the shear force at any place and the pressure drop. No assumption about the type of liquid or flow has been made until this point. This relationship is always valid for any kind of flow and for any type of rheology. Using the shear force-velocity gradient relationship for the fluid concerned (e.g. equation (II.1a) for Newtonian liquids) we obtain after integration an expression for the velocity distribution in laminar flow. From this velocity profile the mean velocity and the volumetric flow rate can be determined as a function of the driving force. We will illustrate this procedure in the following Sections.

II.1.1 Stationary laminar flow between two flat horizontal plates

Consider a liquid flowing between two parallel plates, as shown in Figure II.2. As the control volume we choose the volume between the plane $y = 0$ (centre of flow field) and the plane $y =$ constant between the two cross-sections '1' and '2'. Under steady-state conditions, the momentum balance over this control volume is given by equation (I.7) as:

$$V\frac{d(\rho v_x)}{dt} = 0 = \phi_{v,\text{in}}\rho v_{x,\text{in}} - \phi_{v,\text{out}}\rho v_{x,\text{out}} + \sum F_x$$

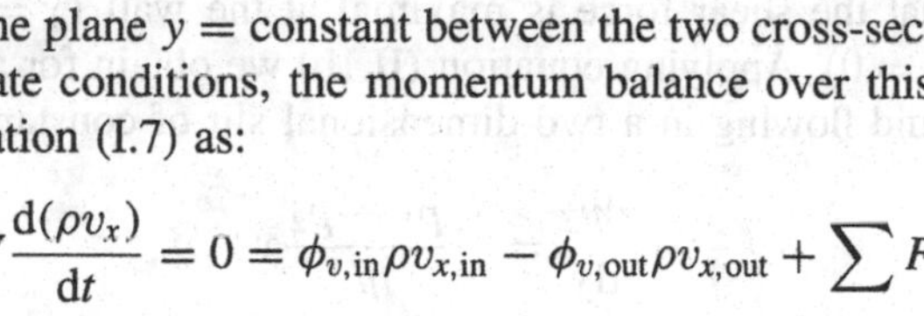

Figure II.2 Flow between two flat plates

Now, since the cross-section available for flow is constant and the liquid is incompressible, the amount of momentum transported by the liquid into and out of the control volume is the same:

$$\phi_{v,\text{in}}\rho v_{x,\text{in}} = \phi_{v,\text{out}}\rho v_{x,\text{out}}$$

and we find that the momentum balance reduces to a force balance $\sum F_x = 0$. This is the case in all situations where compressibility of the liquid can be neglected and where the flow channel has a constant cross-section.

The forces in the x-direction per unit of width acting on the control volume are the following if, at the two cross-sections, the pressures are p_1 and p_2, respectively:

pressure force at '1': $\dfrac{F_x}{\text{width}} = p_1 y$ (positive, because it acts in the positive x-direction)

at '2': $\dfrac{F_x}{\text{width}} = -p_2 y$ (acts in the negative x-direction)

shear force $\dfrac{F_x}{\text{width}} = -\tau_{yx}L$ (negative, because it acts in the negative x-direction)

We can now set up our force balance and we find:

$$\frac{\sum F_x}{\text{width}} = 0 = p_1 y - p_2 y - \tau_{yx}L$$

or

$$\tau_{yx} = \frac{p_1 - p_2}{L}y \qquad (\text{II.2})$$

Thus we see that the shear force is maximal at the wall ($y = \pm d/2$) and zero in the centre ($y = 0$). Applying equation (II.1b) we obtain for a Newtonian non-compressible fluid flowing in a two-dimensional slit of constant width:

$$\frac{dv_x}{dy} = -\frac{p_1 - p_2}{\eta L}y$$

Integrating with the boundary conditions $v_x = 0$ at $y = \pm d/2$ (flow velocity is zero at the wall) we find the velocity distribution:

$$v_x = \frac{p_1 - p_2}{2\eta L}\left(\frac{d^2}{4} - y^2\right) \qquad (\text{II.3})$$

We notice that the velocity profile is parabolic, with the maximum velocity being (at $y = 0$):

$$v_{\max} = \frac{p_1 - p_2}{8\eta L}d^2$$

The flow rate per unit width is found as:

$$\frac{\phi_v}{\text{width}} = \int_{-d/2}^{+d/2} v_x \mathrm{d}y = \frac{p_1 - p_2}{12\eta L}d^3 = d\langle v\rangle \qquad (\text{II.4})$$

and we see that the maximum velocity is given by

$$v_{\max} = \tfrac{3}{2}\langle v\rangle$$

In Figure II.3 the shear force and velocity distribution found for laminar flow in a narrow slit are shown. The reader is asked to work out on his own the solution to a similar problem of a film falling on a vertical plane surface (problem 2 on page 58).

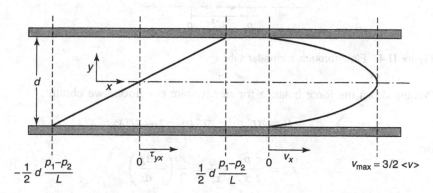

Figure II.3 Shear force and velocity distribution for laminar flow in a narrow slit

We could also have solved the flow problem just discussed by means of the general micromomentum balance developed in I.3. Using equation (I.25) and realizing that there is no flow velocity in the y- and z-directions ($v_y = 0$, $v_z = 0$), no velocity gradient in the x- and z-directions ($\mathrm{d}v_x/\mathrm{d}x = 0$, $\mathrm{d}v_x/\mathrm{d}z = 0$), no shear force in the x- or z-directions ($\mathrm{d}\tau_{xx}/\mathrm{d}x = 0$, $\mathrm{d}\tau_{zx}/\mathrm{d}z = 0$) and that the plates are horizontal (thus $\rho g_x = 0$), we find for steady-state conditions ($\mathrm{d}v_x/\mathrm{d}t = 0$):

$$-\frac{\mathrm{d}(\tau_{yx})}{\mathrm{d}y} - \frac{\mathrm{d}p}{\mathrm{d}x} = 0$$

or, after integration:

$$\tau_{yx} = \frac{\mathrm{d}p}{\mathrm{d}x}y + \text{constant}$$

which is identical with equation (II.2) if we realize that there is no shear force at the centre of the channel (at $y = 0$) and hence that the integration constant must be zero.

II.1.2 Stationary laminar flow through a horizontal circular tube

To solve this problem we can use the same approach as that used in section II.1.1. As a control volume we choose a cylindrical plug of length L and radius r as shown in Figure II.4. The forces acting on the control volume are again pressure forces acting on the cross-sections and a shear force acting at the cylindrical surface in the negative x-direction.

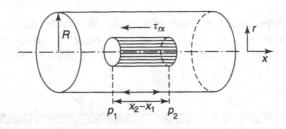

Figure II.4 Flow through a circular tube

Writing down the force balance for steady-state conditions we obtain:

$$\sum F_x = 0 = \pi r^2 p_1 - \pi r^2 p_2 - \tau_{rx} 2\pi r(x_2 - x_1)$$

or:

$$\tau_{rx} = \frac{r}{2} \frac{p_1 - p_2}{x_2 - x_1} = \frac{r}{2} \left(-\frac{\mathrm{d}p}{\mathrm{d}x} \right) \qquad (II.5)$$

This relationship between shear stress and pressure drop is generally valid (there has been no assumption about the type of flow), as was the case with equation (II.2). The shear stress distribution is linear: $\tau_{rx} = 0$ at the centre (where therefore the velocity gradient is also zero) and maximum at the wall.

The velocity distribution for laminar flow of a Newtonian liquid is found by eliminating τ_{rx} in equation (II.5) by using equation (II.1b) and by integrating the differential equation obtained with $v_x = 0$ at the wall ($r = R$). The result found is:

$$v_x = \frac{1}{4\eta} \left(-\frac{\mathrm{d}p}{\mathrm{d}x} \right) (R^2 - r^2) \qquad (II.6a)$$

Thus, again, the velocity distribution is parabolic, with maximum flow velocity at $r = 0$:

$$v_{\max} = \frac{1}{4\eta} \left(-\frac{\mathrm{d}p}{\mathrm{d}x} \right) R^2$$

The flow rate at a given pressure gradient is found from the velocity distribution as:

$$\phi_v = \int_0^R 2\pi r v_x \mathrm{d}r = \frac{\pi R^4}{8\eta} \left(-\frac{\mathrm{d}p}{\mathrm{d}x} \right) = \pi R^2 \langle v \rangle \qquad (II.7a)$$

Combining equations (II.6) and (II.7) we find for the velocity distribution:

$$v_x = 2\langle v \rangle \left(1 - \frac{r^2}{R^2} \right) = \frac{R^2}{4\eta} \left(-\frac{dp}{dx} \right) \left(1 - \frac{r^2}{R^2} \right) \tag{II.8}$$

and we see that in the case of flow through a circular tube:

$$v_{max} = 2\langle v \rangle$$

Equation (II.7) is the Hagen-Poiseuille relationship which (with some corrections) is used in various methods for the measurement of viscosities of Newtonian liquids. In Figure II.5A the shear stress and the velocity distribution found are illustrated.

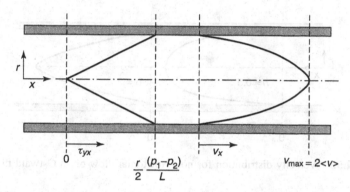

Figure II.5A Shear force and velocity distribution for Newtonian laminar flow in a circular pipe

The velocity gradient at the wall can be found by differentiating equation (II.8) and by determining the value of the first derivative for $r = R$. We obtain:

$$\left. \frac{dv_x}{dr} \right|_{r=R} = -\frac{4\langle v \rangle}{R} = \frac{-R}{2\eta} \left(-\frac{dp}{dx} \right) = -\frac{\tau_w}{\eta} \tag{II.9}$$

where τ_w is the shear stress at the wall. We will need this result later on, when we develop equations to predict pressure drops during flow of liquids in pipes. The reader may check that equation (II.5) could also have been obtained from the appropriate microbalance, equation (I.30), and that equation (II.6) can be derived from equation (I.37).

If we have an Ostwald fluid instead of a Newtonian fluid, the relationship between shear force and velocity gradient reads (see Figure II.1):

$$\tau_{rx} = K \left| \frac{dv_x}{dr} \right|^n$$

as dv_x/dr is positive at all values of r. Inserting this equation into equation (II.5), and integrating with the boundary condition $v_r = 0$ at $r = R$, leads to:

$$v_x = \frac{n}{n+1}\left[\frac{1}{2K}\left(-\frac{dp}{dx}\right)\right]^{\frac{1}{n}}\left(R^{\frac{n+1}{n}} - r^{\frac{n+1}{n}}\right) \tag{II.6b}$$

For $n = 1$ and $K = \eta$, we have the Newtonian case (equation (II.6a)). Velocity distributions for $n = 0.5$ and $n = 2$ are shown in Figure II.5B. From the velocity distribution we find for the flow rate:

$$\phi_v = \frac{\pi n}{3n+1}\left[\frac{1}{2K}\left(-\frac{dp}{dx}\right)\right]^{\frac{1}{n}} R^{\frac{3n+1}{n}} = \pi R^2 \langle v \rangle \tag{II.7b}$$

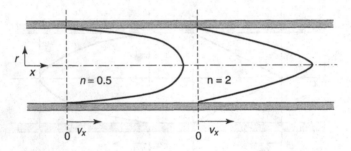

Figure II.5B Velocity distribution for non-Newtonian flow of an Ostwald fluid

The reader is asked here to calculate the relationship between the average and maximum velocities.

A fluid for which $n < 1$ is called a pseudoplastic fluid, while one for which $n > 1$ is called a dilatant fluid.

II.1.3 Stationary laminar flow through a horizontal annulus

In many cases it is not possible to define directly a macroscopic momentum balance as used in the previous two examples. In such cases a microscopic momentum balance must be developed. Consider the annulus shown in Figure II.6. As a control volume we choose a streamline with radius r, thickness dr and length L. We can then set up the following force balance over the control volume:

$$\sum F_x = 0 = 2\pi r\, dr\, p_1 - 2\pi r\, dr\, p_2 + \tau_{rx}2\pi r L|_r - \tau_{rx}2\pi r L|_{r+dr}$$

Dividing by $2\pi L\, dr$ we obtain:

$$\frac{p_1 - p_2}{L}r = \frac{r\tau_{rx}|_{r+dr} - r\tau_{rx}|_r}{dr}$$

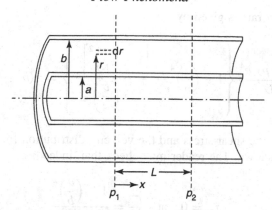

Figure II.6 Laminar flow in an annulus

The right-hand side of this equation is the mathematical definition of the first derivative of $r\tau_{rx}$, so we can write:

$$\frac{d(r\tau_{rx})}{dr} = \frac{p_1 - p_2}{L} r$$

and integrating this equation we obtain:

$$\tau_{rx} = \frac{p_1 - p_2}{2L} r + \frac{C_1}{r} \tag{II.10}$$

Applying Newton's law (equation II.1b) and integrating again we obtain:

$$-v_x = \frac{p_1 - p_2}{4\eta L} r^2 + \frac{C_1}{\eta} \ln r + C_2 \tag{II.11}$$

In this equation the constants C_1 and C_2 still have to be evaluated. We can do this by applying the boundary conditions $v_x = 0$ at $r = a$ and at $r = b$. We then find for the shear stress distribution (from equation II.10):

$$\tau_{rx} = \frac{p_1 - p_2}{2L} \left[r - \frac{1 - \left(\frac{a}{b}\right)^2}{2 \ln \left(\frac{b}{a}\right)} \frac{b^2}{r} \right] \tag{II.12}$$

and for the velocity distribution:

$$v_x = \frac{p_1 - p_2}{4\eta L} b^2 \left[1 - \left(\frac{r}{b}\right)^2 + \frac{1 - \left(\frac{a}{b}\right)^2}{\ln \left(\frac{b}{a}\right)} \ln \left(\frac{r}{b}\right) \right] \tag{II.13}$$

whereas the flow rate is given by:

$$\phi_v = \frac{\pi(p_1 - p_2)b^4}{8\eta L} \left\{ 1 - \left(\frac{a}{b}\right)^4 - \frac{\left[1 - \left(\frac{a}{b}\right)^2\right]^2}{\ln\left(\frac{b}{a}\right)} \right\} = \pi(b^2 - a^2)\langle v \rangle \quad \text{(II.14)}$$

In Figure II.7 the shear stress and the velocity distribution for laminar flow in an annulus are shown. The reader may check that, indeed:

$$\tau_{rx} = 0 \quad \text{at} \quad \frac{r^2}{b^2} = \frac{1 - \left(\frac{a}{b}\right)^2}{2\ln\left(\frac{b}{a}\right)}$$

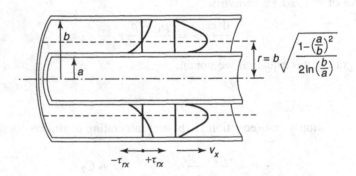

Figure II.7   Shear stress and velocity distribution for laminar flow in a horizontal annulus

and:

$$v_x = v_{max} \text{ at the same value of } r^2/b^2$$

For $a = 0$, equations (II.12), (II.13) and (II.14) reduce to the corresponding relationships for flow through a circular tube. For a very narrow annulus, i.e. for $a = b - d(d \ll a)$ and $r = a + y(y \ll a)$, the corresponding relationships for flow in a narrow slit between two flat plates are obtained (see problem 1, p. 58).

The shear stress and the velocity distribution for flow in a circular pipe can, of course, also be found from equation (II.11) by applying the proper boundary conditions $v_x = 0$ at $r = R$ and $v_x \neq \infty$ at $r = 0$. The last boundary condition leads to $C_1 = 0$, which we would have also found from equation (II.10) by realizing that at $r = 0$, $\tau_{rx} \neq \infty$. The reader may check that equation (II.10) can also be derived from the proper micromomentum balance (equation (I.30)).

Returning to laminar flow in an annulus it is interesting to compare the shear forces at the inner and outer walls. From equation (II.12) we find the forces at the walls to be:

at inner wall: $F_a = -2\pi aL\tau_w|_a = -\pi a(p_1 - p_2)\left[a - \dfrac{1 - \left(\dfrac{a}{b}\right)^2}{2\ln\left(\dfrac{b}{a}\right)}\dfrac{b^2}{a}\right]$

at outer wall: $F_b = 2\pi bL\tau_w|_b = \pi b(p_1 - p_2)\left[b - \dfrac{\left(1 - \dfrac{a}{b}\right)^2}{2\ln\left(\dfrac{b}{a}\right)}\dfrac{b^2}{b}\right]$

The sum of the two forces is then (τ_{ax} points in the negative x-direction!):

$$F_a + F_b = -2\pi aL\tau_w|_a + 2\pi bL\tau_w|_b = \pi(b^2 - a^2)(p_1 - p_2)$$

i.e. the total shear force at both walls equals the pressure force on the cross-section of the annulus, as should be the case.

The ratio of the forces at the inner and outer wall is:

$$\frac{F_a}{F_b} = \frac{-a\tau_{ax}}{b\tau_{bx}} = \frac{\left[\dfrac{b^2/a^2 - 1}{2\ln(b/a)}\right] - 1}{b^2/a^2 - \left[\dfrac{b^2/a^2 - 1}{2\ln(b/a)}\right]}$$

This relationship is shown graphically in Figure II.8. We see that for the limiting case $b/a \longrightarrow 1$ the two forces become equal. The ratio of the force on the inner cylinder to the total force is given by:

$$\frac{F_a}{F_a + F_b} = \frac{-a\tau_{ax}}{-a\tau_{ax} + b\tau_{bx}} = \frac{1}{2\ln(b/a)} - \frac{1}{(b/a)^2 - 1}$$

This relationship is also shown in Figure II.8. If the ratio of the surfaces of the outer to the inner cylinder is 5:1, then 27 % of the total force still acts on the smaller cylinder! Considerations of this kind are of importance during the extrusion of, for example, plastic pipes.

II.1.4 Flow caused by moving surfaces

The foregoing sections have dealt with the flow of liquids caused by a pressure gradient over a length of pipe. Flow can also be caused by moving a surface through a liquid.

Let us consider what would happen if, in an infinitely long horizontal annulus filled with a Newtonian liquid, the inner tube was moving in the x-direction

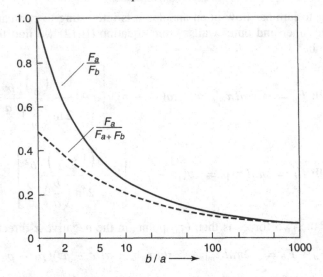

Figure II.8 Forces acting on the walls of an annulus

(Figure II.9) at a constant velocity v_0. Since there is no pressure gradient induced from the outside, the force balance under stationary conditions becomes for this case:

$$\sum F_x = 0 = \tau_{rx} 2\pi r L|_r - \tau_{rx} 2\pi r L|_{r+dr}$$

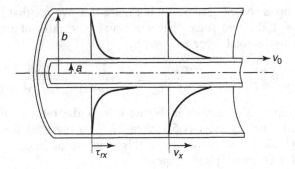

Figure II.9 Horizontal annulus with moving inner cylinder

from which we find:

$$\frac{d(r\tau_{rx})}{dr} = 0$$

Integration, application of Newton's law (equation II.1b), second integration and determination of the two constants of integration via the boundary conditions

$(r = b, v_x = 0$ and $r = a, v_x = v_0)$ leads to:

$$\tau_{rx} = \frac{\eta v_0}{r} \frac{1}{\ln(b/a)} \tag{II.15}$$

and:

$$v_x = v_0 \frac{\ln(b/r)}{\ln(b/a)} \tag{II.16}$$

The first relationship shows that the forces on both surfaces are equal, as should be the case:

$$F_a = \pi a \tau_{ra} L = F_b = \pi b \tau_{rb} L = \frac{\pi \eta v_0}{\ln(b/a)} L$$

In Figure II.10 both relationships are shown graphically. The reader may check that for $b/a \longrightarrow 1$ both expressions reduce to the proper relationships for two parallel flat plates (see also problem 4, p. 59).

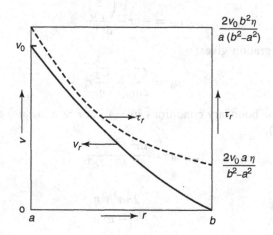

Figure II.10 Annulus with inner moving cylinder

A different situation occurs if the inner cylinder is not moving in the x-direction but is rotating at a rotational speed $n(s^{-1})$. This case can easily be treated by applying cylindrical coordinates (Figure II.11).

A momentum balance for the torque (M) leads to:

$$M = 0 = 2\pi r L r \tau_{r\theta}|_r - 2\pi r L r \tau_{r\theta}|_{r+dr}$$

from which follows:

$$\frac{d(r^2 \tau_{r\theta})}{dr} = 0$$

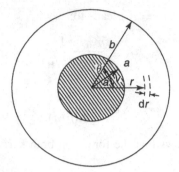

Figure II.11 Shear stress and velocity distribution in an annulus with inner rotating cylinder

Integration of this equation yields, after application of Newton's law, which reads in cylindrical coordinates for non-compressible fluids:

$$\tau_{r\theta} = -\eta r \frac{d\left(\dfrac{v_\theta}{r}\right)}{dr}$$

and second integration gives:

$$v_\theta = \frac{C_1}{2\eta r} - \frac{C_2}{\eta} r$$

and, applying the boundary conditions $v_\theta = v_0$ at $r = a$, $v_\theta = 0$ at $r = b$, we find (using $v_0 = na$):

$$v_\theta = \frac{na^2}{r} \frac{b^2 - r^2}{b^2 - a^2} \tag{II.17}$$

and:

$$\tau_{r\theta} = \frac{2na^2 b^2 \eta}{(b^2 - a^2) r^2} \tag{II.18}$$

As should be the case, the torque at the inner and at the outer wall $M = 2\pi r \tau_r r L$ is the same.

The above situation is present, for example, in friction bearings. A similar situation occurs in a rotary viscosimeter. Here the inner cylinder is immobile and the outer cylinder rotates. The annulus is filled with liquid and the force on the inner cylinder is measured (e.g. by means of the torque of a wire). The reader may check that the viscosity of the liquid in the annulus is given by:

$$\eta = \frac{M}{4\pi n L} \frac{b^2 - a^2}{a^2 b^2}$$

if M is the torque, L the height of the cylinder and if effects caused by the bottom of the annulus are neglected.

The reader is asked (problem 4) to solve by himself the analogous simple problem of two parallel moving plates.

II.1.5 Flow through pipes with other cross-sections

The calculation of velocity profiles for pipes with non-circular cross-sections is difficult but possible. Boussinesq developed analytical solutions for the velocity profiles for Newtonian liquids in pipes with elliptic, triangular and square cross-sections. These results can be summarized by defining a flow coefficient M_0 for which the volumetric flow rate for flow through a slit with the same smallest dimension d has to be corrected. For flow between two parallel plates we have developed equation (II.4), which reads, if B represents the width of the slit:

$$\phi_v = \frac{\Delta p}{12\eta L} B d^3$$

Thus, for other cross-sections the flow rate is given by:

$$\phi_v = \frac{\Delta p}{12\eta L} B d^3 M_0 \qquad (\text{II.19})$$

Values for the correction factor M_0 can be found from Figure II.12.

II.1.6 Non-stationary flow

Until now we have concentrated on stationary (steady-state) situations, i.e. on situations which were independent of the elapsed time. In this section we will consider what happens in the time between starting a flow and reaching the stationary state (transient behaviour) and how long it takes to obtain a steady-state situation.

Consider the situation illustrated in Figure II.13. From $t = 0$ a large flat horizontal plate is moved with a constant velocity v_0. At $t = 0$, the liquid is stagnant ($v_\infty = 0$). At $t > 0$, momentum transport in the $\pm y$-direction takes place, resulting in a velocity distribution which is not only a function of the y-coordinate but also of time.

To analyze this problem we select a control volume parallel to the plate and of height dy. The micromomentum balance over this height dy then reads:

$$\frac{\partial(\rho v_x)}{\partial t} dy = \tau_{yx}|_y - \tau_{yx}|_{y+dy} = -\frac{\partial \tau_{yx}}{\partial y} dy$$

and, after applying Newton's law (equation II.16) we obtain:

$$\frac{\partial v_x}{\partial t} = \nu \frac{\partial^2 v_x}{\partial y^2}$$

The same equation could have been obtained from the general micromomentum balance, equation (I.30). This differential equation is analogous to the equations

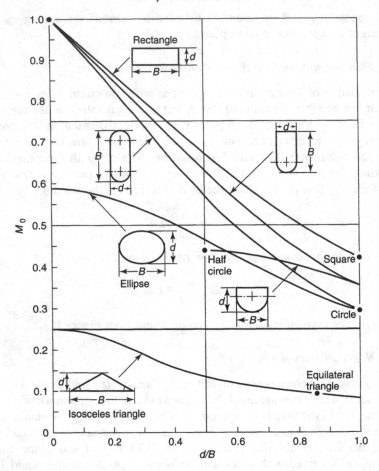

Figure II.12 The flow coefficient M_0 for stationary flow of Newtonian liquids in cross-sections of various geometries

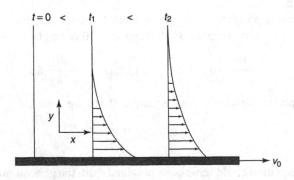

Figure II.13 Horizontal flat plate in an infinite viscous liquid

describing the penetration of heat and mass:

$$\frac{\partial T}{\partial t} = a\frac{\partial^2 T}{\partial x^2} \quad \text{and} \quad \frac{\partial c}{\partial t} = \mathbb{D}\frac{\partial^2 c}{\partial x^2}, \text{ respectively}$$

which will be treated in more detail in Chapters III and IV. There, also, the exact solution of this differential equation will be discussed.

In Figure II.14 the velocity distribution for the system of Figure II.13 is shown for two short times. As can be seen, the distribution can be approximated by straight lines which cut the abscissa at $y = \sqrt{\pi v t}$. This distance, $\sqrt{\pi v t}$, is called the penetration depth, i.e. the distance over which momentum has penetrated. This solution is applicable as long as the penetration depth is smaller than the depth of the liquid.

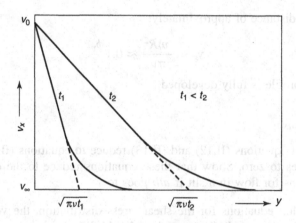

Figure II.14 Velocity distribution for the system of Figure II.13

We can now estimate the velocity gradient as:

$$\frac{\partial v_x}{\partial y} = \frac{v_\infty - v_o}{\sqrt{\pi v t}}$$

or, since $v_\infty = 0$:

$$\tau_w = -\eta\frac{\partial v_x}{\partial y} = \frac{v_o\eta}{\sqrt{\pi v t}} \tag{II.20}$$

A somewhat different situation occurs if, instead of a moving plate, liquid with an initial velocity v_o flows along a fixed plate. This situation will be discussed in Section II.4 (see Figure II.37).

We can now estimate the distance needed for laminar pipe flow to develop its velocity distribution completely (Figure II.15). If we move with the liquid and assume that the velocity distribution will be developed completely when the

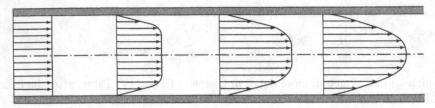

Figure II.15   Development of the velocity profile in a tube

penetration depth at $y = y_s$ is equal to the distance R to the centre of the channel, we can write:

$$R \approx \sqrt{\pi v t} = \sqrt{\pi v \frac{x_s}{\langle v \rangle}}$$

Thus, after a distance of approximately:

$$y_s \approx \frac{\langle v \rangle R^2}{\pi v} \approx 0.1 \frac{\phi_v}{v} \tag{II.21}$$

the velocity profile is fully developed.

II.1.7 Problems

1. Show that equations (II.12) and (II.13) reduce to equations (II.5) and (II.6a) if a/b goes to zero. Show that these equations reduce to the corresponding expressions for flow in a slit if a/b goes to 1.

2. (a) Develop equations for the shear stress distribution, the velocity profile and the volumetric flow rate for a liquid film flowing down on a vertical surface. Use the notation given in Figure II.16 (width of plate $= W$).

   Answer: $\tau_{xy} = -\rho g(d - x)$

   $$v_y = \frac{g\rho}{\eta} \left(dx - \frac{1}{2}x^2 \right)$$

   $$\frac{\phi_v}{W} = \langle v_y \rangle d = \frac{\rho g d^3}{3\eta}$$

   (b) On a vertical plane (width $W = 1$ m) 1200 kg/h water flows downward. What is the mean velocity, the film thickness and the maximum velocity?

   Answer: $\langle v \rangle = 0.725$ m/s

   $$d = 0.46 \times 10^{-3} \text{ m}$$

   $$v_{\max} = 1.09 \text{ m/s}$$

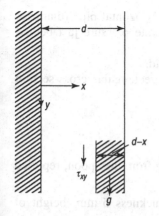

Figure II.16 Laminar film on a vertical plate

*3. In a rotary viscosimeter filled with paint (a Bingham liquid) the shear veloc-
ities are measured as a function of shear force:

Shear force (N/m²)	Shear velocity (s⁻¹)
15	6
18	12
21	18

If this paint is smeared on a vertical wall, how thick can the layer be before
the paint starts to drip ($\rho = 1200$ kg/m³)?

Answer: $d < 10^3$m

4. The slit (width 1 mm) between two horizontal flat plates is filled with a
viscous Newtonian liquid ($\eta = 100$ Ns/m², $\rho = 10^3$ kg/m³). If the surface of
the plates is 10^{-2} m², which force has to be applied under steady-state conditions to one plate in order to move it parallel to the other plate at a velocity
of 1 cm/s? Calculate and draw the shear force and the velocity distribution
between the plates. After what time is the stationary state reached?

Answer: $F = 10$ N, $t \approx 3 \times 10^{-6}$ s

5. The 'viscosity' of a non-Newtonian liquid is approximately given by:

$$\eta = \eta_0 \left| \frac{dv_x}{dr} \right|$$

where η_0 is a constant. Find the volumetric flow rate of this liquid in a
horizontal pipe (diameter $2R$) under laminar and isothermal conditions.

Answer:

$$\phi_v = \frac{2\pi}{7} R^{\frac{7}{2}} \sqrt{\frac{\Delta \rho}{2\eta_0 L}}$$

6. Viscous oil ($\eta = 0.1$ Ns/m^2) is pumped through a horizontal pipe (diameter 1 cm, length 10 m) into an open vessel. The absolute pressure just behind the pump is 3×10^5 N/m^2.
   (a) Determine the volumetric flow rate of the liquid.
   (b) What would be the flow rate if a pipe with rectangular cross-section (0.5 × 2.0 cm) was used?

   Answer: (a) $\phi_v = 4.9 \times 10^{-5}$ m^3/s

             (b) $\phi_v = 3.47 \times 10^{-5}$ m^3/s

*7. Check John's calculations about the syrup dripping from the spoon, reported at the beginning of this section.

   Assume: $\eta = 1000$ cP, $\rho = 2000$ kg/m^3, initial thickness 5 mm, height of spoon 5 cm

Comments on problems

Problem 3

For a liquid with Bingham characteristics the following relationship between shear force and velocity gradient is valid if we use the coordinates shown in the following drawing:

$$\tau_{xy} = \tau_0 - \eta \frac{dv_y}{dx}$$

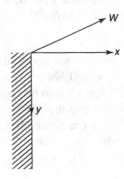

The value of the yield value τ_0 for this paint can be determined from the experimental results as $\tau_0 = 12$ N/m^2.

A force balance will lead to the maximum possible layer thickness before dripping takes place:

$$\rho g h d W < \tau_0 h W$$

and thus:

$$d < \frac{\tau_0}{g\rho}$$

and we find $d < 10^{-3}$ m.

Problem 7 John and his spoon

We are dealing with a situation where a liquid film is flowing down from a vertical surface. An expression for ϕ_v was found in problem 2 of this section:

$$\phi_v = \frac{\rho g d^3 W}{3\eta} = -WH\frac{dd}{dt}$$

Rearrangement leads to:

$$\int_{d=d_0}^{d=d} \frac{dd}{d^3} = \frac{\rho g}{3\eta H}\int_{t=0}^{t=t} dt$$

and after integration:

$$-\frac{1}{2}\left[\frac{1}{d_0^2} - \frac{1}{d^2}\right] = \frac{\rho g t}{3\eta H}$$

or:

$$\frac{d}{d_0} = \sqrt{\frac{1}{1 + \frac{2\rho g t d_0^2}{3\eta H}}}$$

For long times:

$$\frac{2\rho g t d_0^2}{3\eta H} \gg 1$$

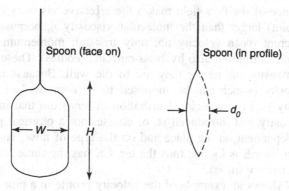

Spoon (face on)   Spoon (in profile)

and we can simplify the above relationship to:

$$\frac{d}{d_0} = \sqrt{\frac{3\eta H}{2\rho g t d_0^2}}$$

For $d/d_0 = 0.01$ we find $t = 1500$ s $= 25$ min. The period to diminish the syrup layer to molecular dimensions is found by substitution of

$$\frac{d}{d_0} = \frac{5 \times 10^{-10}}{5 \times 10^{-3}} = 10^{-7}$$

as 15×10^{12} s.

II.2 Turbulent flow

> *John stood at the edge of a circular sewer channel of 4 m diameter and contemplated that the dead body had been found floating in the channel 7.2 km downstream, in a fishing village, at 11.45 pm. The suspect had a watertight alibi for the time before 9.15 pm. The channel was now half-filled and a piece of wood passed John at a speed of 1 m/s. One of the sewage plant operators had informed him that until the time the body was found the water height had been less than one-quarter of the diameter of the channel. Thinking about one of his college courses, of which you will find a brief outline below, he made a quick calculation and ordered that the suspected man be set free.*

II.2.1 Turbulent flow in pipes

In a well-known experiment in which he injected a filament of dye into a fluid, Reynolds demonstrated that above a certain flow velocity in a pipe the flow changes from laminar into turbulent. In this case, velocity fluctuations occur with components perpendicular to the main direction of the flow.

The turbulence of the flow field makes the effective viscosity (from a macroscopic viewpoint) larger than the molecular viscosity η, because neighbouring layers of different mean velocity not only transmit momentum to each other by molecule interaction but also by cross-currents (eddies). These cross-currents decrease in intensity the nearer they are to the wall. Because of this varying effective viscosity (which can be supposed to be composed of the η and the 'eddy' viscosity ρE) the velocity distribution differs from that in laminar flow. The eddy viscosity just introduced is, of course, not a physical property of the liquid but is dependent on the place and on the type of flow. The dimension of the eddy viscosity ρE is kg/ms; thus the term E has the same dimension as the kinematic viscosity ν (m^2/s).[†]

Figure II.17 shows an example of the velocity profile in a pipe with a circular cross-section for laminar (parabolic) and turbulent flow (more flattened).

[†] The scale of the eddies, which dissipate most of the energy, λ_K, as discussed in Section I.5 (equation (I.48)), differs in this case because turbulence in pipe flow is non-homogeneous (ε differs locally) and not isotropic (directionally invariant).

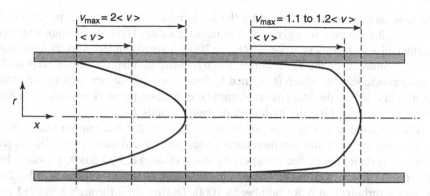

Figure II.17 Laminar and turbulent velocity profiles in a pipe

If we want to know something about the frictional force exerted on the wall by a liquid flowing at a velocity $\langle v_x \rangle$, the shear force τ_w at the wall must be known. Since for incompressible Newtonian liquids:

$$\tau_w = \eta \left.\frac{dv_x}{dr}\right|_{r=R}$$

holds (no turbulence at the wall, ρE at wall $= 0$), we must know the velocity distribution in order to calculate τ_w. For turbulent flow this is not possible by momentum balances, because with turbulence their solution presents insurmountable difficulties. However, on the basis of physical intuition it can be postulated that:

$$\tau_w = f(\rho, \eta, \langle v_x \rangle, D, \text{ geometry})$$

from which, with dimensional analysis, it follows that the quotient of τ_w and $\rho \langle v_x \rangle^2$ for geometrically similar pipes only depends on the Reynolds number $(= \rho \langle v_x \rangle D / \eta)$. This is usually written as:

$$\tau_w = f \tfrac{1}{2} \rho \langle v_x \rangle^2 \qquad \text{(II.22)}$$

With this equation the factor f, the so-called friction factor is defined (compare equation (I.14)); on account of the above it is a function of the geometry and of Re. For turbulent flow this function can be obtained only from experiments on friction losses in pipes and channels. Conversely, these friction losses are, in practice, calculated by using this friction factor f, the value of which is well known for widely divergent conditions (see Section II.4.).

Equation (II.22) can be read as follows: the quotient of τ_w and $\rho \langle v_x \rangle^2$ which is fully determined by the velocity distribution is for geometrically similar pipes only a function of Re. This means that in geometrically similar pipes at the same value of Re, the velocity distribution in corresponding cross-sections must be similar. The two flow fields are then called dynamically similar. Consequently, the flow pattern is characterized by the Re number. The change from laminar to turbulent flow is therefore given by a critical Re value. The flow through a straight circular pipe is turbulent if Re > $\text{Re}_{cr} \simeq 2000$. This also applies to straight, non-circular pipes if, in the Re number, the hydraulic diameter is introduced as the characteristic length, which is defined as four times the surface area of the cross-section divided by the wet circumference (see, further, the next section, and check that for a circular pipe the hydraulic diameter equals D).

The Re number can be interpreted as the ratio of the momentum transport by convection ($\sim \rho v^2$) and the momentum transport by diffusion ($\sim \eta v/D$). Apparently, in turbulent flow the transport by convection, i.e. by eddies, prevails. In a free-falling liquid film these eddies already occur at lower velocities; film flow becomes turbulent at a Re number $\simeq 1000$. During flow through a curved pipe (e.g. a spiral; radius of spiral R) the centripetal forces stabilize the flow and the change to turbulence does not occur until higher Re numbers have been attained. Empirically it was found for the critical Re number:

$$\text{Re}_{cr} = 20\,000 \left(\frac{D}{2R}\right)^{0.32}, \quad \text{for } \frac{2R}{D} \leq 10^3 \tag{II.23}$$

The fact that turbulence decreases towards the wall has led to the concept of the laminar boundary layer. No definite thickness can be assigned to such a layer because the turbulence changes gradually. For a qualitative insight it can be useful to schematize the flow field by a laminar boundary layer, with thickness δ_h, in which the velocity gradient is constant, and an adjacent turbulent flow region, with the mean velocity $\langle v_x \rangle$, in which the velocity gradient is practically zero. The shear force at the wall is then given by:

$$\tau_w = \eta \frac{\langle v_x \rangle}{\delta_h} \tag{II.24}$$

and the thickness δ_h can be estimated with the help of equation (II.22) as soon as the friction factor is known.

With the aid of the laminar boundary layer thickness δ_h we can give a new interpretation to the physical meaning of the Re number. The ratio of the tube diameter and thickness of the laminar boundary layer is found to be:

$$\frac{D}{\delta_h} = \frac{D\tau_w}{\eta \langle v_x \rangle} = \frac{fD\rho \langle v_x \rangle}{2\eta} = \frac{f}{2}\text{Re} \tag{II.25}$$

This means that for pipe flow at Re = 10^5 ($f = 0.0045$) the distance from the wall at which turbulent momentum transport overrides the transport by viscosity is $\delta_h \approx 4 \times 10^{-3}D$.

A velocity distribution as given above is physically impossible, but it can serve as a rough model for treating heat and mass transfers in turbulent flow. In the case of heat transfer, a layer of thickness δ_T is assumed over which the complete temperature drop occurs. Outside this layer the turbulent eddies disperse the heat so efficiently that a uniform temperature exists in the core of the flow. Within the layer, eddies are assumed to be absent. This layer of thickness δ_T is called the thermal boundary layer, analogous to the hydrodynamic boundary layer (thickness δ_h) treated above. It will be clear that both models do not contradict each other as long as $\delta_T \leq \delta_h$.

The question now arises as to whether a relationship exists between the thicknesses of the hydrodynamic and the thermal boundary layers. Some insight into this problem can be gained from the fact that the eddy diffusivity E increases with the third power of the distance from the pipe wall:

$$E = C_1 y^3 \tag{II.26}$$

From the definition of δ_h, this means that eddy momentum transfer overrides viscous transfer if:

$$\rho E = \rho C_1 y^3 \geq \eta \quad \text{or} \quad E = C_1 y^3 \geq \eta/\rho = v$$

Hence:

$$C_1 \delta_h^3 = v \tag{II.27}$$

The diffusivity of heat by eddies is also described by equation (II.26), whereas the diffusivity of heat by conduction only is given by $a = \lambda/\rho c_p$ (m²/s). Hence, the distance δ_T from the pipe wall at which heat transfer by eddies overtakes conductive heat transfer, is given, by a reasoning similar to that applied above, as:

$$C_1 \delta_T^3 = a \tag{II.28}$$

Hence, the sought relationship between δ_T and δ_h is obtained from equations (II.27) and (II.28) and is found to be:

$$\frac{\delta_h}{\delta_T} = \left(\frac{v}{a}\right)^{\frac{1}{3}} = \text{Pr}^{\frac{1}{3}} \tag{II.29}$$

where Pr is named the Prandtl number. For gases, $\text{Pr} \approx 1$ and therefore $\delta_h \approx \delta_T$, for viscous liquids, $\text{Pr} > 1000$ and $\delta_h > 10\delta_T$, and for water at room temperature, $\text{Pr} \approx 7$ and $\delta_h \approx 2\delta_T$.

The reader can imagine that mass transfer between a wall and turbulent pipe flow can be depicted by using a concentration boundary layer of thickness δ_c. The ratio between this thickness and the thickness of the hydrodynamic boundary layer will then be (see also Section IV.3.3):

$$\frac{\delta_h}{\delta_c} = \left(\frac{v}{\mathbb{D}}\right)^{\frac{1}{3}} = \text{Sc}^{\frac{1}{3}} \tag{II.30}$$

where Sc is the Schmidt number. For gases, $Sc \approx 1$ and $\delta_h \approx \delta_c$, and for liquids, $Sc \approx 1000$ and $\delta_h \simeq 10\delta_c$. We see that in these cases the thickness of the hydrodynamic boundary layer is greater than those of the respective thermal and concentration boundary layers, i.e. $\delta_h > \delta_T, \delta_c$.

We have already stated at the beginning of this chapter that it is impossible to derive theoretically a relationship for the velocity distribution during turbulent flow. We can, however, construct a relationship between velocity and the friction factor and see whether we can determine the coefficients in this by experimental results. Analogously to the development of equation (II.22), we assume that the mean velocity is a function of the shear stress at the wall, the density and the kinematic viscosity of the fluid and of the pipe diameter; so we obtain:

$$\langle v \rangle = f(\tau_w, \rho, \nu, D)$$

Via dimensional analysis we find:

$$\frac{\langle v \rangle}{\sqrt{\dfrac{\tau_w}{\rho}}} = f\left(\sqrt{\frac{\tau_w}{\rho}}\frac{D}{\nu}\right) = C\left(\sqrt{\frac{\tau_w}{\rho}}\frac{D}{\nu}\right)^n \tag{II.31}$$

which can be rearranged to give:

$$\frac{\tau_w}{\rho\langle v \rangle^2} = \frac{f}{2} = (C)^{-2/(1+n)}\left(\frac{\nu}{\langle v \rangle D}\right)^{2n/(1+n)} \tag{II.32}$$

Equation (II.32) indicates that in the turbulent region the friction factor will also be a function of the Reynolds number only. Results of friction factor measurements in smooth pipes have been collected and correlated by Blasius as:

$$4f = 0.316\,\text{Re}^{-0.25}$$

This correlation is discussed in more detail in the next section. At the moment we can use it to determine the constant n in equation (II.32). Via:

$$\frac{2n}{1+n} = 0.25$$

we find $n = \frac{1}{7}$.

Repetition of the foregoing procedure for the flow velocity in the x-direction at any place r yields, if we realize that v_x must be zero at $2r = D$, instead of equation (II.31):

$$\frac{v_x}{\sqrt{\dfrac{\tau_w}{\rho}}} = C_1\left[\sqrt{\frac{\tau_w}{\rho}}\frac{(D-2r)}{\nu}\right]^{\frac{1}{7}} \tag{II.33}$$

Thus, maximum flow velocity occurs in the centre of the tube at $r = 0$ and we can write:

$$\frac{v_x}{v_{max}} = \left(1 - \frac{2r}{D}\right)^{\frac{1}{7}} \tag{II.34}$$

In order to find the ratio between the mean and maximum velocities we can calculate $\langle v \rangle$ from equation (II.34) as:

$$\langle v \rangle = \frac{1}{D/2} \int_0^{D/2} v_x \, dr = \frac{v_{max}}{D/2} \int_0^{D/2} \left(1 - \frac{2r}{D}\right)^{\frac{1}{7}} dr$$

Solving this equation we find:

$$\frac{\langle v \rangle}{v_{max}} = \frac{7}{8} = 0.875$$

compared with $\langle v \rangle / v_{max} = \frac{1}{2}$ found for laminar flow in circular tubes.

The velocity distribution (equation (II.33)) gives a rather good description of the actual situation, as illustrated in Figure II.18, where measured values of $(v/v_{max})^7$ are plotted against $2r/D$.

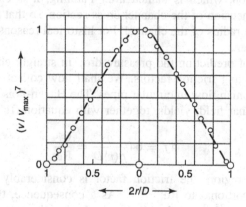

Figure II.18 Measured velocity distribution for turbulent flow

II.2.2 Pressure drop in straight channels

At the end of Section II.1.2 we saw that the pressure drop in a pipe system can only be predicted as a function of the flow rate if the shear force at the wall is known. For laminar flow, this quantity can be predicted directly because the relationship between τ_{xy} and dv_y/dx (e.g. according to Newton, equation (II.1)), characteristic of the fluid, offers sufficient additional information. For turbulent flow we lack this additional information.

In order to be able to predict the flow rate for turbulent flow on the basis of a given pressure drop (or, inversely, the pressure drop for a given flow rate), use is made of graphs and tables of earlier measurement data in which numerical values can be found of the dimensionless groups that are important to this problem. On the basis of dimensional considerations these results can then be used for solving each new but analogous problem. We shall deal in succession with the flow through a straight channel, the flow through fittings and the frictionless flow.

Since in what follows we only speak of a mean velocity and not of velocity distributions, the microbalances are of no use to us in this discussion. The quantities which we shall now come across belong to balances over an entire channel or channel system (i.e. they are macroscopic quantities to be used in macrobalances).

If the shear stress on the wall of a horizontal straight channel (cross-section A, circumference S) is known, we find for the stationary state the momentum balance (see also section II.1.2 and Figure II.4):

$$\sum F_x = 0 = p_1 A - p_2 A - \tau_w S(x_2 - x_1)$$

and, using equation (II.22), we find for the pressure drop:

$$p_1 - p_2 = \tau_w \frac{S(x_2 - x_1)}{A} = f \frac{1}{2}\rho\langle v_x\rangle^2 \frac{S(x_2 - x_1)}{A} \qquad \text{(II.35)}$$

From this equation, which is named after Fanning, it is clear that A/S is the characteristic dimension of the channel cross-section, so that it is obvious to call A/S the hydraulic radius of the channel. For historical reasons, $4A/S$ is called the hydraulic diameter.

The problem of predicting the pressure drop in straight channels can thus be reduced to predicting friction factors. We shall now collect data on the friction factors for Newtonian flow in circular pipes. The Hagen-Poiseuille formula (II.7) valid in the laminar field, yields, together with equation (II.22), for the friction factor:

$$4f = 64\frac{\eta}{\rho\langle v\rangle D} = \frac{64}{\text{Re}} \qquad \text{(II.36)}$$

In the turbulent region, the friction factor is considerably less dependent on Re, notably proportional to $\text{Re}^{-0.25-0}$. As a consequence, the pressure drop is proportional to $\langle v\rangle^{1.75-2}$, so that for a given pipe the pressure drop increases much more strongly with $\langle v\rangle$ than in the laminar field, but the viscosity of the flow has less influence.

For pipes with a smooth wall the experimentally found relationship proposed by Blasius:

$$4f = 0.316\,\text{Re}^{-0.25}\,(4000 < \text{Re} < 10^5) \qquad \text{(II.37)}$$

gives a good description of the dependency of f on Re.

In the case of turbulent flow the roughness of the wall surface also influences the friction losses. When the mean height $\bar{x}$ of these irregularities (roughness) becomes of the same order of magnitude as the thickness of the laminar boundary region (either in the case of great roughness or at high Re numbers), part of the resistance will be caused by direct momentum transmission between the turbulent core of the flow and the protuberances of the wall. In extreme cases (very great roughness or high Re), this causes a pressure loss proportional to the kinetic energy of the flow, so that the friction factor no longer depends on Re. The wall roughness is allowed for by the dimensionless quantity $\bar{x}/D(=$ relative roughness). Consequently, the friction factor is in general a function of Re and $\bar{x}/D$.

In Figure II.19, the relationship between $4f$ and Re is shown for various values of the relative roughness. It appears that in the turbulent region two parts can be distinguished: at high Re numbers, the so-called completely turbulent region where f only depends on the roughness, and at lower Re, a turbulent transition region where both Re and roughness play a part. The mean roughness of a number of materials is also given in Figure II.19.

For pipes with non-circular cross-sections, the friction factor in the laminar region is proportional to $1/$Re. The proportionality constant depends on the form of the cross-section and is known for a number of cases (see Section II.1.5 and Figure II.12).

If the flow is turbulent, approximately the same values of the friction factor apply as in tubes with circular cross-sections if, in the Fanning equation (II.35) and in the Reynolds number, use is made of the hydraulic diameter. Hence the advantage of using the hydraulic diameter is that (only for turbulent flow) with one graph for the relationship between the friction factor and Re, the pressure drop can be calculated for all forms of the cross-section and for all Newtonian fluids (i.e. accurately for circular cross-sections and approximately for other forms). A number of frequently occurring hydraulic diameters are shown in Figure II.20, together with the cross sections available for flow.

The pressure drop in spirals is, for the reasons stated in Section II.2.1, higher than in a straight tube. For a rough estimation of the pressure drop of turbulent flow of water-like liquids, the empirical relationship:

$$\Delta p_{\text{spiral}} = \Delta p_{\text{straight pipe}} \times \left(1 + 3.74 \frac{D}{2R}\right) \tag{II.38}$$

can be used ($D =$ tube diameter, $2R =$ diameter of spiral).

In practice, the chemical engineer often has to determine which flow rate can be attained at a given pressure drop. Thus he has to solve the relationship (II.35):

$$\Delta p = 4f \frac{1}{2} \rho \langle v \rangle^2 \frac{L}{D}$$

for $\langle v \rangle$, which can be done by a trial-and-error procedure. Rearrangement of equation (II.35) yields:

$$4f\text{Re}^2 = \frac{2\Delta p \rho D^3}{L\eta^2} \tag{II.39}$$

For a given problem, the right-hand side of this equation is a known quantity; thus $4f\text{Re}^2$ is known. In Figure II.21, the relationship between $4f\text{Re}^2$ and Re

70

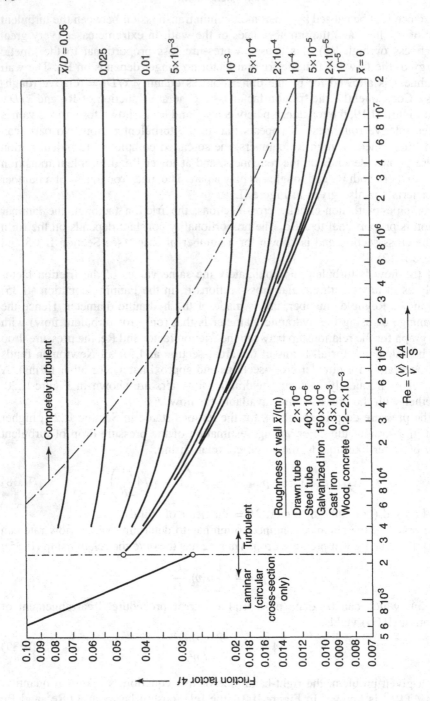

Figure II.19 Friction factor for flow in tubes

Flow situation		Hydraulic diameter $D_h = 4A/S$	A
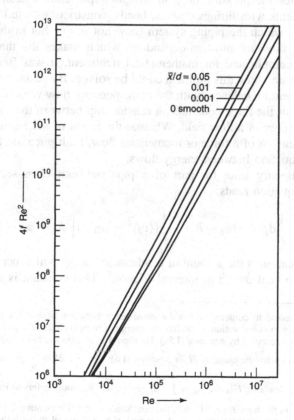 D	Circular pipe	D	$\frac{\pi}{4}D^2$
δ — D_1 D_2	Concentric pipe or slit	$D_2 - D_1 = 2\delta$	$\frac{\pi}{4}(D_2^2 - D_1^2)$
B W	Rectangular pipe	$\frac{2WB}{W+B}$	WB
W H	Open channel	$\frac{4WH}{W+2H}$	WH
$90°$ H	Open channel	$\frac{2H}{\sqrt{2}}$	H^2
D	Half-filled	D	$\frac{\pi}{8}D^2$
δ	Liquid film in a tube	4δ	$\delta\pi D$

Figure II.20   Hydraulic diameters of various channels

Figure II.21   The relationship of $4f\,\text{Re}^2$ and Re for turbulent flow is shown for pipes of differing roughness. With the help of this graph the Re number can be found and consequently the volumetric flow rate.

Now that with f, the pressure drop in the straight horizontal channel is known, it is obvious that the energy dissipation in the channel is also fixed. From the energy equations (I.4a) it follows that the frictional heat released for each transported unit of mass in this case equals $(p_1 - p_2)/\rho$. The energy necessary for pumping the fluid is then:

$$\phi_A = \phi_m \frac{(p_1 - p_2)}{\rho} = \phi_v(p_1 - p_2) = \phi_m \frac{f}{2} \langle v \rangle^2 \frac{S(x_2 - x_1)}{A} \qquad \text{(II.40)}$$

This means that the dissipated mechanical energy per unit of volume equals the undergone pressure difference. Thus, for laminar flow, ϕ_A/ϕ_v is proportional to $\langle v \rangle$ and for turbulent flow to $\langle v \rangle^{1.75-2}$.

II.2.3 Pressure drop in pipe systems

Calculations for the pressure drop in straight pipes cannot easily be extended to piping systems with fittings (valves, bends, constrictions, etc.). The reason is that the forces which the piping system (now not in one but in three directions) exerts on the flow are position-dependent, which makes the three momentum balances too complicated for mathematical treatment. It was Bernoulli who in principle indicated how this problem could be solved. By scalar multiplication of the three momentum balances with the corresponding flow velocities he obtained, after addition of the three equations, a relationship between the various forms of mechanical energy in a flow field. Whereas the terms of the momentum balance have the dimension of a force or momentum flow, multiplication by the velocity leads to an equation between energy flows.

In the stationary state for part of a pipe between cross-sections 1 and 2 Bernoulli's equation reads:

$$0 = -\left[\int_1^2 \frac{1}{\rho} dp + g(h_2 - h_1) + \frac{1}{2}(\langle v_2 \rangle^2 - \langle v_1 \rangle^2) \right] \phi_m + \phi_A - E_{fr}\phi_m \qquad \text{(II.41)}$$

E_{fr} in this equation is the amount of mechanical energy which per unit of mass is converted into heat owing to internal friction[†]. This constant is always positive

---

[†] The reader is asked to contemplate on the consistency between Bernoulli's result, which is an expression of the non-conservation of mechanical energy in stationary situations, and the conservation law of energy as expressed by equation (I.5a). He should start with the basic expression for a point change in the characteristic quantity H in equation (I.6): $dH = T\, dS + \frac{1}{\rho} dp$, which for reversible, isentropic change leads to: $H_{in} - H_{out} = \int_1^2 \frac{1}{\rho} dp$. However, Bernoulli refers to irreversibility, which is accounted for in the term $E_{fr}\phi_m$. Hence, the conclusion is that for isentropic processes the rate of thermal energy production by friction equals the rate of heat removal through the system's walls.

because the dissipation is a result of statistical transport, which according to thermodynamics is attended by degradation of energy (entropy production).

Hence we may conclude from equation (II.41) that owing to friction the total mechanical potential (the first four terms of the equation) decreases in the direction of the flow, so that here we can speak of the law of 'non-conservation' of mechanical energy. The reader should realize that the Bernoulli equation is not a new balance but rather a summary of the three momentum balances.

If the amount of energy added to the system per unit of mass $\phi_A = 0$ and the energy loss due to friction is negligible, equation (II.41) can be written for any point in the flow field as:

$$\frac{1}{\rho}\,dp + g\,dh + v\,dv = 0 \tag{II.42}$$

or, also, if at the same time the density is constant:

$$\frac{p}{\rho} + gh + \tfrac{1}{2}v^2 = \text{constant along streamline} \tag{II.43}$$

These are two forms of the well-known equation of Bernoulli for potential flow[†], which can also be derived in a simple way by using the momentum balance for a frictionless fluid (Newton's first law). The more general equation (II.41), which plays a very important role in the technical flow theory, is further referred to as the extended Bernoulli equation.

On account of what we know about the pressure drop and the energy dissipation in a straight channel we can relate the constant E_{fr} for such a channel (length L) to the friction factor and the mean velocity $\langle v \rangle$ in that channel (equation II.40):

$$E_{fr} = f\tfrac{1}{2}\langle v \rangle^2 \frac{SL}{A} \tag{II.44}$$

Local pressure losses are mainly due to a sudden increase in the cross-section of the pipe. According to Bernoulli's law, in such a diverging flow the velocity downstream decreases and the pressure increases. This pressure gradient may cause the fluid near the wall to flow upstream. This results in systematic eddies (not the same as turbulence!). In these eddies, part of the kinetic energy is destroyed by internal friction. Only if the diameter is carefully widened (e.g. via a diffusor with a vertical angle of at most 8°) do these eddies not occur.

The dissipation per unit of mass in a fitting is described with a friction loss factor K_w which is defined as:

$$E_{fr} = K_w \tfrac{1}{2}\langle v \rangle^2 \tag{II.45}$$

[†] This equation is characterized by the fact that the potential energies can be transformed into kinetic energy without loss.

In practical cases, K_w is only a function of local geometry and is, for not too low Re numbers, independent of Re. For $\langle v \rangle$, the mean velocity in the pipe on the downstream side of the local resistance is usually taken.

To calculate the resistance of a pipe having various pieces of different diameter, bends, valves and other local resistances, both equations (II.44) and (II.45) should be introduced in the extended Bernoulli equation (II.41), via the lost frictional energy E_{fr}. The latter constant integrated over the entire pipe then becomes:

$$E_{fr} = \sum_i \left(f\frac{1}{2}\langle v \rangle^2 \frac{SL}{A} \right)_i + \sum_j \left(K_w \frac{1}{2}\langle v \rangle^2 \right)_j \tag{II.46}$$

where the first sum is taken over all pieces of straight pipe and the second over all local resistances. In air conditioning channels, often short bends have to be made owing to lack of space. However, the K_w values then become unpermissibly high so that frequently guide vanes are built in to lower these values (avoidance of eddies, streamlines).

Empirical correlations between K_w values and the geometry of the pipe system are known for a great number of situations. A number of K_w values are listed in Table II.1. In some cases it is possible to derive K_w values theoretically. Consider the situation shown in Figure II.22. For the calculation of the energy dissipation three equations are available, viz:

the mass balance: $\qquad \phi_m = \rho v_1 A_1 = \rho v_2 A_2$

the momentum balance: $\qquad 0 = p_1 A_1 + \phi_m v_1 - p_2 A_2 - \phi_m v_2 + F_w$

the Bernoulli equation (II.41): $0 = -\left(\int_1^2 \frac{dp}{\rho} + \int_1^2 gdh + \int_1^2 vdv \right)\phi_m$
$$+ \phi_A - E_{fr}\phi_m$$

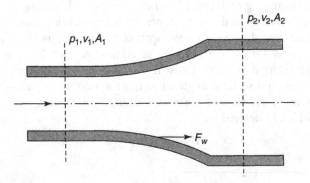

Figure II.22 Flow in a channel with changing cross-section

Table II.1 Some values of the friction loss factor K_w (referring to downstream velocity for Re $> 10^5$)

$$K_w = k\left(1 - \frac{A_1}{A_2}\right)^2$$

Θ	$< 10°$	$10°$	$20°$	$30°$	$40°$	$50°$	$60°$	$70°$	$80°$	$90°$
k	0	0.17	0.41	0.71	0.90	1.03	1.12	1.13	1.10	1.05

Θ	$10°$	$20°$	$30°$	$40°$	$50°$	$60°$	$70°$	$80°$
K_w	0.16	0.20	0.24	0.28	0.31	0.32	0.34	0.35

$$K_w = \left(\frac{A_2}{A_1} - 1\right)^2$$

$$K_w = 0.45\left(1 - \frac{A_2}{A_1}\right)$$

$\dfrac{A_0}{A_1}$	0.1	0.2	0.3	0.4	0.5	0.6	0.7	0.8	0.9	
K_w	226	47.8	17.5	7.8	3.75	1.80	0.80	0.30	0.06	

$\dfrac{A_0}{A_2}$	0.1	0.2	0.3	0.4	0.5	0.6	0.7	0.8	0.9	1.0
K_w	232	51	20	9.6	5.3	3.1	1.9	1.2	0.73	0.48

$A_1 > 20\, A_2$

Θ	$20°$	$40°$	$60°$	$80°$	$90°$	$100°$	$120°$	140
K_w	0.05	0.14	0.36	0.74	0.98	1.26	1.86	2.43

$$K_w = \left(0.131 + 0.163\left(\frac{D}{R}\right)^{3.5}\right)\frac{\Theta}{90°}$$

K_w (referring to downstream velocity for Re $> 10^5$)

Θ	$15°$	$30°$	$45°$	$60°$	$90°$
K_w	0.02	0.11	0.26	0.50	1.20

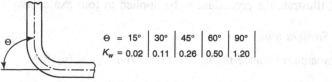

(continued overleaf)

Table II.1 (*continued*)

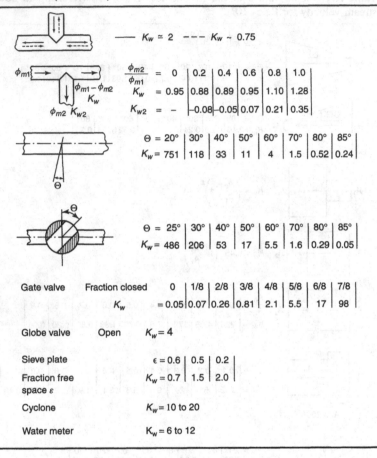

$\dfrac{\phi_{m2}}{\phi_{m1}}$	=	0	0.2	0.4	0.6	0.8	1.0
K_w	=	0.95	0.88	0.89	0.95	1.10	1.28
K_{w2}	=	–	–0.08	–0.05	0.07	0.21	0.35

Θ =	20°	30°	40°	50°	60°	70°	80°	85°
K_w =	751	118	33	11	4	1.5	0.52	0.24

Θ =	25°	30°	40°	50°	60°	70°	80°	85°
K_w =	486	206	53	17	5.5	1.6	0.29	0.05

Gate valve	Fraction closed	0	1/8	2/8	3/8	4/8	5/8	6/8	7/8
	K_w	= 0.05	0.07	0.26	0.81	2.1	5.5	17	98

Globe valve   Open $K_w = 4$

Sieve plate	$\epsilon = 0.6$	0.5	0.2
Fraction free space ε	$K_w = 0.7$	1.5	2.0

Cyclone $K_w = 10$ to 20

Water meter $K_w = 6$ to 12

The latter equation simplifies for horizontal pipes, no energy input and incompressible fluids to:

$$E_{\mathrm{fr}} = \frac{p_1 - p_2}{\rho} + \tfrac{1}{2}(\langle v_1 \rangle^2 - \langle v_2 \rangle^2)$$

Using the above three balances, it is possible to estimate the amount of dissipated energy E_{fr} if we can find an expression for the force the liquid exerts on the wall, F_w. We will illustrate the procedure to be applied in four examples.

Example 1. Straight tube

Here the momentum balance reduces to ($A_1 = A_2$, $\langle v_1 \rangle = \langle v_2 \rangle$):

$$p_1 A_1 - p_2 A_2 + F_w = 0$$

Since the liquid exerts on the wall a shear force only:

$$F_w = -\tau_{w-f} SL = -f \tfrac{1}{2}\rho v^2 SL$$

we find:

$$p_1 - p_2 = f \tfrac{1}{2}\rho v^2 \frac{SL}{A}$$

The Bernoulli equation reads under these conditions ($\langle v_1 \rangle = \langle v_2 \rangle$):

$$E_{\mathrm{fr}} = \frac{p_1 - p_2}{\rho}$$

as we found earlier. Substituting for ($p_1 - p_2$) the expression developed from the momentum balance, we finally find:

$$E_{\mathrm{fr}} = f \tfrac{1}{2}v^2 \frac{SL}{A}$$

as should be the case. Extending the definition of the K_w value to the case of straight pipes:

$$E_{\mathrm{fr}} = K_w \tfrac{1}{2}v^2$$

it follows that for straight pipes:

$$K_w = f \frac{SL}{A}$$

Example 2. Sudden expansion

In the case of sudden expansion, eddy currents will occur, leading to extra energy losses. The momentum balance for this case (see Figure II.23) between

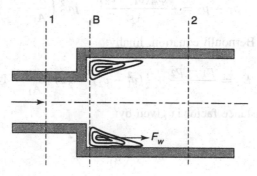

Figure II.23 Sudden expansion of pipe

planes '1' and '2' reads:

$$p_1 A_1 - p_2 A_2 + \phi_m v_1 - \phi_m v_2 + F_w = 0$$

If we neglect wall friction F_w is given by

$$F_w = (A_2 - A_1) p_B$$

where p_B is the pressure just behind the expansion in plane 'B'. This pressure will be $p_1 < p_B < p_2$. Only experimental results can show which pressure has to be used.

A. Let us assume $p_B \approx p_2$, thus $F_w = p_2(A_2 - A_1)$. Then we find from the momentum balance:

$$p_1 - p_2 = \frac{-\phi_m(v_1 - v_2)}{A_1} = -\rho v_2^2 \frac{A_2}{A_1}\left(\frac{A_2}{A_1} - 1\right)$$

We see that the momentum balance predicts correctly that $p_2 > p_1$. Substituting $(p_1 - p_2)$ in the Bernoulli equation:

$$E_{fr} = \frac{p_1 - p_2}{\rho} + \tfrac{1}{2}(v_1^2 - v_2^2)$$

we find after some sorting and applying of the momentum balance:

$$E_{fr} = -\tfrac{1}{2}v_2^2\left(\frac{v_1}{v_2} - 1\right)^2 = -\tfrac{1}{2}v_2^2\left(\frac{A_2}{A_1} - 1\right)^2$$

This is evidently an impossible solution (E_{fr} must be positive, otherwise we would gain energy), thus our estimate of p_B must have been wrong.

B. Repeating the exercise assuming that $p_B \approx p_1$, we find:

$$F_w = p_1(A_2 - A_1)$$

$$p_1 - p_2 = \frac{-\phi_m(v_1 - v_2)}{A_2} = \rho v_2^2\left(\frac{A_2}{A_1} - 1\right)$$

and, with the Bernoulli equation, finally:

$$E_{fr} = \frac{p_1 - p_2}{\rho} + \tfrac{1}{2}(v_1^2 - v_2^2) = \tfrac{1}{2}v_2^2\left(\frac{A_2}{A_1} - 1\right)^2$$

Thus, the resistance factor is given by:

$$K_w = \left(\frac{A_2}{A_1} - 1\right)^2$$

which fits very well with experimental results.

Example 3. Gradual expansion

For the case shown in Figure II.24 we can roughly estimate (neglecting friction losses):

$$F_w = \frac{p_1 + p_2}{2}(A_1 - A_2)$$

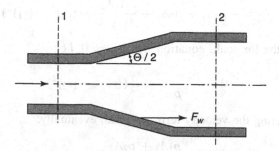

Figure II.24 Gradual expansion of pipe

where $(p_1 + p_2)/2$ is the average pressure. With the momentum balance we find:

$$p_1 - p_2 = -\frac{2\phi_m(v_1 - v_2)}{A_1 + A_2}$$

and, applying the Bernoulli equation, we get:

$$E_{fr} = \frac{p_1 - p_2}{\rho} + \tfrac{1}{2}(\langle v_1 \rangle^2 - \langle v_2 \rangle^2) = \frac{-\phi_m(v_1 - v_2)}{\frac{A_1 + A_2}{2}} + \tfrac{1}{2}(\langle v_1 \rangle^2 - \langle v_2 \rangle^2)$$

Now $(A_1 + A_2)/2$ is the mean cross-section, where the mean flow velocity $(v_1 + v_2)/2$ is present. Thus:

$$\phi_m = \rho \frac{v_1 + v_2}{2} \frac{A_1 + A_2}{2}$$

Substituting ϕ_m in the Bernoulli equation we find $E_{fr} = 0$ i.e. there is no energy loss. E_{fr} has been proved experimentally to be negligible indeed if $\Theta/2 < 8°$. In practical cases of smoothly diverging flow, only small friction losses occur and a practical value of the friction loss factor K_w is 0.05.

Let us now reverse the reasoning. How great is F_w? Knowing that $E_{fr} = 0$, we can derive an expression for F_w. This force acts at a mean cross-section $\overline{A}$, where the flow velocity is given by $(v_1 + v_2)/2$. A mass balance shows:

$$\phi_m = \rho \frac{v_1 + v_2}{2}\overline{A} = \rho v_1 A_1 = \rho v_2 A_2$$

and:

$$\bar{A} = 2\frac{v_1 A_1}{v_1 + v_2} = \frac{2A_1 A_2}{A_1 + A_2}$$

Using these values in the momentum balance, we find:

$$F_w = -p_1 A_1 + p_2 A_2 - v_1 \phi_m + v_2 \phi_m$$

$$= -p_1 A_1 + p_2 A_2 + \frac{A_1 A_2}{A_1 + A_2}\rho(\langle v_2 \rangle^2 - \langle v_1 \rangle^2)$$

and, applying the Bernoulli equation with $E_{\mathrm{fr}} = 0$, i.e.:

$$0 = \frac{p_1 - p_2}{\rho} + \tfrac{1}{2}(\langle v_1 \rangle^2 - \langle v_2 \rangle^2)$$

and by substituting the velocity term, we find, eventually:

$$F_w = \frac{p_1 A_1 + p_2 A_2}{A_1 + A_2}(A_2 - A_1)$$

In the special case where $\Theta/2$ is very small, $A_1 \approx A_2$ and the weighed mean is simplified to:

$$F_w = \frac{p_1 + p_2}{2}(A_2 - A_1)$$

exactly as we guessed at first sight when starting this example.

Example 4. Obstacles in the direction of flow

A problem which often gives rise to false discussions is the question as to what pressure drop is caused by an obstacle in the flow. This problem is relatively simple in a straight channel of uniform cross-section in which, over the range in which the influence of the obstacle makes itself felt, the frictional resistance at the pipe wall is negligible. If over this range the pressure drop is Δp and the cross-section of the channel is A, the momentum balance (for this case $\sum F = 0$) yields $\Delta p\, A = F_{f-w}(=$ the force the channel flow exerts on the obstacle). The Bernoulli equation gives $E_{\mathrm{fr}} = \Delta p/\rho$, so $E_{\mathrm{fr}} = F_{f-w}/\rho A$.

This can be explained as follows. A force F_{f-w} between two media with a mutual relative velocity $\langle v \rangle$ represents an energy flow $F_{f-w}\langle v \rangle$, and hence the energy which is lost per unit of mass flowing through:

$$\frac{F_{f-w}\langle v \rangle}{\rho\langle v \rangle A} = E_{\mathrm{fr}} = F_{f-w}/\rho A$$

If the channel over the length in which the body influences the flow is straight but changes in cross-section (Figure II.25), it is still possible to calculate the pressure drop as a function of F_{f-w}, provided E_{fr} can be written as the sum total of E_{fr} for only the obstacle, and E_{fr} for only the channel $(= E'_{fr})$:

the momentum balance:
$$p_1 A_1 - \langle v_1 \rangle \phi_m - p_2 A_2 - \langle v_2 \rangle \phi_m$$
$$= F_{f-w} + F'_{f-w}$$

the Bernoulli equation:
$$\frac{p_1}{\rho} + \tfrac{1}{2}\langle v_1 \rangle^2 - \frac{p_2}{\rho} - \tfrac{1}{2}\langle v_2 \rangle^2$$
$$= \sum E_{fr} = \frac{F_{f-w}}{\rho(A_1 + A_2)/2} + E'_{fr}$$

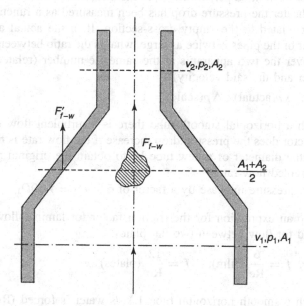

Figure II.25 Obstacle in the direction of flow

From the Bernoulli equation the pressure drop follows at given F_{f-w} and given E'_{fr}, and from the momentum balance it follows at given F'_{f-w}. This method can generally be applied if F_{f-w} is either large or small with respect to

$$\rho E'_{fr}(A_1 + A_2)/2$$

However, as soon as the contributions of the two walls of obstacle and channel can no longer be added, because the one considerably influences the flow near the other, the only solution to the problem is to measure the friction loss factor in a model of the total system.

II.2.4 Problems

1. The hydrodynamic behaviour of a submarine with an underwater speed of 20 knots ($\simeq 10$ m/s) is investigated in a wind tunnel using a scale model (1 : 10). What air velocity has to be used in the wind tunnel?

$$\eta_{water} = 10^{-3} \text{Ns/m}^2 \qquad\qquad \rho_{water} = 10^3 \text{kg/m}^3$$

$$\eta_{air} = 1.6 \times 10^{-6} \text{Ns/m}^2 \qquad \rho_{air} = 1.2 \text{kg/m}^3$$

Answer: 133 m/s

2. An air heater consists of a rectangular box in which a bundle of heating pipes has been fitted perpendicular to the direction of flow. In a scale model of this heater the pressure drop has been measured as a function of the air velocity related to the empty cross-section. If in the actual air heater the diameter of the pipes is twice as large, what is the ratio between the pressure drops over the two apparatus at the same Re number (related to the pipe diameter and the said velocity)?

Answer: Δp(actual) : Δp(scale) $= 1 : 4$

*3. Through a horizontal smooth tube there is a turbulent flow of liquid. By what factor does the pressure drop increase if the flow rate is trebled? What should the diameter of a new tube be to obtain the original pressure drop at the trebled flow rate?

Answer: pressure increase by a factor of 6.8; $D_2 = 1.50 D_1$

4. Develop an expression for the friction factor for laminar flow in a falling film and for flow between two flat plates.

Answer: $f = \dfrac{6}{\text{Re}}$ (film); $\quad f = \dfrac{12}{\text{Re}}$ (plates)

5. Through a smooth horizontal tube 1 kg/s water is forced (Re $\approx 2 \times 10^4$). The pressure difference over the tube is $2 \times 10^5 \text{N/m}^2$. What energy P is needed and what is the temperature rise ΔT of the water if there is no heat exchange with the environment? What will be the values for P and ΔT:
   (a) if the water flow is doubled;
   (b) if 1 kg/s water is forced through a tube 1.5 times as narrow?

   Answer: 200 W, 0.048 °C
   (a) 1350 W, 0.16 °C
   (b) 1372 W, 0.33 °C

6. Calculate the pressure drop over a capillary flow meter with a smooth wall if the gas velocity is 5 m/s, 50 m/s and 100 m/s, respectively.

capillary length 10 cm $\eta_{air} = 2 \times 10^{-5}$ Ns/m^2
capillary diameter 1 mm $\rho_{air} = 1.2$ kg/m^3

Answer: $\Delta p = 3.2 \times 10^2$ N/m^2
6.5×10^3 N/m^2
2.1×10^4 N/m^2

*7. Through a steel tube of 60 m length, 120 m^3/h of water is pumped against a height difference of 25 m. What will be the difference in pressure between the beginning and end of the tube if:
(a) the tube is of circular cross-section ($D = 10$ cm);
(b) the tube has a square cross-section (10×10 cm);
(c) the water flows through an annulus ($D_0 = 14$ cm, $D_i = 10$ cm)?

Answer: (a) 3.4 × 10^5 N/m^2
(b) 3.1 × 10^5 N/m^2
(c) 5.2 × 10^5 N/m^2

8. In a factory there is an 8 in pipe from a trough of a rotating filter to a hollander which is situated 3.75 m lower. Cellulose slurry which flows over the brim of the trough is returned to the system through this pipe. The pipe is 15 m long and has two short 90° bends and one open gate valve. What is the flow rate of the slurry in the pipe and what is the capacity? The viscosity of the slurry is 5×10^{-3} kg/ms (at 20 °C, five times as viscous as water) and the density is 1200 kg/m^3.

Answer: 4.5 m/s; 520 m^3/h

9. A vertical pipe (1 in diameter, 2 m long, smooth) contains one rotameter, three right-angled bends for connecting the pipe to the rotameter and two valves. The K_w value of these valves, when open, is 6. The rotameter, which has approximately the diameter of the pipe, contains a float of 25×10^{-3} kg ($\rho = 2600$ kg/m^3). Calculate the pressure drop over this pipe if water flows through it at an upward velocity of 1 m/s.

Answer: $\Delta p = 28\ 060$ N/m^2

*10. When combustion gases enter a chimney the pressure is 250N/m^2 below atmospheric pressure. The chimney is made of smooth steel plates which are bent to an internal radius of 1.6 m; 15 t/h combustion gases are discharged. The mean temperature of the gases is 260 °C, the outside temperature is 21 °C and the barometer reading is 1000 mbar. What is the height H of the chimney? The density of the burned gas at 25 °C is $\rho = 1.27$kg/m^3 and $\eta = 15.6 \times 10^{-6}$ kg/ms at 260 °C.

Answer: H = 52 m

11. Through a pipe with a 5 cm internal diameter flows 2 l/s of water. An obstacle in the pipe shows a flow resistance of 100 mm water column.
    (a) What is the force the flow exerts on the obstacle?
    (b) How much energy is lost by friction?
    Answer: (a) $= 2$ N
                (b) $= 1.96$ W

*12. A strong fireman sends up a jet of water of 2 m³/min at an angle of 45°. The maximum height he can reach is 20 m. What is the force with which he is pressed against the ground as a result of spouting?
    Answer: 667 N

13. A garden hose ($D_i = 1$ cm) of smooth, inelastic material, 20 m long, is connected to a tap in the ground; the resistance value of the tap related to the downstream velocity is $K_w = 7.5$. The pressure before the tap is 2.65×10^5 N/m². A nozzle ($D_i = 0.25$ cm) can be connected to the spouting end of the hose, in this case $K_w = 0.1$ (related to the velocity in the nozzle). The spouting end is kept 1.5 m above the ground. Calculate the velocity v at which water of 18 °C leaves the garden hose:
    (a) without a nozzle;
    (b) with a nozzle.
    Neglect the water velocity before the tap:
    Answer: (a) 2.3 m/s
                (b) 14.7 m/s

14. At least how high can a pole vaulter jump if he can run at a speed of 10 m/s?
    Answer: 5 m

15. On the Mons St. Nicolas (Luxemburg) lies an artificial lake with a capacity of 7.6×10^6 m³ of water. For twelve hours (mainly during the night) the lake is filled with water from the Our (500 m lower). During the next twelve hours, turbines on the Our are driven with water from the lake in order to generate electric power. The water supply from the lake to the turbines takes place through two parallel circular concrete tubes, 650 m long and 6 m in diameter.
    (a) What power could this station deliver, apart from losses, if it is in continuous operation for 12 out of 24 hours?
    (b) What are the friction losses expressed as a percentage of the theoretically possible power? Let the resistance value for the entry and exit losses, each related to the velocity in the tubes, be 2.5.
    (c) Calculate that, if the pump delivery and the turbine efficiency are 75%, this way of energy storage makes the electric power 1.8 times as expensive; the engineering works, as such, may be considered as being fully written off.
    Answer: (a) 863 MW; (b) 0.65%

16. Two types of machines are to be considered, i.e. prime movers by mechanical energy (e.g. the water powered turbine) and thermodynamic machines. Both are treated by idealized theories, complemented by relatively small correction terms. Describe these theories and identify the two types of corrections required.

Answer: Water powered turbines:

$$-\phi_A = \phi_m \left[\Delta(\tfrac{1}{2}v^2 + gh) - E_{fr} \right]$$

Thermodynamic machines, which work (almost) reversibly and hence perform cycles:

$$\oint (vdv + gdh) \text{ is almost zero.}$$

*17. Show that, in the case discussed at the beginning of this section, John was right in letting the suspected man go.

Comments on problems

Problem 3

(a) An increase in throughput will affect, among others, the Re number and therefore also the friction factor. (See Figure II.19.)
(b) Here again you should not forget the change in f due to the changing Re number.

Problem 7

This problem has been included to show you how widely Figure II.19 can be used, provided that you base your Re number on the proper hydraulic diameter. Don't forget the hydrostatic head pressure!

Problem 10

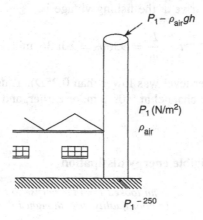

A sketch of the chimney with an expression for the pressures at the bottom and top will help you to find an expression for the height of the chimney by using the Bernoulli equation. It turns out that the losses due to friction are negligible.

Problem 12

With the substitution of the maximum height h in the Bernoulli equation you will find the vertical velocity component v_y of the nozzle. The force in the y-direction can be calculated from the mass flow and v_y. Do you find the same answer when you calculate the weight of water that the poor fellow has to carry?

Problem 17   John and the sewer murder case

The flow conditions in the sewer are turbulent (Re $\simeq 4 \times 10^6$) and we can assume f to be practically constant. Because the driving force $\rho g \Delta h$ is constant we can write:

$$4f \frac{1}{2} \rho v_1^2 \frac{L}{D_{h1}} = 4f \frac{1}{2} \rho v_2^2 \frac{L}{D_{h2}}$$

and we find:

$$v_1 = v_2 \sqrt{D_{h1}/D_{h2}}$$

For a half-filled sewer channel, $D_{h1} = D$, but if the channel is filled to one-quarter of its diameter:

$$D_{h2} = \frac{4\left(\dfrac{\pi R^2 \theta}{360} - \dfrac{R^2 \sin \theta}{2}\right)}{\dfrac{\pi R \theta}{180}} = 0.59D$$

if we use $\cos \theta/2 = (R - \frac{1}{2}R)/R = \frac{1}{2}$.

Since $v_2 = 1$m/s (half-filled channel) we find $v_1 = 0.77$m/s and the time necessary for the body to arrive at the fishing village is:

$$t = \frac{L}{v_1} = 9360\text{s} = 2 \text{ h } 36 \text{ min}$$

(or longer, if the water level was lower than $0.25D$). Thus the corpse must have been dropped into the channel at 9.09 p.m. or earlier, and John came to the right conclusion.

II.3 Flow with negligible energy dissipation

John looked from the corpse to the drum, from which wine (a poor quality rose, thought John) still spouted through the

two 6 mm holes which two of the bullets had left there. The
fat man who had called him had stated that he had been
in this isolated shed with only one other person, who had
run away twenty minutes ago. John thought: the drum of
0.6 m diameter stands on its flat bottom, the holes are 1 m
from its top and a wet rim indicates that the wine level has
decreased by 20 cm. He remembered what is outlined in the
pages to come and arrested the fat man.

For a great number of flow situations the energy dissipation E_{fr} is zero or negligibly small with respect to the amount of mechanical energy which changes from one form into the other (see example 3 of the previous section). In those cases of potential flow the relationship between velocity and driving force is simply given by Bernoulli's law (in the form of equations (II.42) or (II.43)). This equation can also be written in such a way that the terms acquire the dimension of a length:

$$\frac{p}{\rho g} + h + \frac{v^2}{2g} = \text{constant along a streamline} \qquad (II.47)$$

The three terms are then called pressure head, static head and velocity head, respectively; the sum total of these terms remains constant if there is no energy change due to friction or mechanical work. These heads also have physical significance, as can be seen from Figure II.26.

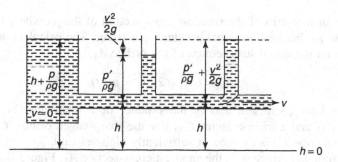

Figure II.26 Static head, pressure head and velocity head

With the left manometer tube only the pressure prevailing in the liquid is measured. The lower opening of the right-hand tube is opposing the stream. Right in front of this opening the velocity is negligible so that all kinetic energy is converted into pressure and causes a thrust equal to $\frac{1}{2}\rho v^2$. This thrust is used when measuring velocities with a Pitot tube.

We will treat here some well-known examples of flow with negligible energy dissipation, the flow from orifices, the flow over a sharp-edged weir and the flow in flowmeters.

II.3.1 Flow of a liquid from an orifice

Friction hardly plays a role in this type of problem. For calculation of the outflow velocity, use is made of the mass balance and the Bernoulli equation. Using this mass balance it appears that some distance before the orifice the velocity is negligible compared with the velocity in the orifice. Next, with equation (II.41) it can be derived that for a sharp-edged orifice (Figure II.27) the volume flow ϕ_v becomes:

$$\phi_v = A_1' \sqrt{2(p_1 - p_0)/\rho}$$

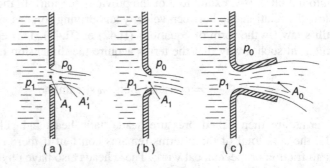

Figure II.27 Flow of liquids through holes

A_1' is the surface area of the smallest cross-section of the jet where the pressure should be p_0, because no radial accelerations occur. The calculations can better be based on the actual surface area of the orifice A_1:

$$\phi_v = CA_1 \sqrt{2(p_1 - p_0)/\rho} \qquad (II.48a)$$

C is the flow coefficient and is the product of a coefficient of contraction $C_c = A_1'/A_1$ and a friction factor C_f. For the sharp-edged orifice, C_f is practically 1, whereas C_c is ca. 0.62 at sufficiently high velocity.

For a rounded orifice with the smallest cross-section A_1, Figure II.27b applies:

$$\phi_v = C_f A_1 \sqrt{2(p_1 - p_0)/\rho} \qquad (II.48b)$$

with $C_f = 0.95 - 0.99$.

If a 'diffusor' (angle not greater than 8° to avoid eddying) is attached to the rounded orifice then:

$$\phi_v = C_f A_0 \sqrt{2(p_1 - p_0)/\rho} \qquad (II.49)$$

where A_0 is the cross-sectional area at the end of the diffusor (Figure II.27c). With the diffusor, more liquid flows through the smallest cross-section than without the diffusor, because in the former case the pressure in the constriction is lower than p_0. The diffusor cannot be lengthened infinitely because then the friction losses become too great. On the other hand, if the pressure in the throat falls below the vapour pressure of the liquid, vapour generation (cavitation) occurs; the given calculations are then no longer valid.

Table II.2 shows the flow coefficients C for various orifices. These coefficients must be used with equation (II.48a).

Table II.2 Flow coefficients for flow through orifices

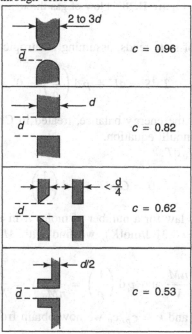

II.3.2 *Flow of gases through orifices*

The approach followed in the last section can also be applied to the flow of compressible gases through orifices. Friction losses can be neglected, and we can assume that the change of state takes place adiabatically. However, due to the adiabatic expansion, the pressure in the jet will be different from the initial pressure p_1 and the final pressure p_0 outside the gas jet (see Figure II.28).

In the stationary state the Bernoulli equation (II.41) reads in its differential form ($g\,dh = 0$, $\phi_A = 0$, $E_{fr} = 0$) for this situation:

$$0 = -\int_1^2 \left\{ \frac{1}{\rho}dp + v\,dv \right\}$$

(II.50)

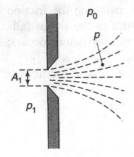

Figure II.28 Flow of gas through an orifice

whereas the equation of state reads, assuming isentropic change:

$$T\,dS = dU + p\,d\left(\frac{1}{\rho}\right) = 0 \qquad (II.51)$$

Under such conditions the energy balance, treated in Chapter I, gives the same information as the Bernoulli equation.
Realizing that $dU = C_v\,dT$:

$$0 = C_v\,dT + p\,d\left(\frac{1}{\rho}\right) \qquad (II.52)$$

Applying the ideal gas law for a number of moles n in a volume V, $pV = nRT$ (R= gas law constant $= 8.31$ J/molK), we find with M = molecular weight of the gas (kg/mol):

$$\rho = \frac{nM}{V} \quad \text{and} \quad p\,d\left(\frac{1}{\rho}\right) = \frac{p}{nM}dV = \frac{RT}{MV}dV$$

With $R = M(c_p - c_v)$ and $\kappa = c_p/c_v$ we now obtain from equation (II.52):

$$0 = \frac{dT}{T} + (\kappa - 1)\frac{dV}{V} \qquad (II.53)$$

Integration of this equation finally yields $TV^{\kappa-1} = $ constant, which can be transformed by means of the ideal gas law to:

$$pV^\kappa = \text{constant} \quad \text{and} \quad p\rho^{-\kappa} = \text{constant}$$

These last equations describe the adiabatic compression and expansion of ideal gases.

Returning to the Bernoulli equation (II.50) for the flow of gases through orifices under steady-state conditions:

$$0 = \frac{1}{\rho}dp + v\,dv$$

we can now replace ρ by $\rho =$ constant $p^{1/\kappa}$, and by integration between p_1, $v = 0$ and p, v we find:

$$\left(\frac{1}{\text{constant}}\right)\left(\frac{1}{\dfrac{-1}{\kappa}+1}\right)(p^{-1/\kappa+1} - p_1^{-1/\kappa+1}) + \frac{1}{2}v^2 = 0$$

$$= \frac{1}{2}v^2 + \left(\frac{\kappa}{\kappa-1}\right)\left(\frac{p}{\rho} - \frac{p_1}{\rho_1}\right)$$

and:

$$v^2 = \left(\frac{2\kappa}{\kappa-1}\right)\frac{p_1}{\rho_1}\left[1 - \left(\frac{p}{p_1}\right)^{(\kappa-1)\kappa}\right] \tag{II.54}$$

So the mass flow rate at any place of the variable cross-section A of the gas jet where the pressure is p is given by:

$$\phi_m = \rho v A = \left(\frac{p}{p_1}\right)^{1/\kappa}\rho_1 v A$$

$$= A\left\{2\left(\frac{\kappa}{\kappa-1}\right)p_1\rho_1\left[\left(\frac{p}{p_1}\right)^{2/\kappa} - \left(\frac{p}{p_1}\right)^{(\kappa+1)/\kappa}\right]\right\}^{\frac{1}{2}} \tag{II.55}$$

The change of the cross-section of the gas stream with pressure is shown schematically in Figure II.29, where:

$$\frac{A\left[2\left(\dfrac{\kappa}{\kappa-1}\right)p_1\rho_1\right]^{\frac{1}{2}}}{\phi_m}$$

has been plotted against p/p_1.

It can be seen that the cross-section of the jet reaches a minimum at

$$\frac{p}{p_1} = \left(\frac{2}{\kappa+1}\right)^{\kappa/(\kappa-1)}$$

At this point the critical conditions A_c, ρ_c, v_c and p_c are present. For the critical velocity v_c we find with equation (II.55):

$$v_c = \left(\frac{2\kappa}{\kappa+1}\frac{p_1}{\rho_1}\right)^{\frac{1}{2}} = \left(\frac{\kappa p_c}{\rho_c}\right)^{\frac{1}{2}} = \left(\kappa\frac{RT}{M}\right)^{\frac{1}{2}} \tag{II.56}$$

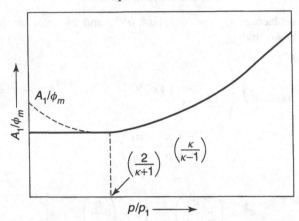

Figure II.29   Flow of gas through an orifice according to equation (II.55)

which is the velocity of sound at p_c and ρ_c (for air at $0\,°\text{C}$, $v_c = 331$ m/s). For gases consisting of two atoms, $\kappa = 1.4$ and consequently $p_c/p_1 = 0.53$.

The expression developed for the mass flow rate (equation (II.55) can be simplified for a number of conditions.

(a) $p_0 \approx p_1$. Under these conditions, the compressibility of gases can be neglected and equation (II.55) simplified to:

$$\phi_m = A\sqrt{2(p_1 - p_0)\rho_1} = CA_1\sqrt{2(p_1 - p_0)\rho_1} \qquad (\text{II.57})$$

where $A =$ cross-section of jet at $p = p_0$, $A_1 =$ cross-section of orifice and $C =$ contraction factor which is approximately $0.6 < C < 0.7$.

(b) $p_c < p_0 < p_1$(for air, $p_c = 0.53 p_1$). Equation (II.55) is valid, A being given by CA_1 with $0.6 < C < 0.7$.

(c) $p_0 \leq p_c(= 0.53 p_1$ for air). When p_0 is so low that at the cross-section A_c the speed of sound is reached, a further decrease of p_0 will not result in a lower pressure at A_c. The reason for this is that pressure waves travel with the velocity of sound; hence as soon as the pressure p_c is reached at A_c the pressure difference $p_1 - p_c$ determines the flow rate completely independently of the value of p_0. Equation (II.55) becomes:

$$\phi_m = \rho_c v_c A_c = A_c \left[\kappa p_1 \rho_1 \left(\frac{2}{\kappa + 1} \right)^{(\kappa+1)/(\kappa-1)} \right]^{\frac{1}{2}}$$

$$= C_K A_c \sqrt{p_1 \rho_1}$$

with:

$$C_K = \left[\kappa \left(\frac{2}{\kappa+1}\right)^{(\kappa+1)/(\kappa-1)}\right]^{\frac{1}{2}}$$ (II.58)

and $A_c = C_c A_1$, where $C_c \approx 0.6$ for sharp-edged orifices and $C_c \approx 0.97$ for rounded orifices. The factor C_K can be calculated for each gas from the known κ (air:$C_K = 0.684$; steam:$C_K = 0.668$).

For flow of gases in pipes or pipe systems the same considerations as those just discussed apply. If $p_0 \approx p_1$, the compressibility can be neglected and the relationships discussed in Section II.2.2 can be applied. If $p_0 < p_1$, the calculation becomes more involved and Figure II.30 can be used for a quick estimation of pressure drop during turbulent flow. This graph is based on air at $17\,^\circ$C. The lines show the calculation of the pressure drop for a mass flow of 0.1 kg/s air at $17\,^\circ$C and 10 atm pressure through a pipe of 25×10^{-3}m diameter and a total friction loss factor of $\sum K_w = 4$. We find $\Delta p = 0.06$atm (for straight pipes $K_w = fSL/A$; for circular pipes at Re $> 10^5$, $K_w \approx 0.02L/D$). We can further read from the graph that under these conditions $v \approx 30$ m/s and $\rho \approx 11$ kg/m^3.

For gases other than air or for other temperatures, the inset displays two correction factors, C_M and C_T. For these conditions, instead of the gas pressure p_0 a corrected pressure $p_0 C_M C_T$ has to be used, and instead of the flow velocity v, the product $v C_M C_T$ is found, as well as $\rho/C_M C_T$ instead of the density ρ. The same applies to the pressure drop determined, which is then $\Delta p C_M C_T$. So, if we had used $CO_2 (M = 44)$ at $100\,^\circ$C, $C_M C_T = 1.3 \times 0.9 = 1.17$, the example drawn in the graph is valid for an initial pressure of 8.5 atm. The flow velocity found is then 25.6 m/s, the density is 12.9 kg/m^3 and the pressure drop would be $\Delta p = 0.051$ atm.

II.3.3 Flow through weirs

The total liquid flow can, in this case, be calculated approximately as a function of the height h of the liquid surface above the edge (see Figure II.31), with Bernoulli's law assuming that mainly potential energy is converted into kinetic energy.

On the upstream side of the weir at a height h, the static pressure is $p_0 + \rho g(h_0 - h)$ (p_0 = ambient pressure), whereas the velocity may be considered negligibly low. The pressure in the overflowing jet is p_0 at all places, whereas the velocity of the streamline coming from height h is termed v_h. Application of the Bernoulli equation yields for the velocity distribution:

$$v_h = \sqrt{2g(h_0 - h)}$$ (II.59)

94

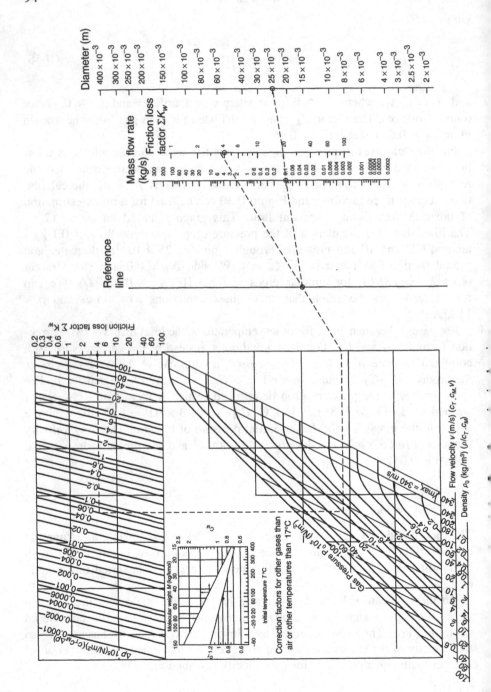

So at $h = 0$ the flow velocity is maximum and $v_h = 0$ at $h = h_0$. Hence the volumetric flow rate in the streamline between h and $h + dh$ is:

$$d\phi_v = v_h W \, dh$$

The total flow rate then becomes (theoretically):

$$\phi_v = W\sqrt{2g} \int_0^{h_0} \sqrt{h_0 - h} \, dh = \tfrac{2}{3}W\sqrt{2gh_0^3}$$

In reality the result must be multiplied by a flow coefficient $C = 0.62$, owing to contraction and some friction loss. Then for $W \gg h_0$ and $h_0 > 0.01\text{m}$, the practical formula becomes:

$$\phi_v = 0.59W\sqrt{gh_0^3} \tag{II.60}$$

A weir can be used for measuring liquid streams with a free surface area. In addition to the rectangular weir there are differently shaped weirs. Figure II.32 gives a survey of the relationships which are applicable to the various weirs.

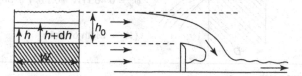

Figure II.31 Flow through a narrow-crested weir

Occasionally a viscous liquid is passed over a narrow-crested weir. In this case, the flow rate is given by:

$$\phi_v = 0.285W\sqrt{gh_0^3}\left\{\frac{gh_0^3}{\nu^2}\right\}^{\frac{1}{2}} \quad \text{if } \frac{\phi_v}{W\nu} < 0.2 \tag{II.61}$$

according to Slocum (*Can. J. Chem. Eng.*, **42**, 1964,).

II.3.4 Flow meters

In order to measure the flow or the flow rate (volume flow rate ϕ_v or mass flow rate ϕ_m) of a fluid one can distinguish between:

A. Displacement meters which divide the gas or liquid stream into known volume units and give the number of these units which have flowed through during

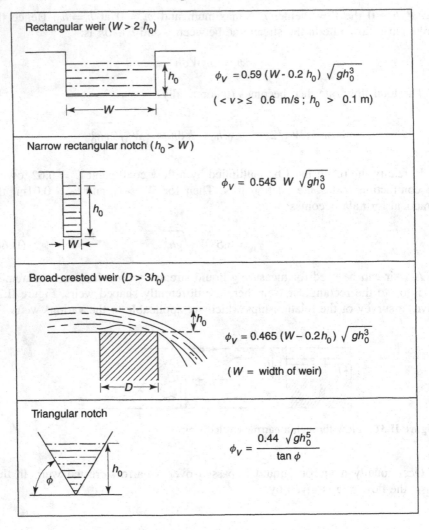

Figure II.32 Flow of water through weirs

a certain time interval (e.g. wet and dry gas meters, various types of petrol gauge).

B. Flow meters with which the mean velocity of the flow can be determined in various ways, e.g.

  (a) dynamic: the indication is based on the flow laws (constriction in the pipe — orifice plate, venturi tube, rotameter and also the weir and the Pitot tube (see the previous section)).

(b) kinematic: the velocity is measured with a mechanical system (small mill, blade wheel, water meter).

(c) thermic: the increase in temperature of the flow as a result of a certain input of electric energy is determinative of the velocity.

We confine ourselves here to treatment of group B(a) and discuss in succession the venturi tube, the orifice plate and the rotameter. These apparatus are designed in such a way that the pressure drop is as low as possible ($E_{fr} \simeq 0$).

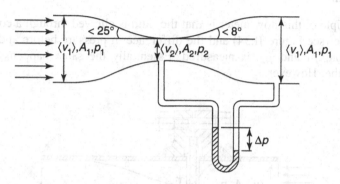

Figure II.33 The Venturi meter

Venturi tube

This tube consists of a gradual constriction (angle 25–30°) and an even more gradual widening (angle maximal 8°) (see Figure II.33). With this construction we can be sure that the stream follows the boundary lines and eddy formation is excluded; p_1 is measured before (or behind!) the constriction and p_2 in the narrowest cross-section (cross-sectional area A_2). For the relationship between ϕ_m and $(p_1 - p_2)$ use is made of the mass balance:

$$\phi_m = \rho \langle v_1 \rangle A_1 = \rho \langle v_2 \rangle A_2$$

and of the Bernoulli equation:

$$p_1 + \tfrac{1}{2}\rho \langle v_1 \rangle^2 = p_2 + \tfrac{1}{2}\rho \langle v_2 \rangle^2$$

from which follows:

$$\phi_m = \left(1 - \frac{A_2^2}{A_1^2}\right)^{-\frac{1}{2}} A_2 \sqrt{2\rho(p_1 - p_2)} \qquad \text{(II.62)}$$

The factor $(1 - A_2^2/A_1^2)^{-1/2}$, if necessary with a correction factor μ_w for the (slight) energy dissipation, is called the flow coefficient α of the Venturi tube.

This factor μ_w is almost 1 (for $0.20 < A_2/A_1 < 0.50$ within 10%, provided the Reynolds number related to the greatest diameter is higher than 2×10^4).

A neat finish of the Venturi tube is a prerequisite, so that it is more expensive than the orifice plate to be discussed below. The pressure loss owing to energy dissipation is, however, small (not more than 20% of the pressure difference measured), so that Venturi tubes are used if extra pressure losses are inadmissible or too expensive (very large pipes, water supply, etc.).

Orifice plate

The principle of this flow meter is that the fluid is allowed through a constriction in the pipe (see Figure II.34) and the difference in pressure before and after the constriction, p_1 and p_2, is measured. Essentially the same happens as in the Venturi tube. However:

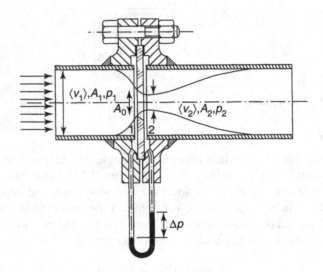

Figure II.34 Sharp-edged orifice meter with corner taps

(a) The smallest cross-section does not coincide with the orifice (cross-section A_0) because the liquid jet after the orifice still constricts to the cross-section $A_2(= \mu A_0)$.

(b) A considerable energy dissipation is given because the flow separates from the wall and eddying occurs on the side of the orifice.

The relationship between the mass flow and the measured pressure difference is expressed by equation (II.62), in which A_2 has been replaced by μA_0. The factor $\mu(1 - \mu^2 A_0^2/A_1^2)^{-1/2}$ times the correction factor for the energy dissipation is called the flow coefficient α for the orifice plate. If necessary, for the transport of gases allowance can be made with the compressibility of the medium by introducing, in addition to α, another correction factor which lies between 0.9 and 1.0 for the usual orifice plates. The flow coefficient α depends on $Re_1 = \langle v \rangle_1 D_1 / \nu$, on the ratio A_0/A_1, on the shape of the orifice and the way in which its edge has been finished and on the places where p_1 and p_2 are measured.

The finish of the orifice has been normalized in many countries.[†]If tabulated values of α are used it should be noted that the prescribed dimensioning is accurately maintained and that the orifice plate is kept at a distance of at least $50 \, D_1$ behind and $10 \, D_1$ before an irregularity (bend, valve, change in diameter, etc.). In the case of careful construction the calibration curve can thus be predicted to within $\pm 1\%$. Orifice plates are preferably dimensioned in such a way that the flow coefficient in the working range is constant (high Re numbers). (See Table II.3.)

Table II.3 Flow coefficient α for orifices with sharp edges and corner taps

A_0/A_1	0.05	0.1	0.2	0.3	0.4	0.5	0.6	0.7
Re	$>10^4$	$>2 \times 10^4$	$>6 \times 10^4$	$>8 \times 10^4$	$>10^5$	$>10^5$	$>2 \times 10^5$	$>4 \times 10^5$
α	0.598	0.603	0.617	0.635	0.660	0.693	0.740	0.805
K_w	1000	240	52	19	8.5	4	2	1

A particular disadvantage of using an orifice plate can be that it causes a considerable pressure loss in the pipe. The kinetic energy of the flow coming out of the orifice is only partly converted into pressure. The resistance value K_w of sharp-edged orifice plates is therefore not zero, as can be seen from Table II.3.

Rotameter

A rotameter is a tapered vertical tube (usually made of glass) containing a float (Figure II.35). The greater the upward fluid flow, the higher the level where the float remains. The rotameter can be considered as an orifice plate with variable free cross-section A_0 (dependent on the position of the float in the tube) and a constant pressure drop which is determined by the dimensions and the weight of the float. The latter can be explained as follows: the pressure on the underside of the float is every where almost p_1 and on the upper surface p_2, their difference balancing the weight of the float (density ρ_f):

$$Vg(\rho_f - \rho) = (p_1 - p_2)A_f$$

[†] Accurate instructions for the design and use of orifice plates can be found in: *V DI-Durchflussmessregeln* (DIN, 1952) and R. F. Streams and others., *Flow Measurements with Orifice Meters*, Nostrand, New York, 1951.

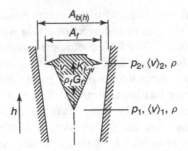

Figure II.35   Rotameter

with A_f being the greatest cross-section of the float, thus making the flow coefficient α of the rotameter dependent on its shape.

II.3.5 Problems

1. A V-shaped weir (aperture 90°) discharges 5.4×10^{-2} m³/s when the water height above its apex is 30 cm. What is its flow coefficient?

   Answer: $C = 0.46$

2. A free-falling liquid jet contracts owing to acceleration in the field of gravitation. For a given jet a 20% contraction is attained at a distance of 8 cm from the outlet opening. Calculate at what distance 20% contraction is found if the outflowing liquid flows 1.5 times as fast.

   Answer: 18 cm

*3. In the bottom of a cylindrical vessel (area of horizontal cross-section $A_1 = 0.5$ m², height $h_0 = 3$ m) there is a sharp-edged round orifice (surface area $A_2 = 5 \times 10^{-4}$ m²). Calculate the time t necessary for the water-filled vessel to drain through orifice A_2. The flow coefficient $C = 0.632$ is assumed to be independent of the water height.

   Answer: 1240 s

4. In the house of Professor T there is a rectangular bath tub. If the tap is fully opened and the plug is put in the drain the bath tub is filled in 20 min. The water height is then 50 cm. When the tap is turned off, the bath filled up to this height empties in 32 min through the drain in the bottom. The flow resistance in the hole is then determinative of the velocity. One day, Professor T wants to take a bath and turns the tap wide open, but in his absent mindedness he forgets to put the plug in the drain. Calculate the eventual height of the water in the bath tub.

   Answer: 32 cm

*5. A gas cylinder has a volume of 40 l and an initial pressure of 200 atm. If we open the main valve (2 mm diameter), how long would we have to wait (at 20 °C) before the pressure is reduced to 10 atm, if the cylinder is filled with:
(a) air;
(b) hydrogen?

Answer: (a) 312 s
(b) 82 s

6. A spray column is applied for the extraction of free fatty acids from edible oils by a solvent, which forms the continuous phase. The oil is supplied under a cylindrical cap (wall thickness 2 mm, height 10 cm), the lower circumference of which is equipped with forty V-notches (equilateral, height 15 mm). Furthermore, the top of the cap has thirty holes (sharp-edged) of 4 mm diameter. What oil flow rate can be handled by this distribution device if the oil height in the notches is allowed to be 10 mm? Which percentage of oil flows through the holes?

$$\rho(\text{oil}) = 800 \text{ kg/m}^3 \quad \rho(\text{solvent}) = 1000 \text{ kg/m}^3$$

Answer: 1.93 m^3/h; 40.2% through holes

*7. Show that in the case presented at the beginning of this chapter, John took the right action when he arrested the fat man.

8. Calculate the approximate dependence of the loss of mechanical energy per unit of time on the flow rate in:
(a) an orifice plate; (b) a rotameter.

Answer: (a) $E_{\text{fr}}\phi_m \sim \phi_v^3$
(b) $E_{\text{fr}}\phi_m \sim \phi_v$

9. (a) Can a rotameter, which has been calibrated for water, be used without renewed calibration for measuring the mass flow of an alcohol stream? If so, calculate the conversion factor.
(b) Can a rotameter destined for measuring a H_2SO_4 flow be calibrated with a glycerol/water mixture? What difficulty is encountered? ($\rho_{\text{alcohol}} = 789$ kg/m^3, $\rho_{\text{float}} = 7000$ kg/m^3)

Answer: (a) 0.89

10. A water supply system contains a device as illustrated in Figure II.36 for measuring the water consumption. In the main pipe (I) an orifice plate M_a with a sharp-edged circular orifice has been incorporated. The parallel pipe (II) contains an orifice plate M_b with a sharp-edged circular opening

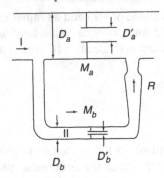

Figure II.36 Water meter

and a rotameter R. The dimensions have been chosen in such a way that the
flow resistance in the parallel pipe is fully determined by the orifice plate.
Calculate the maximum capacity of the rotameter needed at a water flow of
at most 500 m^3/h in the main pipe.

$$D_a = 0.105 \text{ m} \qquad m_a = \left(\frac{D'_a}{D_a}\right)^2 = 0.345$$

$$D_b = 18.8 \times 10^{-3} \text{ m} \quad m_b = \left(\frac{D'_b}{D_b}\right)^2 = 0.065$$

Answer: 2.8 m^3/h

*11. Through a pipe ($D_i = 0.10$ m) flows an amount of water ($T = 20\,^\circ$C) from
12×10^{-3} to 24×10^{-3} m^3/s. The flow must be accurately measured with
a sharp-edged orifice plate with 'corner taps'. The following requirements
should be met:
(i) The flow coefficient of the orifice plate should be constant over the entire
range.
(ii) The pressure difference to be measured should not exceed 68 cmHg.
Determine the limits between which the opening ratio m should be chosen.

Answer: $0.3 \le m \le 0.5$.

Comments on problems

Problem 3

For the volumetric flow rate from the drum, with equation (II.48a):

$$\phi_v = A_1 \frac{dH}{dt} = CA_2 \sqrt{\frac{2(p_1 - p_0)}{\rho}}$$

where A_1 = cross-section of drum, A_2 = cross-section of hole, and H = height of the liquid above the orifice. Furthermore, the liquid pressure is given by:

$$p_1 = p_0 + \rho g H$$

Substitution and integration between $t = 0$, $H = H_0$, $t = t_e$, $H = 0$, yields the effluent time.

Problem 5

For air, $p_c = 0.53 p_1$ and the pressure outside the cylinder (p_0) is 1 atm, whereas the lowest pressure in the cylinder is $p_e(= 10$ atm$)$. Thus $p_c > p_0$, i.e. critical conditions are present and we can use equation (II.58):

$$\phi_m = \frac{d(V\rho_1)}{dt} = V\frac{d\rho_1}{dt} = C_\kappa A_c \sqrt{p_1 \rho_1}$$

Expressing ρ_1 in p_1 with the help of the ideal gas law and integrating between $t = 0$, $p_1 = 200 \times 10^5$ N/m^2 and $t = t_e$, $p_e = 10 \times 10^5$ N/m^2, we can find the effluent time t_e.

Problem 7   John and the barrel of wine

The flow rate of the wine (irrespective of its quality!) from the barrel at any height h is given by:

$$\phi_v = 2CA\sqrt{2gh} = -\frac{\pi}{4}D^2\frac{dh}{dt} \qquad (II.63)$$

where C = flow coefficient ($= 0.82$ according to Table II.2), $A = (\pi/4)d^2$ = cross--section of one hole, D = diameter of barrel and h = height of wine above the holes. Note that we have assumed the barrel to be cylindrical.

Integrating between $t = 0$ and $t = t$ and $h = h_0$ and $h = h$, we find $t \simeq 6.5$ min; in other words, the character running away 20 minutes before John entered could never have fired the fatal shots. For the time being, the behaviour of the fat man is suspicious.

Problem 11

Calculate the Re number for volume flow:

$$Re = \frac{\rho v_1 D_1}{\eta} = \frac{\rho 0.8 v_1 D_1^2}{0.8 \eta D_1} = \frac{\phi_{v_1} \rho_1}{0.8 \eta D_1}$$

$$Re = \frac{12 \times 10^{-3} \times 10^3}{0.8 \times 10^{-3} \times 10^{-1}} = 1.5 \times 10^5 \quad \text{(for minimum flow)}$$

(and for maximum flow, 3×10^5).

From Table II.3, it follows that the upper limit of m is 0.5. In other words, m has to be ≤ 0.5. The lower limit of m is controlled by the maximum pressure that is allowed:

$$\phi_m = \alpha A_0 \sqrt{2\rho \Delta p}$$

$$\phi_m = \alpha m \tfrac{1}{4}\pi D_1^2 \sqrt{2\rho \Delta p}$$

$$m \geq \frac{\phi_m}{\alpha \tfrac{1}{4}\pi D_1^2 \sqrt{2\rho \Delta p}} \geq 0.33$$

The combination of the two conditions to be fulfilled leads to:

$$0.33 \leq m \leq 0.5$$

II.4 Flow around obstacles

> Inspector John relaxed for a moment and read a report of an inventor about a gun which fired almost spherical tear gas grenades at a muzzle velocity of 45 m/s. The report claimed that the grenades could travel about 180 m. John observed that the terminal free-fall velocity of these grenades was 25 m/s and wrote the following text in the margin, before he turned to work again: 'I am afraid that when fired at a distance of 200 m from a crowd, they will drop down just in between the crowd and the police force'. It took his superior one night of studying the following text, before he decided that again John was right.

II.4.1 General approach

Flow around obstacles results in a force F of the flowing medium (with relative velocity v_r) on the obstacle. Generally there are two different mechanisms, which together account for the total force, viz. the friction resistance (a) and the geometric resistance (b).

(a) In the extreme case of flow along a flat plate (Figure II.37), the force is only caused by the friction resistance. This figure shows that the disturbance of the relative flow velocity v_r increases in the y-direction if we move from $x = 0$ to higher values of x. The velocity distribution in this area can be calculated with the aid of the boundary layer theory. This theory shows that the thickness of the hydrodynamic boundary layer is given by:

$$\delta_h = 3\sqrt{\nu \frac{x}{v_r}} \tag{II.64}$$

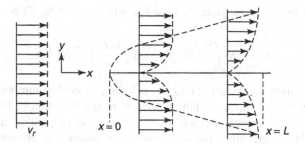

Figure II.37   Flow along a flat plate

This result is somewhat different from what we found in Section II.1.6, when we discussed the penetration of momentum for a stagnant liquid and a moving plate. There we had found that:

$$\delta_p = \sqrt{\pi \nu t} = \sqrt{\pi \nu \frac{x}{v_r}}$$

Apparently, at a given distance, $\delta_h > \delta_p$. This is due to the fact that in the case of the moving liquid there is also a velocity component in the y-direction which increases the thickness of the boundary layer. This velocity component is zero in the case of the moving plate.

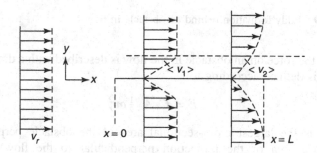

Figure II.38   Mean velocity of flow along a flat plate

The presence of a velocity component v_y in the flowing liquid is illustrated by Figure II.38, where the mean velocities averaged over a constant boundary layer thickness are shown at two values of x. It is evident that $\langle v_1 \rangle > \langle v_2 \rangle$ and therefore some liquid must move upwards at a velocity v_y. The shear stress at the wall is now given by:

$$\tau_{yx}|_{y=0} = \eta \frac{v_r}{\delta_h} = \frac{2}{3} \sqrt{\frac{\nu}{v_r x}} \cdot \frac{1}{2} \rho v_r^2 \qquad \text{(II.65)}$$

and the total force on the plate (both sides, length L, width B) is given by:

$$F = 2B \int_0^L \tau_w \mathrm{d}x = 2.66 \left(\frac{v_r L}{\nu}\right)^{-\frac{1}{2}} BL\tfrac{1}{2}\rho v_r^2 \qquad (\text{II.66})$$

The dimensionless combination $(v_r L/\nu)$ is here a Reynolds number related to the relative velocity v_r and the length dimension in the direction of flow.

(b) If a flat plate with surface area A is placed perpendicular to the flow (Figure II.39) an extreme case of geometric resistance is attained. Before the plate a pressure builds up of the order of magnitude of the thrust $(\tfrac{1}{2}\rho v_r^2)$. At the edge of the plate this thrust has been converted into extra kinetic energy which, however, owing to eddying in the 'dead water' behind the plate, is only partly converted into pressure again. As a result, the order of magnitude of the force on the plate is given by:

$$F = A\tfrac{1}{2}\rho v_r^2$$

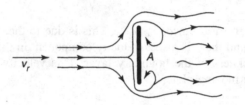

Figure II.39 Eddy formation behind an obstacle in flow

In general, the force on an obstacle in the flow is described with a drag coefficient C_w which is defined according to:

$$F = C_w A_\perp \tfrac{1}{2}\rho v_r^2 \qquad (\text{II.67})$$

where $A_\perp$ is the largest cross-sectional area of the obstacle perpendicular to v_r (= surface area of the projection perpendicular to the flow) and C_w is dependent on $\mathrm{Re} = v_x D/\nu$, with D = characteristic dimension of the obstacle. Hence equation (II.67) has been adapted to the mechanism for the geometric resistance (b).

At high values of Re this effect predominates the friction resistance so that C_w is constant. In Table II.4, a survey is given of the C_w values of a number of obstacles in the turbulent range. In the extreme case that the total momentum flux is used for building up the force on the obstacle, this force is given by

$$F = \phi_v \rho v$$

Comparing this with equation (II.67) we see that always $C_w \leq 2$.

Table II.4 Drag coefficients of various objects ($10^3 <$ Re $< 10^5$)

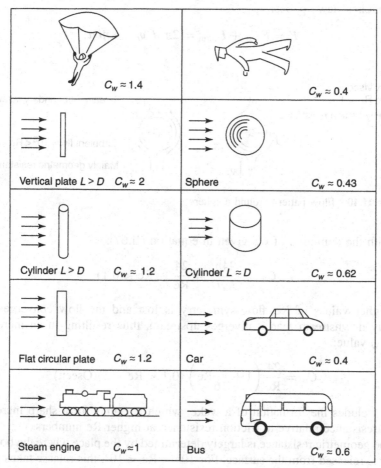

$C_w \approx 1.4$	$C_w \approx 0.4$
Vertical plate $L > D$ $C_w \approx 2$	Sphere $C_w \approx 0.43$
Cylinder $L > D$ $C_w \approx 1.2$	Cylinder $L \approx D$ $C_w \approx 0.62$
Flat circular plate $C_w \approx 1.2$	Car $C_w \approx 0.4$
Steam engine $C_w \approx 1$	Bus $C_w \approx 0.6$

II.4.2 Spherical particles

In engineering we often come across situations where liquid or solid particles are dispersed in a fluid phase. The flow resistance determines the velocity at which they move under the influence of an external force (gravity, centrifugal accelerating force, magnetic force or electric force) with respect to the continuous phase.

The flow field round one single sphere depends on the Reynolds number, which is here defined as Re $= v_r D / \nu$, with D being the external diameter (see Figure II.40). At highly viscous or "creeping" flow (Re < 0.1), no inertia effects occur at all and the streamlines converge behind the sphere as they diverge on inpact, i.e. the streamlines show complete symmetry (see Figure II.40). In this case, Stokes' law applies:

$$F = F_{\text{drag}} + F_{\text{form}} = 2\pi\eta D v_r + \pi\eta D v_r$$

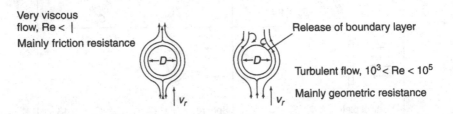

Figure II.40   Flow pattern around a sphere

or, with the definition of C_w given in equation (II.67):

$$C_w = \frac{24\eta}{\rho v_r D} = \frac{24}{\text{Re}} \quad (\text{Re} < 0.1)$$

At higher values of Re, flow symmetry is lost and the flow converges more rapidly downstream than it diverges upstream, thus resulting in an increase in the C_w value:

$$C_w = \frac{24}{\text{Re}} \left(1 + \frac{3}{16}\text{Re}\right) \quad (0.1 < \text{Re} < 5)(\text{Oseen})$$

This preludes the formation of a wake, which accounts for a sharp increase in form resistance relative to friction resistance at higher Re numbers.

The geometric resistance is largely determined by the place where the boundary layer is released from the surface. For $10^3 < \text{Re} < 10^5$, this release takes place a short distance behind the centre of the sphere where C_w is constant and approximately 0.43 (Newton's law of resistance). At $\text{Re} > 2 \times 10^5$, the friction boundary layer becomes turbulent, as a result of which the point where the boundary layer is released shifts to the back so that C_w decreases fairly rapidly. Because of the two mechanisms stated above, the drag coefficient C_w is a continuous function of Re opposite to the friction loss factor f for flow in pipes.

The stationary rate of fall, v_s, of a sphere in a stagnant medium is calculated by assuming its apparent weight to be equal to the resistance force:

$$\frac{\pi}{6}d_p^3(\rho_p - \rho)g = C_w \frac{\pi}{4}d_p^2 \tfrac{1}{2}\rho v_s^2$$

The solution for v_s is found by solving this equation together with the relationship for C_w as a function of Re. This can only take place analytically in the Stokes range (Re < 1):

$$v_s = \frac{(\rho_p - \rho)g d_p^2}{18\eta} \quad (\text{Re} < 1) \tag{II.68}$$

or, in the Newton range ($10^3 < \text{Re} < 10^5$):

$$v_s = 1.76\sqrt{\frac{(\rho_p - \rho)g d_p}{\rho}} \quad (10^3 < \text{Re} < 10^5) \tag{II.69}$$

In the intermediate range ($1 < \text{Re} < 10^3$) which is often encountered, only a numerical solution is possible. The force balance is then written as follows:

$$C_w \text{Re}^2 = \frac{8}{6}\frac{d_p^3 \rho(\rho_p - \rho)g}{\eta^2} \tag{II.70}$$

The right-hand term of this equation is a known constant for a given problem. With the help of Figure II.41 the relevant Re number can now be found, from which the stationary rate of fall can be calculated according to:

$$v_s = \frac{\eta \text{Re}}{\rho d_p}$$

The relationships (II.68), (II.69) and (II.70) are also valid for gas bubbles rising in a liquid if they have a rigid surface. Because of the upward direction of the force, instead of $(\rho_p - \rho)$, of course $(\rho - \rho_b)$ has to be used. Gas bubbles with diameters <0.8 mm are spherical and the drag coefficients for spheres can be applied. Bubbles with diameters $0.8 < d_b < 1.5$ mm form oblate spheroids, the drag coefficient of which is approximately two thirds of the C_w value of a sphere with the same volume.

Gas bubbles in liquids have a rigid surface as long as:

$$d_b < \sqrt{\frac{\sigma}{g\Delta\rho}} \tag{II.71}$$

If surface active agents or dirt are present in the liquid, the mobility of the surface is decreased and equation (II.71) gives too low values for the critical value of d_b. Thus, air bubbles in water are rigid if $d_b < 2.6$ mm. For bubbles with a mobile surface the flow resistance is lower. The stationary rate of rise is then given by:

$$v_s = \sqrt{\frac{2\upsilon}{\rho d_b} + \frac{\Delta\rho}{\rho}\frac{g d_b}{2}} \quad \left(d_b > \sqrt{\frac{\sigma}{g\Delta\rho}}\right) \tag{II.72}$$

Figure II.42 illustrates the rising velocity of air bubbles of various diameters in water. According to equation (II.72), bubbles with diameters of $d_b = 2\sqrt{\sigma/g\Delta\rho}$

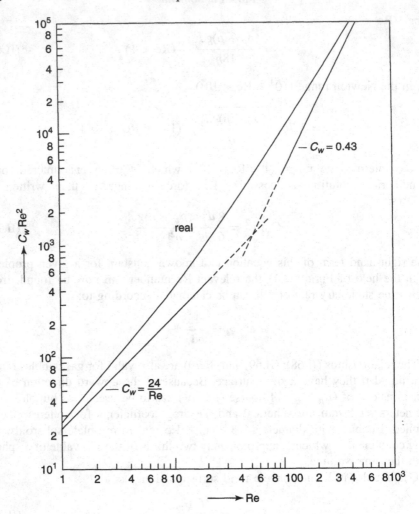

Figure II.41 Drag coefficient and related functions for spherical particles; the real situation compared to the two extremes

have the minimum rate of rise. For very large bubbles (for air in water if $d_b >$ 15 mm), equation (II.72) is simplified to:

$$v_s \approx \sqrt{\frac{\Delta \rho}{\rho} \frac{g d_b}{2}} \quad (\rho g d_b^2 \gg \sigma) \tag{II.73}$$

The rising gas bubbles then have the form of spherical caps. Comparison of equations (II.73) and (II.70), written for half a sphere, shows that for these

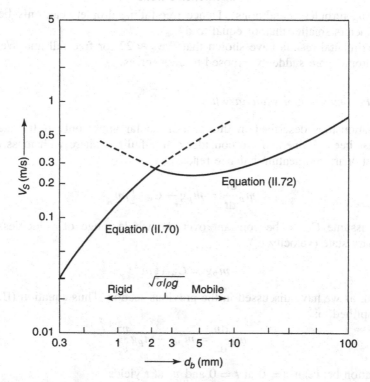

Figure II.42   Velocity of rise of air bubbles in water

conditions $C_w = \frac{4}{3}$ at a Re value of >12000 (air in water). This C_w value is comparable with that of a circular flat plate (see Table II.4).

II.4.3 Free fall of droplets

The friction forces acting on a falling liquid droplet will deform the droplet from a purely spherical to an elongated form until finally the droplet breaks up. The surface tension counteracts this deformation and we can assume roughly that the droplets will no longer break up if this ratio is equal to or smaller than unity:

$$\frac{\text{friction force}}{\text{surface tension}} = \frac{C_w \frac{\pi}{4} d_p^2 \frac{1}{2}\rho_g v^2}{\pi\sigma d_p} = \frac{C_w d_p \rho_g v^2}{8\sigma} \le 1$$

In the dimensionless group we recognize the Weber number, the critical value of which should then be of the order of magnitude of:

$$\text{We}_{cr} = \frac{\rho_g d_{p\,\text{max}} v^2}{\sigma} \approx \frac{8}{C_w} \approx 19 \tag{II.74}$$

(assuming turbulent conditions). Hence free-falling droplets can only be stable for diameters smaller than or equal to $d_{p\,max}$.

Experimental results have shown that $We_{cr} \approx 22$ for free fall and $We_{cr} \approx 13$ if the droplets are suddenly exposed to drag forces.

II.4.4 Particles in non-stationary flow

All relationships described in this section so far apply only if the stationary state has been attained. If we consider a free-falling sphere with mass m_p, the non-stationary momentum balance reads:

$$m_p \frac{dv}{dt} = m_p g - C_w A \tfrac{1}{2} \rho v^2 \tag{II.75}$$

if we assume C_w to be constant over the whole range of velocities. In the stationary state (velocity v_s):

$$m_p g = C_w A \tfrac{1}{2} \rho v_s^2 \tag{II.76}$$

is valid, as we have discussed in the previous section. Thus equation (II.75) can be simplified to:

$$m_p \frac{dv}{dt} = m_p g - m_p g \frac{v^2}{v_s^2}$$

Integration between $v = 0$ at $t = 0$ and v_1 at t yields:

$$\frac{1 + \dfrac{v}{v_s}}{1 - \dfrac{v}{v_s}} = \exp\left(\frac{2gt}{v_s}\right) \tag{II.77}$$

In Figure II.43, v/v_s has been plotted against the dimensionless factor $2gt/v_s$. It appears that the velocity is stationary if $2gt/v_s > 5$. For $2gt/v_s < 1$, the velocity can be represented with good approximation by:

$$\frac{v}{v_s} = \frac{gt}{v_s}$$

i.e. $v = gt$; the relationship applies to free fall in vacuum (no friction losses). This means that up to a time $t < v_s/2g$, friction loss may be neglected, which justifies our assumption of a constant drag coefficient in equation (II.76).

Velocities of free fall are often measured by measuring the time for free fall over a distance L. In this way, the mean velocity $\langle v \rangle = L/t$ is determined. The mean velocity can be calculated as:

$$\frac{\langle v \rangle}{v_s} = \frac{L}{t} \int_0^t \frac{v}{v_s} dt$$

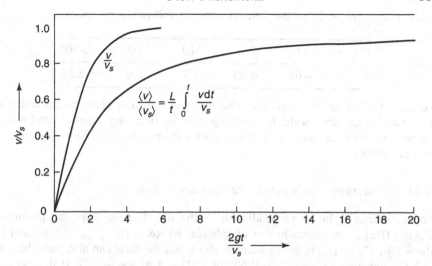

Figure II.43 Non-stationary fall of spherical particles

which yields with equation (II.77):

$$\frac{\langle v \rangle}{v_s} = \frac{v_s}{2gt} \ln \left[\frac{2 + \exp\left(\dfrac{2gt}{v_s}\right) + \exp\left(-\dfrac{2gt}{v_s}\right)}{4} \right] \tag{II.78}$$

This function is also shown in Figure II.43. It can be used to calculate (via trial and error, see problem 4, p. 115) the stationary velocity of free fall from a measured mean velocity.

II.4.5 Rate of sedimentation of a swarm of particles

The rate of sedimentation of a swarm of equally large particles in a liquid is lower than that of one separate particle. In a given system of particles and liquid this rate of sedimentation $(v_s)_s$ appears to depend only on the fraction α of the volume which is taken up by the particles. This constant is also connected with the porosity ε: this is the volume fraction of the continuous phase so that $\varepsilon = 1 - \alpha$. There are various theories about the relationship between $(v_s)_s$ and α or ε. A well-usable empirical relation is that of Richardson and Zaki:

$$(v_s)_s = v'_s \varepsilon^n \tag{II.79}$$

In this equation, v'_s is the rate of settling of one single particle of the system in question $(\varepsilon = 1)$ and n a value dependent on $v'_s d_p / \nu = \text{Re}'_s$ and on the relationship of the particle diameter and the diameter of the vessel D_t in which they settle. For $d_p/D_t < 0.1$, these authors give the following values of n:

Re'_s	<0.1	1	10	100	>500
n	4.65	4.35	3.53	2.80	2.39

Equation (II.79) has the advantage that the exponent n is given as a function of a Re number defined for the free settling velocity of a single particle and not, as happens in most literature, as a function of a Re number based on the unknown swarm velocity.

II.4.6 Cylinders perpendicular to the direction of flow

For an infinite cylinder perpendicular to the direction of flow, the picture of $C_w = f(\text{Re})$ is analogous to that of spheres. At $\text{Re} < 10^{-1}$, $C_w \sim 1/\text{Re}$ and for $\text{Re} > 10^3$, $C_w \simeq 1.2$. If the cylinder is shortened the flow can also pass the ends and C_w becomes lower ($C_w = 0.62$ for $L/D = 1$ at $\text{Re} > 10^3$). If the cylinder is not given a round but a streamlined cross-section, C_w decreases considerably to values between 0.01 and 0.1. If, as in another extreme case, it is flattened (perpendicular to the direction of flow), then $C_w \simeq 2$ for $L/D = \infty$.

In practice, banks of tubes are often used as heating (or cooling) elements in a liquid or gas flow. In flow perpendicular to the tubes, two different ways of arrangement can be distinguished:

(a) in line and (b) staggered (see Figure II.44).

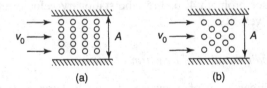

(a)   (b)

Figure II.44 Flow perpendicular to a bank of pipes: (a) in line, (b) staggered

In case (a), most pipes are in the 'wake' of their predecessor, while in case (b) they are not. Owing to this shielding, the resistance of arrangement (a) is lower than that of (b) and so is the heat transfer. For case (b), as a rule of thumb, we can use the fact that the resistance coefficient of a tube (related to the superficial velocity) in the range $30 < \text{Re} < 2000$ is about 1.5 times as high as for a single tube in the flow if the distance between two neighbouring tubes equals the tube diameter.

Naturally, it also occurs quite often that, for example, in the case of heat exchangers, the resistance for flow *along* a bank of tubes must be calculated. For turbulent flow, use is then made of the Fanning equation (II.35) and the hydraulic diameter.

II.4.7 Problems

1. What is the maximum diameter of a rain droplet at 20 °C?

   Answer: $d_p = 6.6$ mm

*2. When driving his car (1.6 m wide, 1.3 m high, 950 kg) at a steady 105 km/h on a highway on a windless day, Klaus notices that, if he pushes the clutch down, it takes 5 s for the car to decelerate to 95 km/h.
   (a) Estimate the maximum value of the drag coefficient.
   (b) Driving steadily, which percentage of the petrol used ($10\ell/100$ km) is converted into mechanical energy, if the heat of combustion is 11000 kcal/kg ($\rho = 700$ kg/m^3)?

   Answer: (a) $C_w < 0.55$
            (b) 15.5%

3. Two metal spheres of equal weight, but of different diameter, fall through air. Calculate the relationship of their stationary rate of fall if the ratio of their diameter is 3 and the flow round the spheres is dynamically similar. Is this supposition correct?

   Answer: $\frac{1}{3}$; yes

4. The 'Euromast' tower in Rotterdam has a height of 104 m. It takes a cherry (diameter 1.7 cm) 6.5 s to fall from the top of the tower to the park below.
   (a) What would be the velocity of stationary free fall?
   (b) What is the drag coefficient?

   Answer: (a) 20 m/s;   (b) $C_w = 0.45$

5. According to a newspaper report, the air in Mexico City (situated at 2200 m above sea level) contains (per volume unit) 25% less oxygen than the air at sea level.
   (a) Ascertain that this statement is correct by calculating from a balance of forces the pressure as a function of the height above sea level.

        Given: Air is an ideal gas.

                 Gas constant $R = 8314$ J/kmol K

                 Air temperature $= 27°$ C

                 'Molecular weight' of air $= 29$ kg/kmol

This lesser air density in Mexico City causes sprinters to meet with lower air resistance.

(b) In what time will a sprinter theoretically be able to cover the 100 m in Mexico City because of this lower air resistance, if he does the 100 m at sea level in 10 s?

   Assume: (i) that both in Mexico City and at sea level he develops the same power.

   (ii) that he runs at constant speed.

   (iii) that the time necessary to pick up speed is negligible.

(c) His actual time in Mexico City is, however, likewise 10 s. What is your conclusion?

   Answer: (b) 9.2 s

   (c) Assumptions (ii) and (iii) are wrong.

6. The gas bubbles formed at the bottom of a glass of beer are released as soon as their diameter is ca. 1 mm.

   (a) Calculate their stationary rate of rise.
   (b) How many times more rapidly would an equally large water drop fall in air?

   Answer: (a) 0.105 m/s; (b) 35 times

*7. A horizontal air duct with rectangular cross-section ($H \times B = 30 \times 30$ cm^2; $L = 10$ m; $\bar{x}/D_h = 2 \times 10^{-3}$) contains an air heating element. This element consists of six rows of fifteen tubes placed transversely to the direction of flow. The rows are installed in a staggered position (Figure II.27b). The external diameter of the tubes is $D = 2$ cm, the mean temperature in the duct is 80 °C and the mean pressure is 1.3×10^5 N/m^2. Calculate the pressure drop in this duct at a mean air velocity of 20 m/s.

   Answer: 2980 N/m^2

*8. Show that John did indeed gave the right answer to the problem of the gas grenade treated at the beginning of this Section.

Comments on problems

Problem 2

Klaus' data collected in a quiet hour on the E10 motorway will help us to calculate a drag coefficient that includes not only air resistance effects but also friction losses. The actual C_w value will therefore be smaller than the value we calculate from the unsteady-state momentum balance:

$$-M\frac{dv}{dt} = C_w A \tfrac{1}{2}\rho v^2$$

Now, from the measured data at $\bar{v} = 100$ km/h:

$$\frac{dv}{dt} = \frac{10 \text{ km/h}}{5 \text{ s}} = 0.556 \text{ m/s}^2$$

and we find:

$$C_{w\,max} = \frac{950 \times 0.556}{2.08 \times 0.60 \times 772} = 0.55$$

If we assume the friction energy loss of the rolling wheels on the road to be of the same order of magnitude as the energy loss due to air friction, we find $C_w \approx 0.30$, which compares well with the value stated in Table II.4.

The total energy content of 10 l of petrol is:

$$10 \times 10^{-3} \times 700 \times 11\,000 \times 4.2 \times 10^3 = 327 \times 10^5 \text{ J}$$

and the mechanical energy for moving the car 100 km:

$$-M\frac{dv}{dt}L = 50 \times 10^5 \text{ J}$$

Thus the percentage transformed into mechanical energy is ~15.5%.

Problem 7

The pressure drop in the channel is caused by friction along the wall and by the flow resistance of the pipes, and we can assume that these two resistances are additive. The pressure drop in the empty channel is then found to be (using $D_h = 0.3$ m, $Re_h \sim 3 \times 10^5$, $4f = 0.024$; equation (II.35)):

$$\Delta p_1 = 200 \text{ N/m}^2$$

The drag coefficient for one single pipe would be (with $Re = 2 \times 10^4$; Table II.4):

$$C_w = 1.2$$

Therefore, for staggered tubes, $C'_w = 1.5C_w = 1.8$. Thus the force on one pipe is:

$$F = C'_w A \tfrac{1}{2}\rho v^2 = 2.66 \text{ N}$$

The force on all pipes together is then $F_{tot} = 6 \times 15F = 239$ N and the pressure drop:

$$\Delta p_2 = 239/0.09 = 2660 \text{ N/m}^2$$

Thus the total pressure drop is:

$$\Delta p_1 + \Delta p_2 = 200 + 2660 = 2860 \text{ N/m}^2$$

Problem 8 John and the tear gas grenades

An unsteady-state momentum balance in the x-direction for this case (see the figure below) reads:

$$M\frac{dv_x}{dt} = -C_w\frac{1}{2}\rho v_x^2 A \tag{1}$$

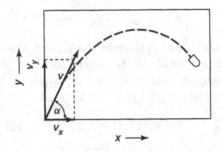

which yields after integration between $t = 0$, v_{x0} and t, v_x:

$$\frac{1}{v_x} - \frac{1}{v_{x0}} = \frac{C_w\rho A t}{2M} \tag{2}$$

Now $v_x = dx/dt$ and after substituting v_x from equation (2) and integration between $t = 0$, $x = 0$ and t_e, x_e we find:

$$x_e = \frac{2M}{C_w\rho A}\ln\left(1 + \frac{C_w\rho A v_{x0}t_e}{2M}\right) \tag{3}$$

Now, for steady-state free fall:

$$Mg = C_w\frac{1}{2}\rho v_s^2 A$$

and thus:

$$2M/C_w\rho A = v_s^2/g \tag{4}$$

Considering now the velocity component in the y-direction (and neglecting friction losses because $g \gg C_w(\rho v^2 A/2M)$) we have:

$$\frac{dv_y}{dt} = -g; v_y = \frac{dy}{dt} = -gt \text{ and } y = v_{y0}t - \frac{1}{2}gt^2$$

So, for $y = 0$:

$$t_e = 2v_{y0}/g$$

Introducing this expression and equation (4) into equation (3) we obtain:

$$x_e = \frac{v_s^2}{g} \ln \left(1 + \frac{2v_{x0}v_{y0}}{v_s^2} \right)$$

Now:

$$v_{x0}v_{y0} = v_0^2 \sin\alpha \cos\alpha$$

which is maximum for $\alpha = 45°$, i.e. $v_{x0} = v_{y0} = 0.71v_0$, and so the maximum distance is:

$$x_{max} = \frac{v_s^2}{g} \ln \left(1 + \frac{v_0^2}{v_s^2} \right) = 90 \text{ m}$$

If you, as did the poor inventor, neglect the air friction, you find for the maximum distance:

$$x_e = \frac{2}{g} v_0^2 \sin\alpha \cos\alpha$$

which yields for $x_{e\,max} = 202$ m!

II.5 Flow through beds of particles

The police doctor rose and stated that the man had died by town gas poisoning not longer than 10 hours ago. 'Remarkable', John thought. The room where they found the body was locked from the inside and there was no smell of gas at all, although concentrations above one-thousandth of the lethal level can be smelt quite easily. The gas valve was open but due to an explosion at the gas works the supply ceased 8 hours before. A small fan in one of the windows pressed air into the room so that the pressure was 80 mm (water column) higher than outside the room. The room had a volume of 80 m^3 and there was about 50 m^2 of wall (30 cm thick), the porosity of the material being approximately 20% and the particles which constituted most of the material having a diameter of ca. 0.3 mm. Lighting a cigarette, John concluded that the man had died at another place.

The flow through a bed of particles is of great importance for all branches of technology; in chemical engineering for filtration, catalytic processes, separation processes in columns with packing materials, etc., and also for petroleum production, soil mechanics and hydraulics. A great deal of theoretical and experimental work has been carried out by various groups. Outwardly, the various formulae we encounter often differ, but on second thoughts they have much in common.

Two types of particle bed are of great technical importance: the fixed bed and the fluidized bed. In the former bed, the particles rest on each other and are not moved by the fluid. If in this bed the velocity of the fluid is increased, a situation is created in which the force on the bed becomes equal to the apparent weight of the particles. The flow lifts the bed, as a result of which the particles become suspended. This is called the fluidized state which is characterized by a thorough mixing, especially of the particles. The result is that temperature- and/or concentration differences in the fluidized bed are rapidly eliminated, so that in the bed uniform conditions prevail which are often desired. If the velocity of the fluid is increased still further, the particles are ultimately blown out of the bed and we enter the field of pneumatic transport. We shall now discuss some essential points of the flow in the two types.

II.5.1 Fixed bed

This bed may be considered as a collection of particles and also as a collection of interconnected winding channels. We stick to the former description in order to calculate the energy dissipation per unit of mass in a bed of spherical particles as a function of the superficial velocity, v_0, related to the cross-section of the tube which forms the boundary of the bed.

The actual velocity $\langle v \rangle$ in the bed is of course higher than v_0, notably $v_0 = \varepsilon \langle v \rangle$ if ε again represents the volume fraction voids (the porosity) of the bed (for beds of spheres, in practice, $0.35 < \varepsilon < 0.45$). The train of thought is now as follows: we calculate the energy which is dissipated around one particle in a flow of velocity $\langle v \rangle$; this energy is multiplied by the total number of particles in the bed in order to obtain the constant $E_{fr}\phi_m$, i.e. the total energy which is dissipated in the bed (see also equation (II.41)). To this end, we base our calculations on the definition equation for C_w (II.67) to determine the force which the flow exerts on one particle:

$$F = C_w \frac{\pi}{4} d_p^2 \frac{1}{2} \rho \langle v \rangle^2$$

The drag coefficient in this relationship is self-evidently different from the one which is found for one sphere in an infinite fluid. From the force F, we can now calculate, by multiplication by $\langle v \rangle$, the energy which is dissipated around one sphere. The total number of spheres in the bed is subsequently calculated from the volume the spheres take up in the bed, $(1 - \varepsilon)AL$, where L is the height of the bed. So:

$$E_{fr}\phi_m = E_{fr}\rho v_0 A = \frac{(1-\varepsilon)AL}{\frac{\pi}{6}d_p^3} C_w \frac{\pi}{4} d_p^2 \frac{1}{2} \rho \langle v \rangle^3$$

or:

$$E_{fr} = \frac{3}{2} C_w \frac{1}{2} v_0^2 \frac{1-\varepsilon}{\varepsilon^3} \frac{L}{d_p}$$

At a sufficiently high velocity in the bed (high Re number) the flow is turbulent and, as we know, C_w is constant. Ergun found for this range $C_w = 2.3$.

At low velocities in the bed, the flow is no longer turbulent, but shows eddy currents. In this region (which is also called 'laminary'), C_w is expected to be inversely proportional to the Re number. The difficulty here is to define the Re number which is characteristic of this flow. The characteristic velocity is, of course, $\langle v \rangle$. However, the characteristic cross-sectional dimension of the flow between the particles is not so much determined by the particle diameter as by the space remaining between the particles. Now it appears that the hydraulic radius A/S of the cavities between the particles can be used for correlating C_w as a function of Re_h ($\equiv 4\rho\langle v \rangle R_h/\eta$) for all packed beds. This hydraulic radius R_h is connected with ε and the specific surface area of the particles S_v:

$$R_h = \frac{A}{S} = \frac{\text{surface of cross-section of cavity}}{\text{circumference cavity}} = \frac{\text{volume cavity}}{\text{wall surface cavity}}$$

$$= \frac{\text{volume cavities/m}^3}{\text{wall surface cavities/m}^3} = \frac{\text{porosity (m}^3/\text{m}^3)}{\text{specific wall surface}} = \frac{\varepsilon}{S_v}$$

The specific surface area is, in turn, again a function of ε and the particle diameter, namely:

$$S_v = \frac{6(1-\varepsilon)AL}{\pi d_p^3}\pi d_p^2 \frac{1}{AL} = \frac{6(1-\varepsilon)}{d_p}$$

so that:

$$\text{Re}_h = \frac{2}{3}\frac{\rho v_0 d_p}{\eta(1-\varepsilon)}$$

Now it appears that for $\text{Re}_h < 1$, $C_w = 152/\text{Re}_h$. This relationship is known as the Carman formula, but is also named after Blake and Kozeny.

According to Ergun, a good description of the flow resistance in the entire Re range is obtained by adding the relationships for laminar and turbulent conditions, with the result:

$$\frac{\Delta p}{\rho L} = \frac{E_{\text{fr}}}{L} = \frac{v_0^2}{d_p}\frac{1-\varepsilon}{\varepsilon^3}\left[170\frac{v}{v_0 d_p}(1-\varepsilon) + 1.75\right] \qquad \text{(II.80)}$$

It appears that the flow may be considered as turbulent if $\text{Re}_h > 700$. The constant $v_0 d_p/v$ is again a Reynolds number and is often used in the case of flow through fixed beds, although somewhat incorrectly. Restrictions in the use of equation (II.80) are that the particles must not deviate too much from the sphere shape and that the diameter of the particles must be considerably smaller than the diameter of the bed ($< \frac{1}{20}$).

For a number of forms of non-spherical particles, E_{fr} has also been determined. In such a case, instead of d_p in equation (II.80) the diameter $\overline{d}_p$ of a sphere having the same volume/diameter ratio as the combined particles in the bed should be used, and the pressure drop so found should be multiplied by a correction factor k.

Table II.5 Factor k with which equation (II.80) underestimates E_{fr} for some agricultural crops (purified products, 'random' arrangement)

Product	k
Peas	1.05
Rapeseed	1.2
Potatoes, beans	1.4
Clover seed, wheat	1.7
Rye, summer barley	2.7
Maize	3.2
Sugar beets, carrots	3.5
Oats	3.8

In Table II.5, the correction factors for a number of agricultural and horticultural products are given.

For asparagus, rather divergent k values are found, just as, for example, for flow through yarn packets and yarn filters. Here it is important to find out whether these cylindrical-shaped bodies lie mainly in the direction of flow ($k \simeq 0.75$) or mainly perpendicular to the flow ($k \simeq 1.1$). In these cases, there is hardly ever a question of 'random' arrangement.

II.5.2 Filtration through a bed of particles

Equation (II.80) can also be used for an idealized description of the filtration process, which is practically always carried out under laminar flow conditions. Rearrangement yields, if V is the amount of filtrate at time t:

$$v_0 A = \phi_v = \frac{dV}{dt} = \frac{\varepsilon^3}{170(1-\varepsilon)^2} \frac{\Delta p A d_p^2}{\eta L} \tag{II.80a}$$

During filtration, the suspended particles will collect on the filter medium (e.g. filter paper, cloth) and form a cake. If we neglect the flow resistance of the filter medium, the term L in equation (II.80a) is the thickness of the filter cake, which is a function of the weight fraction w of solids in the suspension and the amount of filtrate V already produced. A material balance shows:

$$LA(1-\varepsilon)\rho_p = w\rho_1 V \tag{II.81}$$

Substituting L into equation (II.80a) and integrating between $V = 0$ at $t = 0$ and V, t, we find for the amount of filtrate at any time:

$$V = d_p A \sqrt{\frac{\varepsilon^3}{85(1-\varepsilon)} \frac{\Delta p \rho_p}{\eta \rho_1 w} t} \tag{II.82}$$

and for the rate of filtration:

$$\phi_v = \frac{dV}{dt} = d_pA\sqrt{\frac{\varepsilon^3}{340(1-\varepsilon)}\frac{\Delta p \rho_p}{\eta \rho_1 wt}} \qquad \text{(II.83)}$$

The derivation of equations (II.82) and (II.83) was based on the following assumptions:

(a) constant ε, i.e. the filter cake is incompressible;
(b) constant Δp, i.e. the filtration is carried out at constant pressure;
(c) spherical particles;
(d) no flow resistance of filter medium.

In practice, these assumptions are not always permissible. For a given system the influence of the flow resistance of the filter medium can be allowed for by adding, in equation (II.80a), to the resistance of the filter cake which is represented by:

$$R'_c = \frac{170L(1-\varepsilon)^2}{d_p^2\varepsilon^3}, \qquad \text{(II.84)}$$

a resistance R_m for the filter medium. We obtain

$$\phi_v = \frac{dV}{dt} = \frac{\Delta pA}{\eta(R'_c + R_m)} \qquad \text{(II.85)}$$

Substituting L in equation (II.84) from equation (II.81) we find:

$$R'_c = f\frac{V}{A}\frac{\rho}{\rho_p}\frac{170(1-\varepsilon)}{\varepsilon^3 d_p^2} = w\frac{V}{A}R_c$$

With this expression equation (II.85) becomes:

$$\phi_v = \frac{dV}{dt} = \frac{\Delta pA}{\eta\left(w\dfrac{V}{A}R_c + R_m\right)} \qquad \text{(II.86)}$$

Integration with $V = 0$ at $t = 0$ yields:

$$\frac{1}{2}\left(\frac{V}{A}\right)^2 wR_c + \left(\frac{V}{A}\right)R_m = \frac{\Delta p}{\eta}t \qquad \text{(II.87)}$$

or:

$$\left(\frac{A}{V}\right)t = \frac{1}{2}\frac{\eta}{\Delta p}wR_c\left(\frac{V}{A}\right) + \frac{\eta}{\Delta p}R_m \qquad \text{(II.87a)}$$

Thus a plot of $(A/V)t$ versus V/A should yield a straight line with slope $\frac{1}{2}(\eta/\Delta p)wR_c$, whereas the extrapolated value of $(A/V)t$ for $V/A = 0$ yields

$(\eta/\Delta p)R_m$, which enables us to determine the resistances of the filter cake and the filter medium R_c and R_m.

II.5.3 Fluidized bed

Important factors for the fluidized bed are the velocity v_{0i}, at which the fixed bed changes into the fluidized bed, the maximal velocity at which the fluidized bed can only just exist (before pneumatic transport occurs) and the expansion of the bed as a function of the velocity.

The velocity v_{0i} can be calculated from the foregoing (E_{fr} for a fixed bed) if it is borne in mind that at this velocity the force of the fluid on the particles is just equal to the apparent weight of the particles:

$$\Delta pA = g(1 - \varepsilon)(\rho_p - \rho)AL$$

Using the Ergun equation (II.80) it follows that:

$$\frac{v_{0i}^2 d_p^2}{v^2} = \left[170\frac{v}{v_{0i}d_p}(1 - \varepsilon) + 1.75\right]^{-1} \frac{g\rho d_p^3}{\eta^2}(\rho_p - \rho)\varepsilon^3 \qquad \text{(II.88)}$$

The maximal velocity at which the fluidized bed can just exist occurs in a highly dilute bed ($\varepsilon \to 1$) and is equal to the stationary rate of fall v_s of a single particle.

The expansion of the bed as a function of v_0 is more difficult to predict because the flow condition in the fluidized bed is in principle unstable. A gas stream, particularly, tends to move rapidly in large bubbles through the bed (aggregative fluidization). A bed fluidized with liquid (provided its length is not much greater than its diameter) is, as a rule, fairly uniform (particulate fluidization). In the latter case, the bed expansion can be calculated by using the relation of Richardson and Zaki for the rate of settling of a swarm of particles (Section II.5.5). It should be borne in mind that in the fluidized bed the relative velocity between the fluid and the particles is $\langle v \rangle = v_0/\varepsilon$.

The relative velocity between the phases in a settling test (in a vessel with a bottom) must be calculated from the velocity of the swarm $(v_s)_s$. The settling particles displace liquid upwards, so that on settling the relative velocity v_r is equal to the velocity of the swarm plus the velocity of the liquid. Since the volume of particles which settles per unit of time equals the volume of liquid which is displaced per unit of time, the velocity of the liquid on settling is equal to $(1 - \varepsilon)(v_s)_s/\varepsilon$. So, the relative velocity between fluid and particles is:

$$v_r = (v_s)_s + \frac{1 - \varepsilon}{\varepsilon}(v_s)_s = \frac{(v_s)_s}{\varepsilon}$$

Hence the expansion of a uniformly fluidized bed is given by Richardson and Zaki as follows:

$$\varepsilon = \left(\frac{\langle v \rangle}{v_s'}\right)^{1/(n-1)} \quad \text{or} \quad \varepsilon = \left(\frac{v_0}{v_s'}\right)^{1/n}$$

II.5.4 Problems

*1. Equation (II.80) can, apart from numerical factors, also be derived from the Fanning equation (II.35) by regarding the bed as a complex of irregular channels with total length L and hydraulic diameter $4A/S$. Check this statement and calculate that the following applies for the bed: $4fL'/L = 152/Re_h$ for $Re_h < 1$ and $4fL'/L = 2.3$ for $Re_h > 700$.

*2. In a vertical tube ($D_i = 10$ cm, $L = 25$ cm), almost spherical anthracite particles are fluidized with air. The particle size distribution is shown in the following table:

d_p (mm)		Mass (%)
from	to	
1.00	0.84	20
0.84	0.71	29
0.71	0.59	23
0.59	0.50	25
0.50	0.42	3

The density of the anthracite (measured with a pyknometer with mercury) is 1970 kg/m^3, the temperature is 32 °C and the pressure above the bed is 10^5 N/m^2.

(a) At which mass flow rate does the fluidization start?

(b) At which mass flow rate are the particles with 0.42 mm diameter blown out of the bed?

(c) What is therefore the range of mass flow rates suitable for fluidization?

Answer: (a) 0.40 kg/m^{2}s

   (b) 3.5 kg/m^{2}s

   (c) $1 < \phi_m'' < 3$ kg/m^{2}s

3. A suspension is filtered at a constant pressure difference Δp over the filter. At $t = 0$, the filtration begins (the cake thickness is then zero) and after a time t a volume V has been filtered. The following results are found:

V (m^3)	t (s)	Δp (at)
1.00	20	0.2
1.41	40	0.2
1.73	60	0.2

Predict the filtered volume after 20 s if the test is repeated under the following conditions:

(a) at a pressure difference of 0.4 atm;

(b) with a second material, the particles of which are of the same form as the first but with the linear dimensions smaller by a factor of 2 (at $\Delta p = 0.2$ atm).

Answer: (a) 1.41 m³
       (b) 0.5 m³

4. A mixture of oil and bleaching earth is filtered over a plate and frame press of 20 m² filtering surface. After 4 min, 1 m³ and after 10 min, 2 m³ of filtrate are collected. How long does it take to filter one charge of 20 m³?

Answer: t = 460 min.

5. At regular intervals, the filter press in problem 4 must be cleaned. This cleaning operation takes t_0 min, irrespective of the amount of cake to be removed. If the flow resistance of the filter cloth is neglected ($R_m \ll R_c$), how long should the filtration last in order to obtain the maximum filtration capacity over a complete cycle (filtration and cleaning time)?

Answer: t = t_0

*6. Show that John reached the right conclusion in the problem stated at the beginning of this section.

Comments on problems

Problem 1

With equation (II.35) we have:

$$\Delta p = 4f \frac{L'}{D_h} \frac{1}{2} \rho \langle v \rangle^2$$

With:

$$D_h = \frac{4A}{S} = \frac{4\varepsilon d_p}{6(1-\varepsilon)}, \quad \langle v \rangle = \frac{v_0}{\varepsilon} \quad \text{and} \quad Re_h = \frac{2}{3} \frac{\rho v_0 d_p}{\eta(1-\varepsilon)}$$

we find:

$$\Delta p = 4f \frac{L'6(1-\varepsilon)\rho v_0^2}{4\varepsilon d_p 2\varepsilon^2}$$

and:

$$\frac{\Delta p}{\rho L} = \frac{E_{fr}}{L} = 4f \frac{L' v_0^2 (1-\varepsilon)3}{L d_p \varepsilon^3 4}$$

Comparing this with the expression developed in Section II.6.1:

$$\frac{E_{fr}}{L} = \frac{3}{2}C_w\frac{1}{2}v_0^2\frac{1-\varepsilon}{d_p\varepsilon^3}$$

and utilizing the information about C_w given there we find:

$$4f\frac{L'}{L} = \frac{152}{Re_h}(Re_h < 1) \text{ and } 4f\frac{L'}{L} = 2.3(Re_h > 700)$$

Problem 2

Since we are dealing here with a mixture of particles of different size we must be careful which diameter to use in equation (II.88). The derivation of this and of the Ergun equation was based on the specific wall surface S_v given as:

$$S_v = \frac{\text{surface of cavity}}{\text{volume of particles}} = \frac{6(1-\varepsilon)}{d_p}$$

For a mixture of particles of different size we must therefore use:

$$\overline{d}_p = \frac{6 \times \text{volume of particles}}{\text{surface of particles}} = \frac{\sum n_id_i^3}{\sum n_id_i^2} = \frac{\sum M_i}{\sum \dfrac{M_i}{d_i}}$$

where M_i = mass fraction of particles with diameter d_i. Thus we find:

$$\overline{d}_p = \frac{1.0}{\dfrac{0.20}{0.92} + \dfrac{0.29}{0.78} + \dfrac{0.23}{0.65} + \dfrac{0.25}{0.55} + \dfrac{0.03}{0.46}} = 0.69 \text{ mm}$$

Estimating $\varepsilon \approx 0.4$, we now find from equation (II.88) ($\eta = 17 \times 10^{-6}$ Ns/m^2, $\rho = 1.3$ kg/m^3):

$$\phi''_{m0i} = \rho v_{0i} = 0.40 \text{ kg/m}^2\text{s}$$

From equation (II.70) we find for the steady-state velocity of free fall of the smallest particles, $v = 2.65$ m/s; thus $\phi''_{max} = 3.5$ kg/m^{2}s.

Problem 6 John and the body in the locked room

From equation (II.80) we find the flow rate of air leaking through the wall to be ($\Delta p = 800$ N/m^2, laminar flow):

$$\phi_v = v_0A = 7.9 \times 10^{-3} \text{ m}^3/\text{s}$$

Assuming the gas in the room to be perfectly mixed the material balance:

$$V\frac{dc}{dt} = -c\phi_v$$

leads, after integration between $t = 0$, c_0 and t, c to:

$$\ln \frac{c}{c_0} = -\frac{\phi_v}{V}t$$

Thus, after 8 h, the gas concentration should still be one-seventeenth of the lethal level and John should be able to smell the gas.

II.6 Stirring, mixing and residence time distribution

> *John saw that the only man on duty was very busy trying to speed up the circulation of the people past the crashed car. He saw a parallel to a mechanical stirrer and remembered that increasing the speed by only 25% would increase the frictional power by a factor of two. Therefore, he told him not to pay too much attention to the crowd's speed, but, now it had started moving, to concentrate on keeping it flowing smoothly.*

Mixing and stirring are two conceptions which partly overlap. Mixing is the uniform distribution to a certain scale of inhomogeneities over a certain space. Mixing can be attained by stirring. The object of stirring can also be to promote heat or mass transfers or to make a suspension or emulsion. Three transport mechanisms generally occur during mixing:

(a) convection, i.e. the material present is moved;
(b) dispersion, i.e. different parts of the material move at different speeds;
(c) diffusion.

II.6.1 Types of stirrer and flow patterns

Figure II.45 shows a few types of stirrer which are applied in practice. Types (a) and (b) will induce radial flow, type (c) axial flow, whereas type (d) will induce flow in both directions. Figure II.46 shows the flow patterns obtained for two types of impellers.

In the circulation mainstream of radial flow type impellers, four different zones can be distinguished:

(i) the radial liquid flow leaving the stirrer blades;
(ii) the vertical flow along the wall;
(iii) the horizontal flow back to the stirrer shaft;
(iv) the axial flow along the shaft to the stirrer.

Measurements have shown that in the radial liquid flow leaving the stirrer blades:

$$r\bar{v}_r = \text{constant}$$

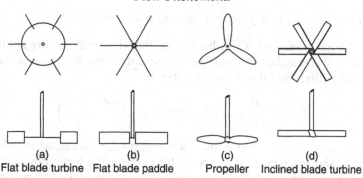

(a) (b) (c) (d)
Flat blade turbine Flat blade paddle Propeller Inclined blade turbine

Figure II.45 Various types of impellers

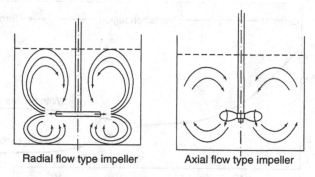

Radial flow type impeller   Axial flow type impeller

Figure II.46 Flow patterns of impellers

with r being the distance to the stirrer shaft and $\bar{v}_r$ the radial velocity of the liquid flow at position r, averaged over the time and stirrer height.

The flow pattern of axial flow type impellers is less distinct. The downward directed liquid flow (these stirrers are usually made to pump downwards) changes direction at the bottom and flows upwards along the wall of the vessel. Here again, a circulation pattern can clearly be distinguished in the flow pattern. A propeller stirrer, however, causes not only axial but also rather strong radial and tangential flow in the vessel. The flow pattern of a propeller stirrer is therefore less systematic compared with that of a radially pumping stirrer. When a draft tube is placed around a propeller stirrer the flow pattern in the vessel becomes more systematic.

When the vessel has not been fitted with baffles, vortexing occurs at high velocities so that gas can be sucked into the liquid. Baffles counteract this effect. Sufficient activity of the baffles is attained when the product of number times width of the baffles divided by the vessel diameter is ≈ 0.4. This situation (mostly reached by applying four baffles of $0.1\ D$ width) is called 'fully baffled'.

II.6.2 *Power consumption, pumping capacity and mixing time*

With dimensional analysis it can be derived that for geometrically similar situations the power number Po must be a function of the Reynolds number Re, related to the stirrer and the Froude number Fr:

$$Po = \frac{P}{\rho n^3 d^5} = f(\text{Re Fr}) = f\left(\frac{\rho n d^2}{\eta}; \frac{n^2 d}{g}\right) \tag{II.89}$$

The Froude number only plays a role when vortexing occurs at the liquid surface. Figure II.47 shows a typical power curve for a stirrer. The lower power consumption in the tank without baffles is caused because air is sucked in, as a result of which the density of the mixture decreases.

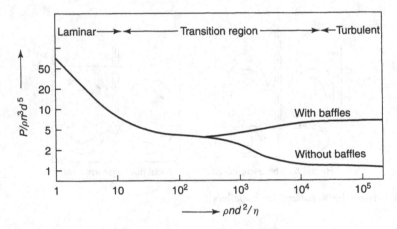

Figure II.47 Power consumption curve for turbine stirrer

The shape of the power curve of baffled systems may be physically interpreted as follows. If we consider one of the impeller blades moving through a liquid, then the force exerted on an object in a stream of fluid moving at mean velocity $\bar{v}$ is given by:

$$F = C_w A \tfrac{1}{2} \rho \bar{v}^2$$

where A is the cross-sectional area of the object and C_w the drag coefficient. Since $A \sim d^2$, and $\bar{v}$ corresponds to the tip speed of the blade ($\pi n d$) multiplied by a slip factor, the force on the blade is proportional to $\rho n^2 d^4$. The power needed to overcome this force is equal to force × arm × number of revolutions per unit time, or $\sim Fnd$. Therefore, the power number $Po = P/\rho n^3 d^5$ is proportional to the drag coefficient C_w. The shape of the power curve is similar to that of the drag coefficient versus the Reynolds number. For Re $> 10^4$, the power number Po, becomes a constant. In this turbulent region Po is only a function of the geometry of the system. At values of Re < 10, Po is proportional to $1/Re$; thus the power consumption is given by:

$$P = Po\rho n^3 d^5 \quad (Re > 10^4)$$

$$P = Po'\eta n^2 d^3 \quad (Re < 10)$$

Table II.6 gives a survey of some approximate values for Po and Po' for various impellers.

Table II.6 Rough estimations of Po and K_v

Impeller	Re < 10 Po'	Re > 10^4 Po	Re > 10^4 K_v
Flat blade turbine			
six blades, width/diameter — 1/5	70	6	1.3
four blades, ditto	70	4.5	0.6
Flat blade paddle			
six blades, width/diameter — 1/6–1/8	70	3	1.3
Propeller, three blades			
pitch = diameter	42	0.3⎫	0.4–0.5
pitch = two times diameter	42	1.0⎭	
Inclined blade turbine			
six blades, width/diameter — 1/8	70	1.5	0.8

The energy consumption for suspending particles in a vessel by means of a stirrer depends on the minimal stirring speed n^* at which the particles no longer settle.

In the turbulent region (Re > 10^4) and for particles larger than the Kolmogoroff scale (see Section I.5), viscosity plays a minor role. From dimensional analysis, it follows that for the average squared velocity on the scale of a particle with diameter d_p:

$$\langle v_p^2 \rangle \sim \varepsilon^{2/3} d_p^{2/3}$$

where ε is the energy dissipation per unit mass. For a constant power number one can show easily that:

$$\varepsilon \sim n^3 d^2$$

and thus:

$$\langle v_p^2 \rangle \sim n^2 d^{4/3} d_p^{2/3}$$

Equilibrating this velocity with the falling velocity of the particle in a turbulent flow, results in:

$$\frac{\rho^{\frac{1}{2}} n^* d^{2/3}}{g^{1/2}(\rho_p - \rho)^{1/2} d_p^{1/6}} = \text{constant} \tag{II.90a}$$

This equation can be compared with an empirical correlation from Zwietering (*Chem. Eng. Sci.*, 1958, **8**, 244) which after some rearranging can be written as:

$$\frac{\rho^{1/2}n^*d^{2/3}}{g^{1/2}(\rho_p - \rho)^{1/2}d_p^{1/6}} = C\mathrm{Re}^{-0.1}\left(\frac{d_p}{d}\right)^{0.05}w^{0.14} \tag{II.90b}$$

where C is a geometry-dependent constant which can be calculated from Zwietering's results and w is the weight fraction of the particles. The constant in equation (II.90a) turns out to be a weak function of the Re number and d_p/d. As equation (II.90a) was derived for very low concentrations of particles (particle/flow and particle/particle interactions being negligible), the dependence on w is not therefore surprising.

The power consumption then is:

$$P = Po\bar{\rho}n^{*3}d^5$$

where $\bar{\rho}$ is the mean density of the suspension.

If we consider only the biggest velocity component of the liquid flowing from the impeller, the pumping capacity is given by the product of the mean velocity of the liquid leaving the impeller and the area described by the blade tips. For the propeller this leads to (d = diameter of impeller):

$$\phi_v = \bar{v}\frac{\pi}{4}d^2$$

whereas for a turbine we find (W = width of blade):

$$\phi_v = \bar{v}\pi dW$$

If we assume $\bar{v}$ to be proportional to the tip speed of the blades (πnd) we find for a given geometry (for turbines $W \sim d$):

$$\phi_v = K_v nd^3 \tag{II.91}$$

In the turbulent region, K_v is a constant for a given impeller geometry. Some values of K_v are listed in Table II.6.

In many chemical engineering operations, e.g. the blending of gasoline, the mixing of reactants in a chemical reactor, etc., knowledge of the time necessary to reach homogeneity may be very important. Usually the mixing time is defined as the time needed for homogenization to the molecular scale. Since measurements on this scale are beyond experimental capabilities, an investigator is only able to measure the terminal mixing time, this being the time required to attain homogeneity on the scale of observation. In the turbulent region, for many types of impellers the relationship (t_m = mixing time):

$$nt_m = \text{constant}$$

holds, for geometrically similar situations.

The mixing time is about four times the circulation time of the vessel contents. Thus, if no dead spaces are present:

$$t_m \approx 4\frac{V}{\phi_v}$$

is a good estimation of the mixing time.

II.6.3 Residence time distribution: the F and E functions

> *The sergeant on duty in front of the supermarket informed John that the man they were looking for had entered the supermarket together with his wife 15 minutes ago. He said that they seemed to be doing their shopping for the next week and the man looked relaxed. John, knowing from his own experience that shopping for the week in this supermarket took on the average about 30 minutes, made a swift calculation. He decided that he had a probability of 60% to catch the man if he searched the supermarket directly with his squad. He therefore ordered them accordingly, but they failed to find the man. That night, he read the following section and came to the conclusion that the probability of an event and the actual course of an event are, as he should know, two completely different things.*

Flow through an apparatus practically always results in a certain residence time distribution. Fluid elements entering the apparatus simultaneously will generally leave it at different times. This distribution in residence time can be described quantitatively with two distribution functions which are closely related, namely the F and the E functions.

Although residence time distribution is of great importance to the design of process equipment, we will restrict our discussion to the principle. A more comprehensive treatment of non-ideal flow, including its influence on the design for chemical reactions, can be found elsewhere.[†]

The F function

Consider the continuous flow system shown in Figure II.48. From a time $t = 0$ on, we replace the feed stream by a fluid containing a tracer material (e.g. colouring agent) at a concentration c_0. If we measure the mean concentration of the tracer material in the effluent stream we will find that the tracer concentration (and also $\langle c \rangle / c_0$, its dimensionless concentration) increases with time, as illustrated in Figure II.49. Now $F(t)$ is defined as the volumetric part of the effluent stream which had a residence time in the apparatus of less than the time t. Since we

---

[†] see, for example, K. R. Westerterp, W. P. M. Van Swaay and A. A. C. M. Beenackers, *Chemical Reactor Design and Operation*, John Wiley & sons, 1987.

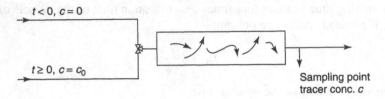

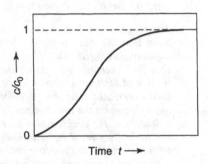

Figure II.48   Measurement of the *F* curve

Figure II.49   The *F* curve

know that all untraced fluid in the effluent stream has been in the apparatus for a time t, $F(t)$ is represented by the fraction of traced fluid ($\langle c \rangle / c_0$) in the effluent.

The *F* function (and also $\langle c \rangle$) is averaged over a small volume taken from the flow (mean cup values). For measuring, part of the flow is collected, mixed and subsequently analyzed. If concentration gradients or velocity profiles are present over the cross-section (as is the case during laminar flow), we must be careful to measure the correctly averaged value.

The *F* function has a number of properties, the most important of which is:

$$\lim_{t \to \infty} (1 - F(t)) = 0$$

(i.e. all fluid originally present in the apparatus will have ultimately left it). Furthermore, from a mass balance we find:

$$\int_0^\infty \phi_v (c_0 - \langle c \rangle) \mathrm{d}t = V c_0$$

and using $F = c/c_0$ we obtain:

$$\int_0^\infty (1 - F) \mathrm{d}t = \frac{V}{\phi_v} = \tau \tag{II.92}$$

The mean residence time can thus be determined from experimentally found $F(t)$ curves, as illustrated by Figure II.50. A line at $t = t_1 = $ constant, creates the areas A_1, A_2 and A_3 such that:

$$t_1 = A_1 + A_3$$

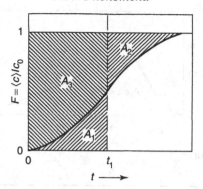

Figure II.50 Determination of τ

Because:

$$\tau = \int_0^\infty (1 - F)\mathrm{d}t = A_2 + A_3$$

This line represents the mean residence time τ if A_1 equals A_2. As a check on the measurements, τ can be calculated as $\tau = V/\phi_v$ from measured values of V and ϕ_v.

Often, instead of the time t, a dimensionless time parameter $\theta = t/\tau$ is used. However, in this present computer-dominated era, for bad reasons such simple checks as those described above are often forgotten.

The E function

The residence time distribution function of the fluid in the apparatus is given by the E function, which therefore represents the age distribution of the fluid leaving the vessel. Figure II.51 shows a typical E curve. The fraction of material in the exit stream with a residence time between θ and $\theta + \mathrm{d}\theta$ is given by $E\,\mathrm{d}\theta$, from which follows that:

$$\int_0^\infty E\,\mathrm{d}\theta = 1$$

The fraction of effluent material 'younger' than θ_1, which is also represented by the F function, is then given by:

$$F(\theta_1) \equiv \int_0^{\theta_1} E\,\mathrm{d}\theta \tag{II.93}$$

The E curve can be measured by injecting a tracer material into a flow system for a very short time and by measuring its concentration as a function of time at a downstream point. A typical plot of these concentrations versus time is shown in Figure II.52. The area under the curve:

$$\int_0^\infty \langle c \rangle \mathrm{d}t \approx \sum c\Delta t \tag{II.94}$$

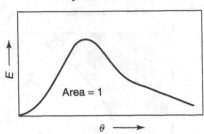

Figure II.51   The *E* curve

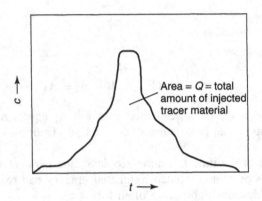

Figure II.52   Response curve to a tracer injection signal

represents the total amount of tracer substance injected, Q. In order to find $E(t)$, this area must be unity; thus all concentration readings must be divided by the total amount of tracer injected:

$$E(t) = \frac{\langle c \rangle}{\int_0^\infty c \, dt} = \frac{\langle c \rangle}{Q} \approx \frac{\langle c \rangle}{\sum \langle c \rangle \Delta t} \qquad (\text{II.95})$$

The mean residence time can then be calculated via:

$$\tau = \int_0^\infty t E(t) dt \approx \sum_0^\infty t E(t) \Delta t = \frac{\sum t \langle c \rangle \Delta t}{\sum \langle c \rangle \Delta t} \qquad (\text{II.96})$$

Since $\theta = t/\tau$ and $E(\theta) = \tau E(t)$ we are now able to construct the actual E curve in dimensionless parameters.

As we will see later, it is often desirable to know the variance of the distribution curve, which can be calculated according to:

$$\sigma^2 = \int_0^\infty (\theta - 1)^2 E \, d\theta \approx \sum \theta^2 E \Delta \theta - 1 = \frac{\sum t^2 \langle c \rangle \Delta t}{\tau^2 \sum \langle c \rangle \Delta t} - 1 \qquad (\text{II.97})$$

II.6.4 Simple applications of F and E functions

Perfect mixer. A perfect mixer is a vessel in which stirring is so effective that the composition of its contents is identical at all places. Naturally the effluent from this vessel has the same composition as its contents.

Keeping in mind that $c_{out} = c$, a mass balance over this vessel reads (see also Section I.1):

$$\phi_v c_0 - \phi_v c = V \frac{dc}{dt} \qquad (II.98)$$

For the determination of the F curve from time $t = 0$ on, the tracer concentration of the feed stream is changed from $c_{in} = 0$ to $c_{in} = c_0$. Integration of equation (II.98) therefore yields:

$$F = \frac{c}{c_0} = 1 - e^{-\theta} \qquad (II.99)$$

Plug flow. During plug flow, all parts of the fluid move at the same speed and there is no axial dispersion and no residence time distribution. Therefore, the F curve is described by:

$$F = 0 \text{ for } t < \tau$$
$$F = 1 \text{ for } t > \tau$$

and for the E curve:

$$E = 0 \text{ for } t \neq \tau$$

but E = tracer input signal at $t = \tau$!

Laminar flow in a circular tube. Here, the residence time distribution is caused by the differences in velocity and by diffusion. If we neglect diffusion, we can write for any streamline with velocity v and residence time t:

$$\frac{v}{\langle v \rangle} = 2 \left(1 - \frac{r^2}{R^2} \right) = \frac{\tau}{t} = \theta^{-1} \qquad (II.100)$$

Because the maximum velocity at the centre is twice the mean velocity, both the F and the E functions are zero for:

$$\frac{v}{\langle v \rangle} < 2 \text{ and } t/\tau < \frac{1}{2}, \text{ respectively}$$

For times $> \tau/2$, we find for the F curve:

$$F = \frac{\int_0^r v 2\pi r \, dr}{\langle v \rangle \pi R^2} = 1 - \left(1 - \frac{r^2}{R^2} \right)^2 = 1 - \frac{\tau^2}{4t^2} = 1 - \left(\frac{1}{2\theta} \right)^2 \qquad (II.101)$$

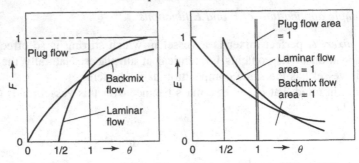

Figure II.53 F and E curves for three simple cases

and for the E curve:

$$E = \frac{\mathrm{d}F}{\mathrm{d}\theta} = \frac{4}{(2\theta)^3}$$ (II.102)

The F and E functions derived for the three cases treated here are shown in Figure II.53.

II.6.5 Continuous flow models

The analysis applied in the last chapter to three idealized cases of flow is not applicable to most practical flow situations. It is therefore desirable to describe actual flow conditions by models. Two of these model systems, the 'dispersion model' and the 'tanks in series model', are widely applied, and we will analyze both models in the following sections.

Dispersion model. The dispersion model is based on the plug flow of a fluid with a certain amount of intermixing in the axial direction. This intermixing can be due to diffusion by Brownian motion or by turbulence (Figure II.54).

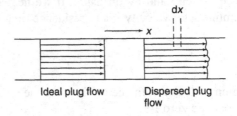

Figure II.54 Dispersed plug flow

If we assume the mass flow rate for dispersion in the x-direction to be proportional to the concentration gradient, a mass balance over a length $\mathrm{d}x$ shows:

$$A\,\mathrm{d}x\frac{\partial c}{\partial t} = \left(v_x cA - DA\frac{\partial c}{\partial x}\right)\bigg|_x - \left(v_x cA - DA\frac{\partial c}{\partial x}\right)\bigg|_{x+\mathrm{d}x}$$

from which follows:

$$\frac{\partial c}{\partial t} + v_x \frac{\partial c}{\partial x} = D \frac{\partial^2 c}{\partial x^2} \qquad \text{(II.103)}$$

Here, D is the dispersion coefficient.

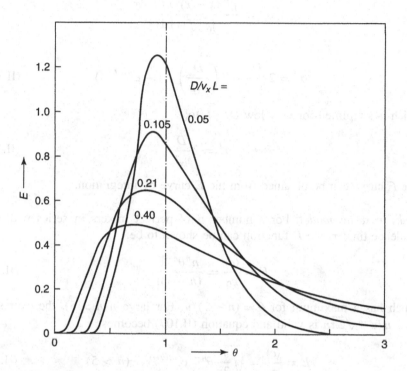

Figure II.55 Dispersion model, E curves

For the case of an instantaneous tracer injection (at time $t = 0$, place $x = 0$) this differential equation can be solved analytically. Accordingly, Figure II.55 shows E curves for various values of the parameter D/v_xL. For $D/v_xL \ll 1$, the complete analytical result can be simplified:

$$E = \frac{c}{c_0} = \frac{1}{2\sqrt{\pi \dfrac{D}{v_xL}}} \exp\left[-\frac{(1-\theta)^2}{4\dfrac{D}{v_xL}}\right]; \quad \frac{D}{v_xL} \ll 0.01 \qquad \text{(II.104)}$$

which is the Gaussian distribution. For low values of D/v_xL, the maximum of E occurs at $\theta = 1$. For higher values of $D/v_xL (> 0.01)$, the maximum occurs at $\theta < 1$.

In order to fit an experimentally determined E curve to one of the possible theoretical curves, the variances of the curves can be compared. For a flow system with negligible entrance effects (i.e. for small values of D/v_xL), the variance can be calculated from equation (II.104) via:

$$\sigma^2 = \frac{\int_0^\infty (x - \bar{x})^2 f(x)\,dx}{\int_0^\infty f(x)\,dx}$$

to be:

$$\sigma^2 = 2\frac{D}{v_xL} - 2\left(\frac{D}{v_xL}\right)^2 (1 - e^{-v_xL/D}) \tag{II.105}$$

which is simplified for very low D/v_xL to:

$$\sigma^2 = 2\frac{D}{v_xL} \tag{II.106}$$

The F curve can be obtained from the E curve by integration.

Tanks in series model. For a number n of perfect mixers in series with total residence time τ, the E function can be shown to be:

$$E = \frac{\langle c \rangle}{c_0} = \frac{n^n \theta^{n-1}}{(n-1)!}e^{-n\theta} \tag{II.107}$$

which has a maximum for $\theta = (n-1)/n$. For large $n(n > 5)$, the expression $n! \simeq n^n e^{-n}\sqrt{2\pi n}$ is valid and equation (II.107) becomes:

$$E = \frac{\langle c \rangle}{c_0} = \sqrt{\frac{n}{2\pi}}\theta^{n-1}e^{-n(\theta-1)} \quad (n > 5) \tag{II.108}$$

Figure II.56 shows E curves according to equations (II.107) and (II.108) for various numbers of stirred tank reactors in series.

The variance of the E curve for n stirred vessels in series is given by:

$$\sigma^2 = \frac{1}{n} \tag{II.109}$$

Comparing equations (II.106) and (II.109) makes it possible to comment on the same phenomenon in terms of both models:

$$n = v_xL/2D \tag{II.109a}$$

F curves for different values of n are shown in Figure II.57.

The analysis of experimentally determined F curves was more complicated than that of E curves before fast computers became available.

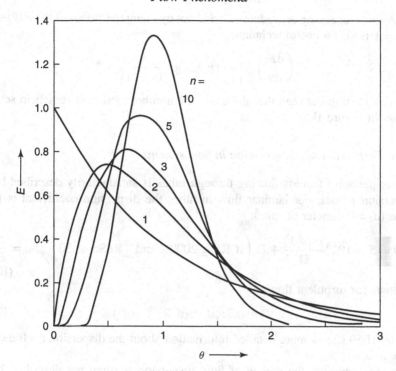

Figure II.56 Tanks in series model, E curves

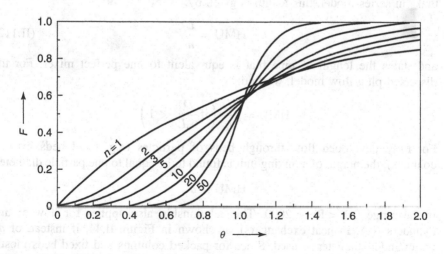

Figure II.57 Tanks in series model, F curves

However, checking procedures carried out by using the slopes of the $F(\theta)$-curve at $\theta = 1$ is still a useful technique:

$$\left(\frac{dF}{d\theta}\right)_{\theta=1} = (E)_{\theta=1} = \frac{n^n}{(n-1)!}e^{-n}$$

The relationship between this slope and the number of stirred vessels in series is shown in Figure II.58.

II.6.6 Dispersion and segregation in flow systems

The dispersion of fluids flowing through tubes is satisfactorily described by the dispersion model. For laminar flow in tubes the dispersion coefficient is found to be ($d_t = $ diameter of tube):

$$D = 5 \times 10^{-3}\frac{(v_x d_t)^2}{\mathbb{D}} + \mathbb{D} \left(\text{if Re} < 2000 \quad \text{and} \quad \text{ReSc} \ll 30\frac{L}{d_t}; \text{Sc} = \frac{v}{\mathbb{D}}\right)$$
(II.110)

whereas for turbulent flow:

$$D \approx 0.2v_x d_t \quad (\text{if Re} > 10^5)$$
(II.111)

Figure II.59 shows more detailed information about the dispersion coefficient for flow in tubes.

In the literature, the degree of fluid dispersion is often not described by the dispersion coefficient but by stating the height of a mixing unit, (HMU). For the tanks in series model, this length is given by:

$$\text{HMU} = \frac{L}{n}$$
(II.112)

and states the length of pipe that is equivalent to one perfect mixer. For the dispersed plug flow model, we find:

$$\text{HMU} = \frac{2D}{v_x} \left(\text{if}\frac{D}{v_x L} \ll 1\right)$$

For fully developed flow through beds of particles (e.g. fixed beds, packed columns), the height of a mixing unit is found to be equal to one particle diameter:

$$\text{HMU} \approx d_p$$

in the range $0.1 < \text{Re} < 2000$. This relationship also applies for flow around cylinders (e.g. in heat exchangers), as shown in Figure II.44, if instead of d_p the cylinder diameter is used. Since for packed columns and fixed beds mostly $d_p \ll L$, the number of mixers in series is so big that plug flow is approximated.

A knowledge of dispersion coefficients, and hence of appropriate models for residence time distributions, enables us to analyze the consequence of uneven

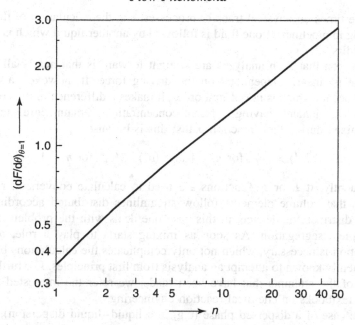

Figure II.58 Tanks in series model; $dF/d\theta$ at $\theta = 1$

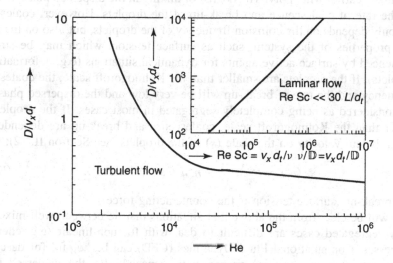

Figure II.59 Dispersion number for flow in tubes

residence times on physical transfer processes, e.g. the calculation of the degree of mixing in pipelines, if one fluid is followed by another liquid which is miscible with the first.

The reason that such analyzes are straight forward is that nearly all physical exchange is linearly dependend on the driving force. If, however, a chemical reaction occurs which is not of first order, it makes a difference in the conversion rate whether elements having different concentrations remain segregated (apart) or are mixed during their reaction paths; this is because:

$$\langle c^n \rangle > \langle c \rangle^n \text{ for } n > 1 \text{ and } \langle c^n \rangle < \langle c \rangle^n \text{ for } n < 1$$

Consequently, if E or F functions are used to calculate conversion rates, one assumes that volume elements follow streamlines distributed according to the E or F distributions. Hence, in this case one deals with the problem as a case of complete segregation. As soon as mixing starts to play a role, additional information is necessary, which not only complicates the calculations but is also insufficiently known to attempt an analysis from first principles. For further information of the attempts that have been made, we refer the interested reader to various textbooks on chemical reaction engineering.

In the case of a dispersed phase (e.g. in a liquid–liquid dispersion), the situation becomes even more difficult, because the degree of segregation has to be known for **both** phases. For the continuous phase, the same arguments as discussed earlier will apply. The degree of mixing in the dispersed phase depends on the rate of coalescence and break-up of the droplets. However, coalescence not only depends on the collision frequency of the droplets, but also on the physical properties of the system, such as surface tension, which may be strongly influenced by surface active agents for dynamical situations (e.g. deformation of droplets). If the droplets are smaller than the Kolmogoroff scale, the coalescence frequency and the rate of break-up will be very low and the dispersed phase can be considered as being completely segregated in most cases. If the droplets are larger than the Kolmogoroff scale, coalescence and break-up are dependent on the turbulent velocity on the scale (s) of the droplets (see Section II.6.2):

$$\langle v_s^2 \rangle \sim \varepsilon^{2/3} s^{2/3} \sim n^2 d^{4/3} s^{2/3}, \text{ for } s > \lambda \tag{II.113}$$

For break-up, surface tension is the counteracting force.

It will be clear from the above that situations that lie between well-mixed and fully segregated cases are difficult to deal with for non-lineair (e.g. chemical) processes. Computational Fluid Dynamics (CFD) can be helpful for describing these processes but remains approximate, especialy for the dispersed phase, because basic physical knowledge is still lacking. In practice, the extremes for the phases being well-mixed or completely segregated can be described with good accuracy in most cases, both for the continuous phase and for the dispersed phase.

II.6.7 Problems

1. A catalyst precipitation process has been developed on a pilot plant scale (reactor diameter 0.25 m). It appeared that using a standard flat blade turbine (six blades, fully baffled tank), a minimum stirring speed of 700 rpm was required in order to obtain a satisfactory product (ρ liquid $\approx$ 1300 kg/m^3).

    A geometrically similar factory scale reactor (0.6 m diameter) has been designed. Calculate which stirring speed should be applied and what the power consumption would be if we keep the following constant:
    (a) the power input per unit volume;
    (b) the pumping capacity per unit volume;
    (c) the mixing time;
    (d) the size of the particles that are only just not suspended in the liquid.

    Answer: (a) 391 rpm, 685 W
            (b) 700 rpm, 4000 W
            (c) 700 rpm, 4000 W
            (d) 391 rpm, 685 W

2. Show that John reached the right conclusion in the problem described at the beginning of this section.

3. Develop an expression for the E and F functions of two perfect mixers in series (each with a residence time $\tau/2$) and show that the result is in agreement with equation (II.107).

*4. The feed stream to a continuous deodorizer (a vacuum steam stripper applied for rendering edible oils tasteless) is at time $t = 0$ changed from bean oil to coconut oil and the refractive index of the effluent stream is recorded. From the refractive index the percentage of coconut oil is determined:

Time (min)	Coconut oil (%)	Time (min)	Coconut oil (%)
30	0	60	70.5
40	5	65	83
45	16.5	70	92
50	34.5	80	99
55	52		

(a) What is the mean residence time of the oil?
(b) How many perfect mixers in series would yield a corresponding residence time distribution?
(c) What would be the dispersion factor if we wanted to describe the residence time distribution by the dispersion model?

Answer: (a) 55.1 min
        (b) 25 vessels
        (c) 0.02

5. Calculate the height of a mixing unit for air and water flowing through a circular pipe at Re numbers of 1100 and 10 000.

	ρ (kg/m^3)	$\mathbb{D}$(m^2/s)	η(Ns/m^2)
Air	1.3	2.5×10^{-5}	1.7×10^{-5}
Water	10^3	2×10^{-9}	10^{-3}

Answer:

	Re		
HMU	1	100	10 000
Air	d_t	$0.5\,d_t$	d_t
Water	$5\,d_t$	$500\,d_t$	d_t

6. A relatively small waste-water stream, mixed with an innoxious flow of $0.1 \text{ m}^3/\text{s}$, is discharged into the environment. If the waste-water stream is stopped, but not the main flow, the concentration of the waste is halved every 100 min. If now 1 kg of waste, uniformly distributed over a period of 10 min, is dumped into the main stream, which is then the maximal concentration of this material in the exhaust?

Answer: 1.12 g/m^3

*7. In order to determine the mixing efficiency in a cascade of six stirred vessels, the following experiment is carried out. A continuous stream of water is passed through and at time $t = 0$ an amount of acid is injected into the first reactor. At regular intervals, samples from the effluent of the last reactor are taken and analyzed:

Time (min)	Concentration (g/l)	Time (min)	Concentration (g/l)
0	0	30	3.00
5	0.10	35	2.12
10	1.63	40	1.39
15	3.23	50	0.51
20	3.96	60	0.10
25	3.71	70	0

(a) What is the mean residence time in the cascade?
(b) How many perfect mixers in series would yield the same residence time distribution?
(c) What would be the dispersion factor if we wanted to describe the residence time distribution by the dispersion model?

(d) How many grams of acid were injected if the fluid flow rate was 100 l/min?

Answer: (a) 25.5 min
  (b) 5.5 vessels
  (c) 0.10 (= Pe^{-1})
  (d) 10.2 kg

8. Show that equation (II.109a), which relates the model of a cascade of mixers to the model of flow with axial dispersion and which is based on the criterion of equal variance, also predicts that when both models are matched in this way:

the E values at $\theta = 1$ are the same for both models and, hence, the slopes $dF/d\theta$ at $\theta = 1$ are equal.

*9. Show that in the problem stated at the beginning of Section II.6.3, John calculated the correct probability of catching the man in the supermarket.

Comments on problems

Problem 4

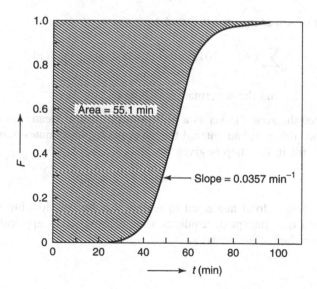

A plot of the F values given versus time (see the figure) yields, after graphical integration, $\tau = 55.1$ min. At τ, we find graphically for the slope of the line:

$$\left(\frac{dF}{dt}\right)_{t=\tau} = 0.0357 \text{ min}^{-1}$$

or:

$$\left(\frac{dF}{d\theta}\right)_{\theta=1} = \left(\frac{dF}{dt}\right)_{t=\tau} \quad \tau = 1.96$$

From Figure II.58, this leads to $n \approx 25$ perfect mixers in series. The dispersion number is then found via

$$\sigma^2 = 2\frac{D}{v_x L} = \frac{1}{n}$$

Problem 7

With equations (II.96) and (II.97), we find:

$$\tau = \frac{\sum tc\Delta t}{\sum c\Delta t} = \frac{2600}{101.8} = 25.5 \text{ min}$$

$$\sigma^2 = \frac{\sum t^2 c\Delta t}{\tau^2 \sum c\Delta t} - 1 = 0.181$$

Using equation (II.105), we find by trial and error:

$$\frac{D}{v_x L} = 0.10$$

The amount of tracer injected is found as:

$$\sum c\Delta t = 101.8 = \frac{M}{\phi_v}; \quad M = 10.2 \text{ kg acid}$$

Problem 9 John and the supermarket problem

John regarded the supermarket as a perfect mixer with a mean residence time of 30 min. Since the man had entered the supermarket 15 minutes ago, the chance that he was still in the shop is given by

$$1 - F = e^{-\theta} = e^{-0.5} = 0.606$$

or 60%. However, John neglected to realize that he was looking for one man only, whereas the concept of residence time distribution is applicable only to a very great number of elements!

CHAPTER III

Heat Transport

In this chapter the transport of energy is studied on the understanding that the changes in heat content are much greater than the changes in mechanical energy. In this case the law of conservation of energy is written as:

$$\frac{dE_t}{dt} = \phi_{min}u_{in} - \phi_{mout}u_{out} - \phi_H + \phi_A \qquad (III.1)$$

where $E_t = \rho c_p TV$ and $u = c_p T^{\dagger}$ (as long as no phase transition occurs), ϕ_H = heat that is exchanged with the surroundings and ϕ_A = mechanical energy added. With this law of conservation, all problems of heat transport in a system can be solved; the following therefore does nothing but ascertain the consequences of this balance.

The transport of heat plays an important role in many processes in engineering. It may be that we wish to promote the transport from one medium to the other (generation of steam, heating or cooling of liquids, removal or supply of heat of reaction, cooling of transmitter tubes, removal of heat generated in nuclear reactors, etc.) or to suppress the transport to restrict loss of heat or cold (e.g. insulation of pipes, heated vessels and cold stores).

Three different mechanisms according to which heat can be transported (invariably from higher to lower temperature) may be distinguished, namely:

(a) *Heat conduction*

In non-moving media in the presence of a temperature gradient, heat is transported by the molecular movement (from high to low temperature). According to Fourier, the heat flux ϕ_H'' (J/m² s) occurring at this temperature gradient is given by:

$$\phi_H'' = -\lambda \frac{dT}{dx} \qquad (III.2)$$

---

† We have taken here for the heat capacity the value which is correct if the process proceeds at constant pressure. Fundamentally, this is not quite accurate, because the presence of a flow means that there are pressure differences. In most cases this is not a great problem, because either c_p and c_v are almost equal to each other (e.g. when solid substances and incompressible liquids are involved) or the process is almost isobaric. If we want to proceed in a precise manner, we should read, instead of c_p, $c_v + [T(\partial p/\partial T)_v - p](\partial V/\partial T)_p$.

Since ϕ_H'' is positive in the direction of decreasing temperature, the minus sign has been added.

In many correlations for heat transport, the heat conductivity λ is found in the combination $\lambda/\rho c_p$ which, because of its analogy with the diffusion coefficient, is called thermal diffusivity $a(m^2/s)$. In Table III.1 approximate values for the heat conductivity and for the thermal diffusivity are collected.

Table III.1   Approximate values of λ and a for various materials

	$\lambda(W/m\,°C)$	$a(m^2/s)$
Gases	0.02	20×10^{-6}
Liquids	0.2 (water 0.6)	0.1×10^{-6}
Solids (non-metallic)	2	1×10^{-6}
Metals	20–200	$5 - 50 \times 10^{-6}$

(b) *Flow (convection)*

In this case, heat is entrained with the moving conducting medium. The flow may be the result of external causes (forced flow or convection) or spontaneous, e.g. as a result of density differences caused by temperature differences (thermal, free flow or convection), or of bubble formation (boiling).

(c) *Radiation*

This is the transfer of heat by electromagnetic waves between surfaces of various temperatures which are separated by a medium transparent to heat radiation (infrared light). At high temperatures, the transfer by radiation becomes important, but it can also play an appreciable role at room temperature.

These three mechanisms often occur in combination. In addition, for the heat transport by (b) the heat conductivity λ is a very important property, for a flowing medium can only absorb heat from a hot wall if it is conducting. Temperature-levelling in moving media takes place by convection (mixing of cold and hot) as well as by conduction.

Just as in the theory of flow phenomena, for cases of stationary heat conduction knowledge of the relationship between the heat flow and the driving force is sufficient to rapidly solve one-dimensional problems. In the theory of flow, Newton's law was the starting point; in this case it will be Fourier's law.

III.1 Stationary heat conduction

> *John realized that the man found in the storage room for deep frozen foods was dead. Although he wore protective clothing his body was quite cold and had started to freeze.*

The doctor informed John that one hour ago the temperature measured at the arm had still been 8 °C. John concluded, realizing that the storage temperature was −30 °C, that the man must have died approximately $3\frac{1}{2}$ hours ago and started questioning his colleagues.

III.1.1 Heat conduction through a wall

In the case of stationary heat conduction through a wall (thickness d, other dimensions infinite, temperature difference over the wall $\Delta T = T_1 - T_0$), the heat flow through any plane parallel to the end faces will be of the same magnitude (i.e. no accumulation of heat in the wall takes place because of its stationary state). Now that the heat flux ϕ_H'' appears to be constant (independent of x), $\lambda dT/dx$ will have to be constant (Fourier); so if λ is constant in the temperature range from T_1 to T_0, the temperature distribution over the wall will be linear. If the heat conductivity λ of the material is known, the heat flux at a given temperature difference can be calculated:

$$\phi_H'' = \lambda \frac{T_1 - T_0}{d} \quad \text{(W/m}^2\text{).} \tag{III.2a}$$

Methods for measuring the heat conduction of solid substances are based on this equation. If the wall consists of more layers (Figure III.1), for the same reasons as in the foregoing, ϕ_H'' must be constant in every cross-section in every layer and so in every layer dT/dx must be constant (and inversely proportional to the conductivity of that layer!). If for every layer we write equation (III.2a) as follows:

$$\phi_H'' \left(\frac{d_i}{\lambda_i} \right) = \Delta T_i = T_i - T_{i+1} \tag{III.2b}$$

and add up all these relationships for all layers we get:

$$\Delta T_{tot} = T_1 - T_4 = \phi_H'' \left(\frac{x_2 - x_1}{\lambda_1} + \frac{x_3 - x_2}{\lambda_2} + \frac{x_4 - x_3}{\lambda_3} \right) \tag{III.3}$$

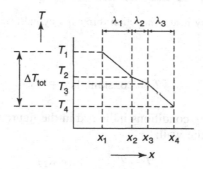

Figure III.1 Stationary heat conduction through walls

The terms $(x_2 - x_1)/\lambda$, etc., can be considered as heat resistances, and the above equation says that the total temperature difference $(T_1 - T_4)$ equals the heat flux multiplied by the sum total of the resistances placed between x_1 and x_4. (Compare this with Ohm's law in electricity.) The sum total of all partial resistances, which are also called the partial heat resistances, is described as the total heat resistance for which usually (and also here) the symbol $1/U$ is used. 'U' is then the total heat transfer coefficient.

III.1.2 Heat conduction through cylindrical walls

The next example is the stationary heat conduction in the thick wall of a cylindrical tube (see Figure III.2). In this case the heat flows, ϕ'_H (per unit of tube length) through the surfaces $r = R_1$ and $r = R_2$ and through all surfaces in between must be equal:

$$\phi'_H = -\lambda \frac{dT}{dr} 2\pi r = \text{constant} \tag{III.4}$$

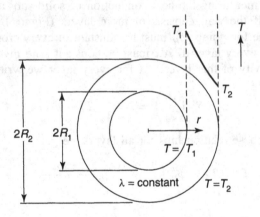

Figure III.2 Stationary heat conduction through a cylindrical wall

Integration yields:

$$T = \text{constant } \ln r + C_2$$

and with the boundary conditions indicated in the figure we find for the temperature distribution in the wall:

$$\frac{T - T_2}{T_1 - T_2} = \frac{\ln(r/R_2)}{\ln(R_1/R_2)} \tag{III.5}$$

Subsequently, the heat flow per unit of length can be calculated as a function of the applied temperature difference:

$$\phi_H' = -2\pi r\lambda \frac{dT}{dr} = \frac{2\pi\lambda(T_1 - T_2)}{\ln R_2/R_1} \qquad (III.6)$$

which is indeed independent of r.

Equation (III.6) can be used to calculate the optimum thickness of thermal insulations of a pipe. There are two optimization criteria. The first is purely technological. With increasing diameter of the insulation the heat transport resistance increases and so does the outer surface of the insulation. Due to this a situation may arise where, with increasing insulation thickness, the radial heat loss per metre length of pipe, ϕ_H', goes through a maximum. If the heat transfer coefficient (by free convection and radiation) at the outer surface is called h_0, we find from equation (III.6) after addition of the transport resistances in the insulation and at the outer surface (T_3 = air temperature):

$$\phi_H' = \frac{(T_1 - T_3)}{\dfrac{\ln R_2/R_1}{2\pi\lambda} + \dfrac{1}{2\pi R_2 h_0}} \qquad (III.7)$$

From this equation it can be calculated that the maximum heat loss is obtained for $h_0 R_2/\lambda = 1$. For values of $h_0 R_2/\lambda$ higher than 1, ϕ_H' decreases monotonously with increasing outer radius R_2 of the insulation. For most insulation materials, $\lambda = 0.05$ W/m °C and h_0 is of the order of magnitude of 10 W/m² °C (see also Section III.6). In these cases, the critical value of R_2 at which the above phenomenon occurs is $R_2 = 5 \times 10^{-3}$ m. With such insulation materials and under such conditions there is therefore no risk that fitting an insulation jacket increases the loss of heat. A different situation arises when, for example, for the insulation of laboratory apparatus, a material is used for which $\lambda = 0.15$ W/m °C; in that case, the critical value of $R_2 = 1.5 \times 10^{-2}$ m!

The second optimization criterion is connected with the consideration that for technical installations the choice of the insulation thickness is determined for economic reasons. At greater thickness the loss of heat decreases but the costs of insulation rise. By writing down the equation for the total capital loss as a function of the insulation thickness (loss of heat, interest and depreciation), and by differentiating this equation to the insulation thickness, the optimal insulation thickness can be calculated.

III.1.3 Heat conduction around a sphere

As the last example of stationary conduction we shall treat the heat transfer from a sphere to a stationary medium (sphere diameter $2R$, sphere temperature T, ambient temperature T_∞, constant λ of the medium around the sphere). Here it also applies that the heat flow through each surface $r(=$ constant) must have the same value:

$$\phi_H = -\lambda \frac{dT}{dr} 4\pi r^2 = \text{constant (independent of } r)$$

from which follows the temperature distribution in the medium around the sphere:

$$\frac{T - T_\infty}{T_1 - T_\infty} = \frac{R}{r} \qquad \text{(III.8)}$$

This result is again sufficient for calculating the heat flow which is transferred between sphere and medium at a given temperature:

$$\phi_H = 4\pi R\lambda(T_1 - T_\infty) \qquad \text{(III.9)}$$

Newton described the cooling of bodies of any form with a pragmatic cooling law in which he made the heat flow proportional to the outer surface of the body and to the prevailing temperature difference $(T_1 - T_\infty)$. The proportionality constant, which is a purely phenomenological quantity and contains all that we do not know of the transfer process, he called the heat transfer coefficient h.

For the sphere from the previous example this cooling law is therefore formulated as follows:

$$\phi_H = h4\pi R^2(T_1 - T_\infty) \qquad \text{(III.10)}$$

It follows that in this case $h = \lambda/R$. In our jargon, this is summarized to $\text{Nu} = 2$, where the Nusselt number Nu is defined as $h2R/\lambda$. The Nu number represents the relationship between the heat resistance which is estimated on account of a characteristic dimension of the body (in the above example $2R/\lambda$) and the actual heat resistance $(1/h)$. We may also consider this quantity as the relationship between the dimension of the body $(2R)$ and the thickness of the layer over which the drop in temperature takes place (λ/h).

III.1.4 General approach for the calculation of temperature distributions

The general approach for the calculation of temperature distributions is a microbalance which here will be given for a Cartesian volume element (λ and c_p constant). The approach is completely analogous to that which we followed for the calculation of velocity distributions during laminar flow. Here again is the restriction that as soon as the eddies take over the heat transport from the molecular transport, exact calculation is no longer possible, because we can describe the molecular transport of heat mathematically (Fourier's law), but not the eddy transport. This means that pure conduction problems and transport to laminar flows can be dealt with exactly, whereas heat transfer to turbulent flows cannot.

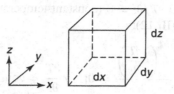

Figure III.3 The Cartesian volume element

The microbalance is as follows. In the x-direction, an amount of heat equal to $(-\lambda(\partial T/\partial x) + \rho c_p v_x T)\,dy\,dz$ flows in at position x. The net transport in the volume element through the walls x and $x + dx$ is therefore:

$$\left(\lambda\frac{\partial^2 T}{\partial x^2} - c_p\frac{\partial \rho v_x T}{\partial x}\right) dx\,dy\,dz$$

Analogous expressions can be written for the y- and z-directions. The heat content of the volume is $\rho c_p T\,dx\,dy\,dz$ and the production of heat per unit of time is $q\,dx\,dy\,dz$ (hence q is production per unit of time and per unit of volume).

The balance, reading:

$$\frac{\text{accumulation}}{\text{unit of time}} = \text{flow in} - \text{flow out} + \frac{\text{production}}{\text{unit of time}}$$

now becomes:

$$c_p\frac{\partial \rho T}{\partial t} + c_p\frac{\partial \rho v_x T}{\partial x} + c_p\frac{\partial \rho v_y T}{\partial y} + c_p\frac{\partial \rho v_z T}{\partial z} = \lambda\frac{\partial^2 T}{\partial x^2} + \lambda\frac{\partial^2 T}{\partial y^2} + \lambda\frac{\partial^2 T}{\partial z^2} + q$$

If we use the continuity equation once to modify this heat balance (just as with the micromomentum balance) the result is:

$$\rho c_p\left(\frac{\partial T}{\partial t} + v_x\frac{\partial T}{\partial x} + v_y\frac{\partial T}{\partial y} + v_z\frac{\partial T}{\partial z}\right) = \lambda\left(\frac{\partial^2 T}{\partial x^2} + \frac{\partial^2 T}{\partial y^2} + \frac{\partial^2 T}{\partial z^2}\right) + q \quad \text{(III.11)}$$

In cylindrical coordinates this equation reads:

$$\rho c_p\left(\frac{\partial T}{\partial t} + v_r\frac{\partial T}{\partial r} + \frac{v_\theta}{r}\frac{\partial T}{\partial \theta} + v_z\frac{\partial T}{\partial z}\right) = \lambda\left(\frac{1}{r}\frac{\partial\left(r\dfrac{dT}{dr}\right)}{\partial r} + \frac{1}{r^2}\frac{\partial^2 T}{\partial \theta^2} + \frac{\partial^2 T}{\partial z^2}\right) + q$$

$$\text{(III.12)}$$

The reader may check that the simple cases treated in the first three parts of this section can be derived from these equations. In the case of conduction through a cylindrical wall treated in section III.1.2, for example, $q = 0$, $v_x = v_y = v_z = 0$, $\partial T/\partial t = 0$ (stationary state), $\partial T/\partial \theta = 0$ (constant temperature in all radial directions) and $\partial T / \partial z = 0$ (constant temperature in axial direction). So we find from equation (III.12):

$$\frac{1}{r} \frac{\partial \left(r \dfrac{\partial T}{\partial r} \right)}{\partial r} = 0; \quad r \frac{\partial T}{\partial r} = \text{constant}$$

which is identical with equation (III.4).

III.1.5 Temperature distribution in a cylinder with uniform heat production

An electric current flowing through a wire will, because of the electric resistance of the material, produce an amount of heat q per unit of volume (W/m^3). By heat conduction the heat produced is transported to the surface of the wire which has a constant temperature T_0. From the surface of the wire there is a heat flow to the surroundings which, in steady-state conditions, equals the amount of heat produced. Thus:

$$\phi_H|_{r=R} = q\pi R^2 L \tag{III.13}$$

In order to find the temperature distribution in the wire we set up a microenergy balance over a cylindrical shell of thickness dr (Figure III.4). For steady-state conditions, equation (III.1) can be written as:

$$\frac{dE_t}{dt} = 0 = 2\pi r L \phi_H''|_r - 2\pi r L \phi_H''|_{r+dr} + 2\pi r\, dr L q$$

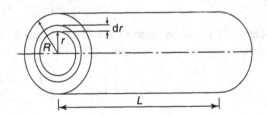

Figure III.4 Temperature distribution in a cylinder

Dividing by $2\pi L\, dr$ we obtain:

$$0 = \frac{r\phi_H''|_r - r\phi_H''|_{r+dr}}{dr} + rq = -\frac{d(r\phi_H'')}{dr} + rq$$

Integrating this equation we find:

$$\phi_H'' = \frac{q}{2}r + \frac{C_1}{r}$$

The integration constant C_1 must be zero because, at $r = 0$, ϕ_H'' cannot be infinite. Thus, applying Fourier's law (equation III.2):

$$\phi_H'' = \frac{q}{2}r = -\lambda\frac{dT}{dr}$$

A second integration (assuming λ to be constant) yields:

$$T = -\frac{q}{4\lambda}r^2 + C_2$$

and, using the boundary condition $T = T_0$ for $r = R$, we finally obtain:

$$T - T_0 = \frac{q}{4\lambda}(R^2 - r^2) = \frac{qR^2}{4\lambda}\left(1 - \frac{r^2}{R^2}\right) \tag{III.14}$$

Thus, the temperature distribution is parabolic, the maximum temperature being at the centre ($r = 0$):

$$T_{max} - T_0 = \frac{qR^2}{4\lambda} \tag{III.15}$$

The average temperature is given by:

$$\langle T\rangle = \frac{\displaystyle\int_0^R 2\pi r T\,dr}{\displaystyle\int_0^R 2\pi r\,dr} = \frac{\displaystyle\int_0^R T_0 r\,dr + \int_0^R \frac{qr}{4\lambda}(R^2 - r^2)\,dr}{\frac{1}{2}R^2}$$

$$\langle T\rangle - T_0 = \frac{q}{8\lambda}R^2 = \frac{1}{2}(T_{max} - T_0) \tag{III.16}$$

For the heat flow at the surface we find:

$$\phi_H = -2\pi RL\lambda\left.\frac{dT}{dr}\right|_{r=R} = \pi R^2 Lq$$

which is in agreement with the overall balance equation (III.13) set up for steady-state conditions.

It is interesting to compare the results just obtained with the equations developed for the shear force and flow velocity distribution for laminar flow in a circular pipe (sections II.1.2 and II.1.3). The microenergy balance applied here is analogous to the micromomentum balance of section II.1.3 and the correlations for laminar flow found in section II.1.2 can be translated into the temperature distributions just developed if we use the conversions given in Table III.2. Apparently, the processes of laminar flow through pipes and of uniform heat production in a cylinder are analogous.

Table III.2 Analogy between laminar flow and uniform heat production

Laminar flow in cylindrical pipe		Cylinder with uniform heat production	
τ_{rx}	(N/m^2)	ϕ_H	(W/m^2)
v_x	(m/s)	$T - T_0$	$(°C)$
η	(Ns/m^2)	λ	$(W/m°C)$
$p_1 - p_2/L = -\dfrac{dp}{dx}$	(N/m^3)	q	(W/m^3)

III.1.6 Problems

1. An oven wall consists in succession of a layer of firebricks ($\lambda_f = 1.21$ J/m °C s), a layer of lagging bricks ($\lambda_1 = 0.080$ J/m °C s) and a layer of bricks ($\lambda_b = 0.69$ J/m °C s). Each layer is 10 cm thick. The temperature on the inside of the wall is 872 °C, on the outside 32 °C.
   (a) If the surface area of the wall is 42 m², calculate how much heat is lost by conduction every 24 h.
   (b) Calculate the temperature T_m in the middle of the layer of lagging bricks.

   Answer: (a) 2.1×10^9 J; (b) 470 °C

2. The *Chemisch Weekblad* of 30 September 1966 gave the following news item:

   > In Elsa, a mining settlement in the Yukon district, situated only 290 km south of the polar circle, where the temperature sometimes falls to −70 °C, the water supply was seriously hampered by freezing. This problem has been solved by putting a floating pumping engine in the nearby lake. The construction is such that the water inlet tube is well below the level of the ice. The 12.5 cm thick pipe connecting the pumping engine with the storage tank in town — a distance of ca. 5 km — was insulated with a 5 cm thick layer of urethane foam. The flow rate of the water in the pipe is on an average 675 l/min. Even at extremely low temperatures, water of 2.5 °C can be pumped from the lake to the town without heating being necessary.

   What follows from these data for the heat conductivity of urethane foam?

   Answer: 0.03 W/m °C

3. A research fellow at Corning glass works struggled with reports on the overall heat transfer coefficients (U) for double-glazed windows, separated by a distance d. He considered that two processes occur simultaneously in the encapsulated air layer: transverse conduction and longitudinal flow in the cell, i.e. up on the warm and down on the cold side. In order to summarize the results of the various reports, he considered that a characteristic width

(Δ) for these windows is:

$$\Delta = \left(\frac{av}{\beta \Delta T g}\right)^{1/3}$$

with the symbols having their usual meaning and β being the temperature expansion coefficient of air.

(a) Give arguments to support your collegue here.

The researcher proved to be able to summarize the available reports as follows:

$$\frac{U\Delta}{\lambda} = 0.16 \quad \text{for} \quad \frac{d}{\Delta} > 12$$

$$\frac{U\Delta}{\lambda} = \frac{\Delta}{d} + 5 \times 10^{-4}\frac{d^2}{\Delta^2} \quad \text{for} \quad \frac{d}{\Delta} < 12$$

(b) Describe the regimes so found in qualitative terms (using such terms as 'conduction', 'convection', 'laminar', 'turbulent').

(c) At which value of d is the overall heat transfer the least?

(d) What is the influence of the ambient temperature on the optimal value of d (compare for instance, ambient temperatures of 5 and 20 °C), if $\Delta T = $ constant.

(e) Sketch a graph of $U\Delta/\lambda$ versus d/Δ for $5 < d/\Delta < 20$.

Answer: (c) $\dfrac{d}{\Delta} = 10$

(d) $\dfrac{d(5\,^\circ C)}{d(20\,^\circ C)} = \left(\dfrac{273}{293}\right)^{4/3} = 0.93$

4. A length of steam pipe ($R_1 = 0.05$ m) has to be insulated ($R_2 - R_1$) under circumstances so that the outside heat transfer coefficient is much higher than λ/R_2. It is established that the annual cost of heat loss is $C_1/\ln(R_2/R_1)$ (compare equation (III.7)). The annual cost of the insulation is $C_2(R_2/R_1 - 1)$. If $C_2/C_1 = 2.86$, what is the most economical optimal insulation?

Answer: $[\ln(R_2/R_1)]^{-1} + 2.86(R_2/R_1 - 1)$ has to be minimal, or $\dfrac{R_2}{R_1} = 1.6$,

$R_2 = 0.08$ m, $R_2 - R_1 = 0.03$ m, if $h_0 > 6$ W/m^2 K for $\lambda = 0.05$ W/mK

5. The rate of heat production of a Newtonian liquid per unit of volume by viscous dissipation in a cylindrical pipe is given by $q = \eta(dv/dr)^2$ [W/m^3].

(a) If the pipe has a constant wall temperature T_w, develop an expression for the steady-state radial temperature distribution during laminar flow.

(b) Show that for most practical cases the radial temperature gradient caused by viscous dissipation can be neglected.

Answer: (a) $T - T_w = \dfrac{\eta\langle v\rangle^2}{\lambda}\left(1 - \dfrac{r^4}{R^4}\right)$

*6. The hydrogenation of edible oils is carried out with a suspended catalyst. What is the maximum temperature difference between the catalyst surface and the oil?

Data: spherical particles, $d_p = 2.5 \times 10^{-6}$ m
apparent density $= 1750$ kg/m^3
0.2 wt% of catalyst applied
hydrogenation rate $r = 1.3 \times 10^{-3}$ kmol/m^3 s
heat of reaction $\Delta H = 25$ kcal/mol
molecular weight of the oil $M = 880$ g/mol
$\rho_{oil} = 900$ kg/m^3; $\lambda_{oil} = 0.2$ kcal/m h °C

Answer: 3×10^{-4} °C

*7. A copper pipe (length 0.5 m, outside diameter 2.5 cm, wall thickness 2 mm) has a thick heat insulation and the ends of the pipe are being kept at 0 °C ($\lambda_{copper} = 400$ W/m °C). Calculate the temperature distribution in the pipe wall if an electric current flows through the pipe giving 20 W in the stationary state.

Answer: $T = 346 \left(\dfrac{L^2}{4} - x^2 \right)$ °C

*8. Show that John drew the right conclusion in the case of the deep frozen body, reported at the beginning of this Section.

Comments on problems

Problem 3

An expression for Δ may be derived in two intuitive ways, namely dimensional analysis or a more physical approach. In the latter case, the laminar flow situation of Section II.1.1 is translated into the present case with a pressure gradient $\beta \rho \Delta T g$:

$$d = \left(\frac{12\eta \phi_v'}{\beta \rho \Delta T g} \right)^{1/3}$$

Next, convective transport with a temperature difference ΔT_L is compared to conduction with a temperature difference ΔT_d:

$$\rho c_p \phi_v' \Delta T_L = \lambda \Delta T_d \text{ or } \phi_v' = a \frac{\Delta T_d}{\Delta T_L}$$

from which the expression for Δ results if $\Delta T_d / \Delta T_L$ is a constant, this being the case when the convective flow breaks up in cells.

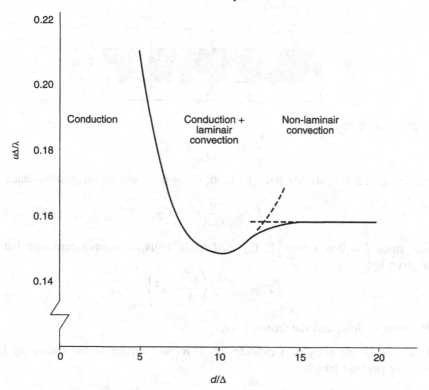

Sketch of $U/\Delta\lambda$ versus d/Δ

Problem 6

A heat balance over one catalyst particle reads:

$$\phi_H = \frac{r\Delta H}{n} = hA\Delta T$$

where n = number of catalyst particles per m³ oil = 1.25×10^{14}. Now for these small particles Nu ≥ 2; thus $h \geq 2\lambda/d$ and ΔT can be calculated.

Problem 7

A heat balance over a length of pipe dx reads:

$$\phi_H = -A\lambda \left.\frac{dT}{dx}\right|_x + \frac{q}{L}dx = -A\lambda \left.\frac{dT}{dx}\right|_{x+dx}$$

or:

$$\frac{d^2T}{dx^2} = -\frac{q}{\lambda LA}$$

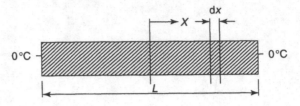

Integration yields:

$$\frac{dT}{dx} = -\frac{q}{\lambda LA}x + C_1$$

but since, at $x = 0$, dT/dx must be zero, $C_1 = 0$. A second integration leads to:

$$T = \frac{-q}{2\lambda LA}x^2 + C_2$$

and, since $T = 0$ at $x = \pm\frac{1}{2}L$, $C_2 = qL/8\lambda A$. Thus the temperature distribution is given by:

$$T = \frac{q}{2\lambda LA}\left\{\frac{L^2}{4} - x^2\right\}$$

Problem 8   John and the frozen body

If we regard the body as a cylinder $(L \gg R)$ we can set up the following heat balance per unit length:

$$\phi_H' = \frac{T - T_a}{\dfrac{\ln R_2/R_1}{2\pi\lambda} + \dfrac{1}{2\pi R_2 h_0}} = -\frac{Mc_p}{L}\frac{dT}{dt}$$

or, collecting all constant values into one constant C:

$$\int_{T_0}^{T} \frac{dT}{T - T_a} = -C\int_0^t dt; \quad \ln\frac{T - T_a}{T_0 - T_a} = -Ct$$

The information that $T_0 = 8\,°C$ at $t = 0$ and $T = 0\,°C$ at $t = 1$ h at $T_a = -30\,°C$ allows us to calculate the value of the constant C as $C = 0.236\ \text{h}^{-1}$.

So, assuming the poor fellow had no fever, his temperature was $37\,°C$ when still alive but $0\,°C$ after t hours and we find, using the same equation, that death must have occurred 3.4 h ago.

III.2 Non-stationary heat conduction

When John entered the empty room, he found a hot iron on an asbestos plate of 1 cm thickness. He lifted the plate which proved still to be cold on the underside. He decided

*that somebody had been in the room less than five minutes
ago and ordered his men to search the apartment.*

If we want to describe the variation of the mean temperature of a body with
time caused by heat conduction we have to start by setting up a microbalance.
From this balance we can calculate the temperature as a function of time and
place. This temperature distribution then enables us to calculate the mean temperature of the body. This complicated procedure is necessary because during heat
conduction large temperature gradients occur in the medium that necessitate a
careful averaging procedure to find the mean temperature.

We have already come across the micro heat balance in the previous Section
(equations (III.11) and (III.12)). For pure conduction without heat production
($v_x = v_y = v_z = 0, q = 0$) equation (III.11) is simplified to:

$$\rho c_p \frac{\partial T}{\partial t} = \lambda \left\{ \frac{\partial^2 T}{\partial x^2} + \frac{\partial^2 T}{\partial y^2} + \frac{\partial^2 T}{\partial z^2} \right\} \qquad \text{(III.17)}$$

In order to solve this equation we generally need to know:

the geometric boundaries of the conducting medium;
the temperature distribution at a certain time, e.g. at $t = 0$ (initial condition);
the conditions for the temperature (or temperature gradient) at the boundaries
of the medium (boundary conditions).

The analytical solutions of equation (III.17) are known for a great number of
problems and can be found in the literature (see e.g. H.S. Carslaw and J.C. Jaeger,
Conduction of heat in solids, 2nd edition, Oxford, 1959).

For non-stationary heat conduction, the duration of the heat penetration process
is of great importance. If this time is so short that the heat penetrating into one
side of a body has not yet reached the other side we speak of heat penetration
into a semi-infinite medium. Accordingly, if the penetrating heat has reached
the geometric boundaries of the medium we speak of penetration into a finite
medium.

III.2.1 Heat penetration into a semi-infinite medium

Let us consider a solid body with a flat boundary at $x = 0$ and an infinite length
in the positive x-direction, as shown in Figure III.5. At $t = 0$, the temperature of
the medium is T_0 in all places (for all values of x; initial condition). At $t \geq 0$,
the wall at $x = 0$ is brought to a constant temperature T_1 (boundary condition).
A second boundary condition is given by the assumption that the medium is
semi-infinite, which means that at $t > 0$ and $x = \infty$, $T = T_0$ (no penetration to
other wall). For this problem equation (III.17) is simplified to:

$$\frac{\partial T}{\partial t} = a \frac{\partial^2 T}{\partial x^2}; \quad a = \frac{\lambda}{\rho c_p} \qquad \text{(III.18)}$$

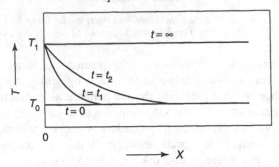

Figure III.5 Penetration of heat into a semi-infinite medium

with:

$$\text{initial condition:} \qquad T = T_0 \quad \text{for} \quad t = 0, x \geq 0$$
$$\text{boundary condition 1:} \quad T = T_1 \quad \text{for} \quad t \geq 0, x = 0$$
$$\text{boundary condition 2:} \quad T = T_0 \quad \text{for} \quad t > 0, x = \infty$$

Equation (III.18) can now be solved by anticipating that the temperature distribution can be described by two dimensionless variables[†]:

$$\theta = \frac{T - T_0}{T_1 - T_0} \quad \text{and} \quad \xi = \frac{x}{\sqrt{4at}}$$

via:

$$\frac{\partial \theta}{\partial t} = \frac{\partial \theta}{\partial \xi} \frac{d\xi}{dt} = -\frac{1}{2} \frac{x}{\sqrt{4at^3}} \frac{d\theta}{d\xi}$$

and:

$$\frac{\partial^2 \theta}{\partial x^2} = \frac{d^2 \theta}{d\xi^2} \left(\frac{\partial \xi}{\partial x} \right)^2 + \frac{d\theta}{d\xi} \frac{\partial^2 \xi}{\partial x^2} = \frac{d^2 \theta}{d\xi^2} \frac{1}{4at}$$

Equation (III.18) is simplified to:

$$\frac{d^2 \theta}{d\xi^2} + 2\xi \frac{d\theta}{d\xi} = 0 \quad \text{with} \quad \left. \begin{array}{ll} \theta = 0 & \text{for} \quad \xi = \infty \\ \theta = 1 & \text{for} \quad \xi = 0 \end{array} \right\} \qquad (\text{III}.19)$$

Integration yields:

$$\frac{d\theta}{d\xi} = C_1 e^{-\xi^2} \qquad (\text{III}.20)$$

[†] The factor '4' is introduced to avoid extra numerical factors in the further analysis.

and:

$$\theta = C_1 \int_0^\xi e^{-\xi^2} \, d\xi + C_2$$

With the boundary conditions, we find $C_2 = 1$ (since $\int_0^0 e^{-\xi^2} \, d\xi = 0$) and:

$$\theta = 0 = C_1 \int_0^\infty e^{-\xi^2} \, d\xi + 1 = \tfrac{1}{2} C_1 \sqrt{\pi} + 1$$

Thus $C_1 = -2/\sqrt{\pi}$, and we find as the final solution:

$$\theta = 1 - \frac{2}{\sqrt{\pi}} \int_0^\xi e^{-\xi^2} \, d\xi \qquad \text{(III.21)}$$

or:

$$1 - \theta = 1 - \frac{T - T_0}{T_1 - T_0} = \frac{T_1 - T}{T_1 - T_0} = \frac{2}{\sqrt{\pi}} \int_0^{x/\sqrt{4at}} e^{-\xi^2} \, d\xi \qquad \text{(III.22)}$$

The right-hand term of this equation is called the error function, which is often represented by the symbol erf, defined by:

$$\text{erf } y = \frac{2}{\sqrt{\pi}} \int_0^y e^{-\xi^2} \, d\xi$$

In Table III.3 a few values of this integral are listed; complete tables can be found in the various handbooks.

Table III.3 The error function

y	erf y
0	0
0.1	0.11
0.2	0.22
0.4	0.43
0.6	0.60
0.8	0.74
1.0	0.84
1.2	0.91
1.5	0.97
∞	1.00

Equation (III.22) enables us to calculate the temperature distribution in the medium at any time t and in any place x. The type of curve found is illustrated in Figure III.6, which shows the temperature penetration into a semi-infinite slab of stainless steel ($a = 7.5 \times 10^{-6}$ m^2/s) at $t_1 = 0.1$, $t_2 = 1$ and $t_3 = 40$ s for

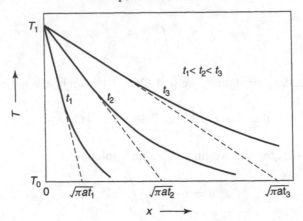

Figure III.6   Temperature distribution in a one-dimensional semi-infinite medium

$T_1 - T_0 = 10\,°C$. It appears that, for small penetration depths x, the temperature curves can be represented by straight lines. The slope of these lines is found from equation (III.22) to be:

$$\frac{\partial T}{\partial x}\bigg|_{x=0} = (T_1 - T_0)\,\frac{\partial \theta}{\partial x}\bigg|_{x=0} = \frac{(T_1 - T_0)}{\sqrt{4at}}\,\frac{\mathrm{d}\theta}{\mathrm{d}\xi}\bigg|_{\xi=0}$$

Using equation (III.20) we find:

$$\frac{\partial T}{\partial x}\bigg|_{x=0} = -\frac{2}{\sqrt{\pi}}\,\frac{T_1 - T_0}{\sqrt{4at}} = -\frac{T_1 - T_0}{\sqrt{\pi at}} \tag{III.23}$$

showing that the tangent to the temperature curve at $x = 0$ goes through the point $T = T_0$, $x = \sqrt{\pi at}$. The distance $x = \sqrt{\pi at}$ is called the penetration depth, which represents the distance x over which the temperature difference $T_1 - T_0$ at $x = 0$ has dropped to 20% of its original value. The last statement can be checked by substituting $x = \sqrt{\pi at}$ into equation (III.22).

The heat flux through the wall of the medium ($x = 0$) is found, using equation (III.23), as:

$$\phi_H'' = -\lambda\,\frac{\partial T}{\partial x}\bigg|_{x=0} = (T_1 - T_0)\frac{\lambda}{\sqrt{\pi at}} = (T_1 - T_0)\sqrt{\frac{\lambda \rho c_p}{\pi t}} \tag{III.24}$$

Equation (III.24) is one of the most important results in technology; it is known as the penetration theory. The number of successful applications of equation (III.24) is enormous, because it is often possible to define a characteristic time for the heat penetration process. Take, for instance, the heating or cooling of particles in a rotating drum. The contact time of the particles with the wall follows from the volume of particles and the rotational speed of the drum. In most cases, the penetration depth ($\sqrt{\pi a t}$) is smaller than half the smallest dimension ($D/2$) of the particles, so that the medium can indeed be regarded as semi-infinite.

The condition for semi-infiniteness of the medium:

$$\sqrt{\pi a t} \ll \frac{D}{2}$$

is often written as:

$$\text{Fo} = \frac{at}{D^2} \ll 0.1 (< 0.05) \tag{III.25}$$

where the dimensionless number Fo is the Fourier number. The physical meaning of this number is:

$$\text{Fo} = \frac{at}{D^2} = \left(\frac{\text{penetration depth}}{\text{thickness of medium}} \right)^2$$

The heat flux through the plane $x = 0$ (equation (III.24)) can also be described with a heat transfer coefficient h:

$$\phi_H'' = h(T_1 - T_0) = (T_1 - T_0) \sqrt{\frac{\lambda \rho c_p}{\pi t}} \tag{III.26}$$

Thus:

$$h = \sqrt{\frac{\lambda \rho c_p}{\pi t}} \quad \text{or} \quad \text{Nu} = \frac{hD}{\lambda} = 0.57 \, \text{Fo}^{-\frac{1}{2}} \tag{III.26a}$$

If the penetration process lasts a time t_e, the mean heat transfer coefficient is given by:

$$\langle h \rangle = \frac{1}{t_e} \int_0^{t_e} h \, dt = \sqrt{\frac{\lambda \rho c_p}{\pi t_e^2}} \int_0^{t_e} \frac{1}{\sqrt{t}} \, dt = 2 \sqrt{\frac{\lambda \rho c_p}{\pi t_e}} \tag{III.27}$$

Applications of this formula are numerous; see, for example, problems 2, 4, 6 and 7 at the end of this Section.

III.2.2 Heat penetration into a finite medium

If the penetration depth is greater than the dimensions of the body (Fo > 0.1), the penetration theory proves inadequate, so another description must be found. The situation then is as illustrated in Figure III.7. After a certain period of time (in the figure for $t > t_3$, which corresponds with Fo > 0.1), the consecutive temperature distributions remain geometrically similar, which means:

$$T - T_0 = (T_1 - T_0)[1 - A(t)f(x/R)] \tag{III.28}$$

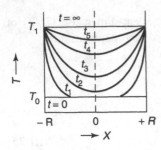

Figure III.7 Heat penetration into a finite medium (for $0 < t < t_2$, penetration theory; for $t_3 < t < \infty$, constant h)

The function $A(t)$ only depends on the time; it decreases with increasing time $(A(\infty) \rightarrow 0)$. The temperature distribution is further given as a function of the place by the time-independent function $f(x/R)$ which must satisfy the following conditions:

$$f(\pm 1) = 0 \quad \text{and} \quad \mathrm{d}f/\mathrm{d}x|_{x=0} = 0$$

The non-stationary temperature distributions in a body of any form should therefore satisfy this general relationship for larger times (Fo > 0.1). We shall now ascertain what the consequences will be by further analyzing the case of the flat plate. If we apply the general relationship for the temperature distribution of the microbalance (equation (III.18)) to the flat plate, the result after a slight adaptation will be:

$$\frac{1}{f(x/R)} \frac{\mathrm{d}^2 f}{\mathrm{d}(x/R)^2} = \frac{R^2}{a} \frac{1}{A(t)} \frac{\mathrm{d}A}{\mathrm{d}t} \tag{III.29}$$

The left-hand term is only a function of the place, while the right-hand term is only a function of the time. Therefore the two terms should be constant, e.g. equal to $-\beta^2$ (a real, negative number). Only then can the two terms for all x and all t be equal to each other and only then will $A(t)$ for $t \rightarrow \infty$ asymptotically approach zero. It follows that:

$$A \sim \exp(\beta^2 at/R^2) \quad \text{and} \quad f(x/R) \sim \cos(\beta x/r)$$

Thus the differential equation for the flat plate and the boundary conditions are satisfied, but not, however, the starting condition, which after all could be expected of a solution which applies only to large times. Now the value of β^2 has still to be determined. It follows from the consideration that only cosine functions of certain wavelengths neatly fit the range $-R < x < R$ (just as when vibrating a string the permissible wavelengths are determined by the length of the string). The highest possible wavelength is here apparently equal to $4R$, with further possibilities in this problem being (considering the boundary conditions $f(\pm 1) = 0$) $4R/3$, $4R/5$, etc. The relevant β values are: $\pi/2$, $3\pi/2$, $5\pi/2$, etc. (as the reader may establish for himself).

These β values, which indicate which of the cosine functions fit, are called the 'eigenvalues' of the problem. The lowest β value (here $\pi/2$) is the most important for long times, for they need an A function which decreases less rapidly with time than the A functions belonging to the other β values. So, of all of the possible solutions (which, for example, have to be summed up with certain weighting factors if also the initial condition has to be satisfied) for long times, only that with $\beta = \pi/2$ remains:

$$\frac{T_1 - T}{T_1 - T_0} \sim \exp\left(-\frac{\pi^2 at}{4R^2}\right)\cos\left(\frac{\pi x}{2R}\right) \quad \text{(flat plate)} \quad \text{(III.30)}$$

The proportionality constant (which is of the order of 1 and here equals $4/\pi$) cannot be found in this way. A complete analysis is necessary (it is this weighting factor which reminds us of the previous history of $t = 0$). It is, however, sufficient for practical purposes to know that it is almost 1; the cumbersome full analysis is seldom required. For short times (Fo < 0.05) it yields a poorly converging series of cosine terms as a solution. A better usable solution for short times is already available (see the previous section).

For longer times we are also often interested not so much in the actual temperature distribution as in the heat transfer coefficient h_i (internal heat transfer coefficient), which we define on the basis of the mean temperature difference:[†]

$$h_i(T_1 - \langle T \rangle) = \phi_H'' = -\lambda \left.\frac{dT}{dx}\right|_{x=R} \quad \text{(III.31)}$$

From equation (III.30) we find for the temperature distribution:

$$\frac{\partial T}{\partial x} = -C_1(T_1 - T_0)\exp\left(-\frac{\pi^2 at}{4R^2}\right)\frac{\pi}{2R}\sin\frac{\pi x}{2R} \quad \text{(III.32)}$$

and at $x = R$:

$$\left.\frac{\partial T}{\partial x}\right|_{x=R} = -C_1(T_1 - T_0)\exp\left(-\frac{\pi^2 at}{4R^2}\right)\frac{\pi}{2R} \quad \text{(III.32a)}$$

The mean temperature $\langle T \rangle$ is given by:

$$\langle T \rangle = \frac{1}{R}\int_0^R \left\{T_1 - C_1(T_1 - T_0)\exp\left(-\frac{\pi^2 at}{4R^2}\right)\cos\frac{\pi x}{2R}\right\}dx \quad \text{(III.33)}$$

which yields for the mean temperature difference:

$$\langle T \rangle - T_1 = -C_1\frac{2}{\pi}(T_1 - T_0)\exp\left(-\frac{\pi^2 at}{4R^2}\right) \quad \text{(III.33a)}$$

[†] $\langle T \rangle = \dfrac{1}{R}\displaystyle\int_0^R T\,dx$

Thus we find with equation (III.31):

$$h_i = \frac{\lambda \frac{\pi}{2R}}{\frac{2}{\pi}} = \frac{\pi^2}{4} \frac{\lambda}{R} \tag{III.34}$$

or:

$$\mathrm{Nu} = \frac{h_i 2R}{\lambda} = \frac{\pi^2}{2} = 4.93 (\mathrm{Fo} > 0.1) \tag{III.34a}$$

Due to the fact that for $\mathrm{Fo} > 0.1$ the shapes of the temperature distributions do not change with time, the Nu values become constant. The value of the Nusselt number depends only on the form of the body. For a flat plate, we find $\mathrm{Nu} = 4.93$, and for a sphere $\mathrm{Nu} = 6.6$. For bodies which can be enclosed by a sphere we can estimate the Nu numbers by applying $\mathrm{Nu}_{\mathrm{sphere}} = \alpha D_1 / \lambda = 6.6$ and taking for D_1 the diameter of the sphere having the same volume as the body. For a cube (with side D) we find:

$$V = D^3 = \frac{\pi}{6} D_1^3; \quad \mathrm{Nu}_{\mathrm{cube}} = \frac{h_i D}{\lambda} = \frac{6.6}{D_1} D = 6.6 \sqrt[3]{\frac{\pi}{6}} = 5.3$$

(exact solution 4.9). For a cylinder with $L = D$ we find analogously:

$$\mathrm{Nu}_{\mathrm{cyl}} = 6.6 \sqrt[3]{\frac{2}{3}} = 5.8$$

(exact solution 5.8). Since for $\mathrm{Fo} > 0.1$ the Nu and h values are constant, a heat balance over the total body becomes:

$$\rho c_p V \frac{d\langle T \rangle}{dt} = h_i A (T_1 - \langle T \rangle) \tag{III.35}$$

Integrating, with $\langle T \rangle = T_0$ at $t = 0$, we find:

$$\ln \frac{T_1 - \langle T \rangle}{T_1 - T_0} = -\frac{h_i A t}{\rho c_p V} = -\frac{h_i D}{\lambda} \frac{\lambda}{\rho c_p} \frac{A}{VD} t = -\mathrm{Nu} \frac{at}{D^2} \tag{III.36}$$

This solution is identical to the relationship found with equation (III.33a). Both relationships show that a plot of the relative temperature equalization $(T_1 - \langle T \rangle)/(T_1 - T_0)$ against $\mathrm{Fo} = at/D^2$, should yield a straight line for $\mathrm{Fo} > 0.1$. This is indeed the case, as shown by Figure III.8. An analogous graph can be produced for the relative temperature equalization in the centre of the body, as shown in Figure III.9. Similar information is given in Table III.4.

Since $\phi_H'' = h_i(T_1 - \langle T \rangle)$ the y-axis of the first graph also represents:

$$\frac{T_1 - \langle T \rangle}{T_1 - T_0} = \frac{\phi_H''}{h_i(T_1 - T_0)} \tag{III.37}$$

Figure III.8 can also be used to find the heat flux into the medium at any time.

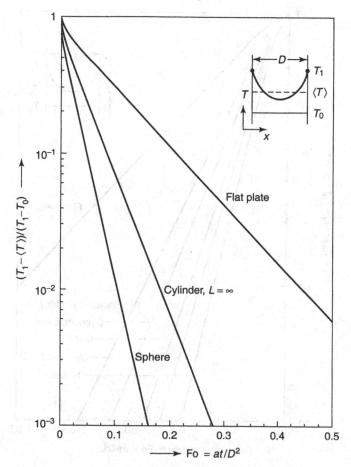

Figure III.8 Mean temperature during non-stationary heat conduction

III.2.3 Influence of an outside heat transfer coefficient

In Sections III.2.1 and III.2.2 we have considered cases where the heat transfer coefficient outside the medium is very large ($h_e \to \infty$) compared with the heat conduction in the medium. This situation is, for example, present if a cold piece of wood (poor heat conductivity) is placed in a steam atmosphere (good heat transfer) with a saturation temperature T_1.

The opposite situation is realized if a piece of metal (good heat conduction) of temperature T_0 is placed in air (poor heat transfer) with a temperature T_1. In this case, the heat transport outside the medium is so slow that temperature equalization in the metal is practically complete; the temperature T inside the

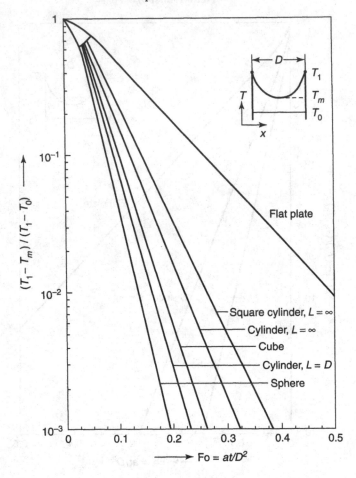

Figure III.9 Centre temperature during non-stationary heat conduction

medium is the same in all places and the rate of heating or cooling is independent of the heat conductivity of the material. A macroscopic heat balance for this case leads to an exponentially decreasing temperature equalization such as equation (III.36), but with the outside heat transfer coefficient h_e instead of h_i

$$\ln \frac{T_1 - T}{T_1 - T_0} = -\frac{h_e A t}{\rho c_p V} \qquad (III.38)$$

If both heat transport resistances, the outside ($1/h_e$) and the inside ($1/h_i \sim D/\lambda$), are important, it is possible to construct graphs similar to Figures III.8 and III.9 for a series of values of h_e. These graphs can be found in a number of handbooks

Table III.4 Temperatures and Nusselt numbers for Fo > 0.1 in finite media during non-stationary heat conduction

Geometry	$\dfrac{T_1 - T_m}{T_1 - T_0}$	Nu_m
Flat plate	$\dfrac{4}{\pi}\exp(-\pi^2\,\text{Fo})$	3.14
Cylinder ($L = \infty$)	$1.60\exp(-23.13\,\text{Fo})$	2.50
Sphere	$2\exp(-4\pi^2\,\text{Fo})$	2.00

Geometry	$\dfrac{T_1 - \langle T\rangle}{T_1 - T_0}$	Nu
Flat plate	$\dfrac{8}{\pi^2}\exp(-\pi^2\,\text{Fo})$	4.93
Cylinder ($L = \infty$)	$0.692\exp(-23.13\,\text{Fo})$	5.78
Sphere	$\dfrac{6}{\pi^2}\exp(-4\pi^2\,\text{Fo})$	6.58

(e.g. *VDI Wärmeatlas*)[†]. We can, however, approximate the exact solution by assuming that also during non-stationary conduction the heat resistances are additive (as is the case during stationary heat conduction).

For the region with constant Nu values (Fo > 0.1, i.e. long penetration times) we can then write instead of equation (III.36):

$$-\ln\frac{T_1 - \langle T\rangle}{T_1 - T_0} = \frac{At}{\rho c_p V}\left\{\frac{1}{h_i} + \frac{1}{h_e}\right\}^{-1} = Nu\frac{at}{D^2}\left\{1 + \frac{h_i}{h_e}\right\}^{-1} \qquad (\text{III.39})$$

or, if we call f the function, shown in Figure III.8:

$$\ln\frac{T_1 - \langle T\rangle}{T_1 - T_0} = f\left\{\frac{at}{D^2}\left(1 + \frac{h_i}{h_e}\right)^{-1}\right\} \qquad (\text{III.40})$$

Apparently, we can also use Figure III.8 to solve problems with an outside heat resistance if we use, instead of at/D^2, the corrected parameter:

$$\frac{at}{D^2}(1 + h_i/h_e)^{-1}$$

i.e. the times necessary to reach the same relative temperature equalization with and without an external heat resistance form a ratio of $(1 + h_i/h_e) : 1$.

In the literature one sometimes finds the so-called Biot number which is defined as:

$$Bi = \frac{h_e D}{\lambda}$$

[†] *VDI Wärmeatlas*, 8th edition, Springer-Verlag, 1997

This number gives an indication of the ratio of the internal and external heat transfer resistances. If $Bi < 1$, the external resistance dominates, while if $Bi \gg 1$ the internal resistance is the most significant. The Biot number should not be confused with the Nusselt number; in the latter the heat transfer coefficient and the heat conductivity relate to the same phase.

In equations (III.39) and (III.40), T_1 is the outside temperature and not the surface temperature T_s of the medium. Because of the resistance $1/h_e$, T_s is lower than T_1. The surface temperature can be calculated from a heat balance:

$$h_i(T_s - \langle T \rangle) = h_e(T_1 - T_s) \tag{III.41}$$

Equation (III.41) appears to be in good agreement with the known exact solutions. The heating or cooling times calculated with the above procedure are within 10% of the exact solutions. In the literature, other cases are also known where an exact analysis showed that place- or time-dependent resistances in series are not exactly additive, but that only a small error is made by assuming additiveness. We can therefore extend the approach applied to cases where the internal heat resistance is time-dependent, i.e. to short penetration times, $Fo < 0.1$. The times needed to reach a certain temperature equalization with and without external resistance form a ratio of $(1 + \langle h_i \rangle / h_e) : 1$. In this equation, $\langle h_i \rangle$ is the mean internal heat transfer coefficient averaged over the time of the heat transfer process t_e (equation (III.27)).

We can, of course, also use the relationships:

$$\frac{1}{U} = \frac{1}{\langle h_i \rangle} + \frac{1}{h_e} \quad \text{and} \quad \phi''_H = U(T_1 - T_0) \tag{III.42}$$

where $\langle h_i \rangle$ is the mean inside heat transfer coefficient, in order to calculate the heat flux into the medium.

III.2.4 Problems

1. A packet of butter of 250 g has a temperature of 20 °C. It is put in a refrigerator at 4 °C. After 120 min the centre of the packet appears to have a temperature of 6 °C.
   (a) After how many minutes (from the beginning) is this temperature 5 °C?
   (b) How long does it take for the interior of a 500 g packet (equal in geometry to the former) to reach a temperature of 6 °C?

   Answer: (a) 150 min;   (b) 190 min

2. Check that if two conducting media a and b with uniform temperatures T_a and T_b are brought into contact with each other, immediately a constant temperature T_c (contact temperature) prevails at the surface, which is

given by:

$$\frac{T_a - T_c}{T_c - T_b} = \sqrt{\frac{(\lambda \rho c_p)_b}{(\lambda \rho c_p)_a}}$$

*3. A cooling drum (external diameter $D_u = 1$ m) has a surface temperature of 20 °C and rotates at an angular velocity of 2π min^{-1}. A bituminous product with a temperature of 80 °C (at this temperature the bitumen is just liquid) is brought as a layer on to the drum and after a half-revolution is scraped off again. The requirement is that the temperature of the cooled bitumen does not exceed 25 °C. Properties of bitumen: $c_p = 920$ J/kg °C; $\rho = 10^3$ kg/m^3; $\lambda = 0.17$ W/m °C.

(a) What is the maximum thickness of the layer which still fulfils this requirement under the conditions mentioned?

(b) By which factor is the capacity of the drum increased if the angular speed of 6π/ min is chosen?

Heat exchange between the layer and the surroundings is neglected. On cooling, no latent heat is liberated.

Answer: (a) 2.2 mm; (b) 1.73

4. Objects are submerged in a hot fluidized bed to warm them up. The particles (diameter 1 mm, $\lambda = 1$ W/m °C, $\rho c_p = 4 \times 10^6$ J/m^3 °C) remain on an average for 0.3 s on the surface of these objects and ca. 60% of the surface is occupied by particles. Calculate the heat transfer coefficient between the fluidized bed and the objects. Assume that every particle remains on the surface equally long, that the gas with which the particles are fluidized does not contribute to the heat transfer and that the objects are submerged in the bed much longer than 0.3 s.

Answer: 2400 W/m^2 °C.

5. Elsevier's *Culinary Encyclopedia* recommends the following boiling times for obtaining hard-boiled, non-crumbly eggs from various birds:

Bird	Egg weight (g)	Boiling time (s)
Turkey	95	850
Goose	85	780
Duck	70	620
Chicken (category 2)	60	500
Guinea fowl	40	420
Partridge, peewit	36	390
Pheasant	30	350
Pigeon	24	300

How can these data be summarized scientifically?

6. A solid glass sphere with a diameter of 0.2 m is suspended in a room which at the beginning has the same temperature as the sphere. The ambient temperature suddenly increases by 10 °C.
   (a) If the heat transfer coefficient is $h = 8$ W/m^2 °C, how long does it take for the centre of the sphere to adopt the temperature change for $\frac{1}{4}$? And if $h = \infty$?
   (b) Answer the same questions for the mean temperature of the sphere. Explain the difference between these answers and those to question (a). In which case is the difference the greatest and why?
   $\lambda_{glass} = 0.8$ W/m °C; $a_{glass} = 44 \times 10^{-8}$ m^2/s

   Answer: (a) 5.5 and 1.26 h; (b) 2.7 and 0.63 h

7. A pan with custard must be cooled from 100 to 20 °C. To this end, the pan is put in a sink filled with water at 10 °C to the same height as the custard in the pan. Water is added to keep the temperature around the pan at 10 °C. There are two ways of cooling:
   (a) Allow the pan with custard to cool gently.
   (b) Stir the pan regularly and use a pan scraper to scrape the custard off the cylindrical pan wall and mix it with the bulk. Each part of the wall surface should be scraped once every minute.
   Calculate the times necessary to cool the custard down to a mean temperature of 20 °C, according to both methods.

   Data: Thermal diffusivity of the custard is 10^{-7} m^2/s and $\lambda = 0.5$ W/m °C over the entire temperature range. The custard is so viscous that free convection is fully suppressed. Heat transport through the bottom is negligible. The pan diameter is 20 cm.

   Answer: (a) 11 h; (b) 32 min

8. If water droplets (temperature 20 °C) fall from a given height on to a hot plate, the droplets will no longer moisten the plate above a certain plate temperature (T_p) (Leidenfrost phenomenon). In the following table, this temperature is given for four materials. Determine whether this phenomenon occurs at a contact temperature characteristic for the droplet.

Material of plate	λ(W/m °C)	$\rho c_p[10^6(\text{J/m}^3 \text{ °C})]$	T_p(°C)
1	382	3.51	199
2	140	2.44	206
3	72	1.45	219
4	14	2.05	244

Answer: $T_c = 190 \pm 2$ °C

*9. Tins filled with spinach (diameter and height 12 cm) are put in a room filled with steam at 120 °C to sterilize the spinach.

(a) The heat transfer coefficient of the steam to the tin is 10 000 W/m² °C. After how many seconds does the heat resistance in the tin determine the rate of heat transfer?

(b) Make a (schematic) drawing of the temperature distribution in and around the tin at the moment the temperature of the centre of the tin begins to rise.

(c) If the temperature of the tins is originally 20 °C, for how long should the tins be put in the room with steam, so as to ensure a spinach temperature of at least 110 °C everywhere in the tin?

Properties of spinach: $\lambda = 0.7$ W/m °C; $\rho = 1200$ kg/m³; $c_p = 4000$ J/kg °C

Answer: (a) 1 s; (c) 2.5 h

*10. Show that John took the right steps in the case of the empty apartment discussed at the beginning of this section ($a_{asbestos} = 10^{-7}$ m²/s).

Comments on problems

Problem 3

The situation is as drawn in the left-hand picture; there is non-stationary heat conduction. The figure on the right shows the temperature distribution in the film.

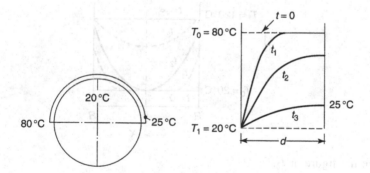

(a) The temperature equalization after a half-revolution (at time $t = \pi/2\pi = 0.5$ min $= 30$ s) is:

$$\frac{T_1 - T_m}{T_1 - T_0} = \frac{20 - 25}{20 - 80} = 0.0833$$

and Figure III.9 yields a corresponding value (flat plate, since heating is only on one side, $2d$ must be used instead of D) for the Fourier number of:

$$Fo = 0.27 = \frac{at}{D^2} = \frac{at}{4d^2}$$

from which we find $d = 2.2 \times 10^{-3}$ m.

(b) Fo stays constant, but $t_2 = \frac{1}{3}t_1$; thus:

$$d_2 = d_1\sqrt{\frac{1}{3}} = \frac{d_1}{1.73}$$

and the capacity:

$$\phi_{m2} = \frac{3}{\sqrt{3}}\phi_{m1} = 1.73\phi_{m1}$$

Problem 9

The figure shows the temperature distribution in the spinach tins during non-stationary heating.

(a) The heat transfer resistance in the tin after a short time is $\delta/\lambda_{spinach}$, with $\delta = \sqrt{\pi a t}$. We can assume that the inside resistance determines the overall heat transfer resistance as soon as:

$$\frac{\delta}{\lambda} \approx 10\frac{1}{h_e}$$

which is the case after:

$$t = 1\ \text{s}$$

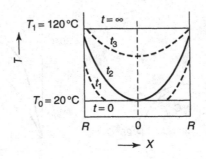

(b) See the figure at t_2.
(c) Temperature equalization:

$$\frac{T_1 - T_m}{T_1 - T_0} = 0.1$$

and Figure III.9 yields:

$$\text{Fo} = 0.09 = \frac{at}{D^2}$$

and we find:

$$t = 2.5\ \text{h}.$$

Problem 10 John in the empty apartment

The penetration depth of heat into the asbestos plate is $\delta = \sqrt{\pi a t} < 10^{-2}$ m, and therefore, with $a = 10^{-7}$ m^2/s, $t < 5$ min, meaning that somebody must have been in the apartment less than 5 minutes before.

III.3 Heat transfer by forced convection in pipes

> *It was near midnight when John entered his flat. An open pot of water was boiling in the kitchen. John thought: 'The pot has a diameter of 20 cm and could have contained at most 6 l of water. The bottom of the pot is covered with a 3 mm layer of scale and the temperature controller of the heating plate is put at 150°C'. He remembered the pages to come and looked for other signs of whether somebody was at home or not.*

In this section we will treat, subsequently, heat transfer during laminar pipe flow and during turbulent flow, and heat transport from one fluid to another through a solid wall.

III.3.1 Heat transfer during laminar flow in pipes

If a fluid moves through a pipe as a plug, the penetration theory discussed in Section III.2 can be applied for calculating heat flows if one substitutes $t = x/\langle v \rangle$. However, if radial velocity distributions occur, the penetration theory cannot be applied as such, because the flowing fluid transports heat as well.

Consider the situation shown in Figure III.10. A viscous liquid passes as a laminar flow through a circular tube. At $x < 0$, the liquid and the pipe wall have uniform temperatures T_0. At $x \geq 0$, the temperature of the pipe wall is $T_w (T_w > T_0)$. Because of this temperature gradient there will be a heat flux q from the wall into the liquid by conduction in the radial direction. Since there is also a temperature gradient in the x-direction, there will be conductive heat transport in the x-direction as well and, furthermore, heat is produced by viscous dissipation and heat is transported in the x-direction by the flowing liquid. An energy balance over a ring-shaped control volume $2\pi r\, dx\, dr$ consists of the following expressions:

conduction in radial direction: $+q_r|_r 2\pi r\, dx - q_r|_{r+dr} 2\pi (r + dr)\, dx$

conduction in axial direction: $+q_x|_x 2\pi r\, dx - q_x|_{x+dx} 2\pi r\, dx$

heat produced by viscous dissipation: $-2\pi\, dr\, dx \tau_{rx} \dfrac{\partial v_x}{\partial x} = \eta \dfrac{\partial^2 v_x}{\partial x^2} 2\pi r\, dr\, dx$

energy transported by fluid: $\rho c_p v_x (T - T_0)|_x 2\pi r\, dr - \rho c_p v_x (T - T_0)|_{x+dx} 2\pi r\, dr$

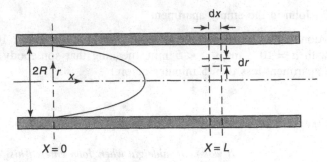

Figure III.10   Heat transport during laminar flow

Equating heat input and heat output over the control volume, neglecting heat conduction in the x-direction and heat production by viscous dissipation (which is allowed in the majority of cases), we find:

$$\frac{(rq_r)_r - (rq_r)_{r+dr}}{dr} + r\rho c_p v_x \frac{(T)_x - (T)_{x+dx}}{dx} = 0$$

or:

$$\frac{\partial(rq_r)}{\partial r} + r\rho c_p v_x \frac{\partial T}{\partial x} = 0 \tag{III.43}$$

Applying Fourier's law (equation (III.2)):

$$q_r = -\lambda \frac{\partial T}{\partial r}; \quad \frac{d(rq_r)}{dr} = -\lambda \frac{\partial}{\partial r}\left(r\frac{\partial T}{\partial r}\right)$$

we find from equation (III.43):

$$-\lambda \frac{\partial}{\partial r}\left(r\frac{\partial T}{\partial r}\right) + r\rho c_p v_x \frac{\partial T}{\partial x} = 0 \tag{III.44}$$

The velocity distribution for laminar flow in a circular pipe is given by (equation (II.8)):

$$v_x = 2\langle v\rangle\left(1 - \frac{r^2}{R^2}\right)$$

Introducing this expression into equation (III.44) we find:

$$-\lambda \frac{\partial}{\partial r}\left(r\frac{\partial T}{\partial r}\right) + 2r\rho c_p\langle v\rangle\left(1 - \frac{r^2}{R^2}\right)\frac{\partial T}{\partial x} = 0 \tag{III.45}$$

This differential equation (the so-called Graetz equation) has been solved for a number of boundary conditions. For the conditions constant wall temperature and uniform fluid temperature at the inlet, the solution is, for large distances from the pipe entrance:

$$Nu = \frac{hD}{\lambda} = 3.66 \quad \text{for} \quad \frac{ax}{\langle v \rangle D^2} > 0.1 \qquad (\text{III.46a})$$

and for the region near the pipe entrance:

$$Nu = \frac{hD}{\lambda} = 1.08 \left(\frac{ax}{\langle v \rangle D^2} \right)^{-\frac{1}{3}} \quad \text{for} \quad \frac{ax}{\langle v \rangle D^2} < 0.05 \qquad (\text{III.46b})$$

The mean Nu number over the entrance region is accordingly given by:

$$\langle Nu \rangle = \frac{1}{x_e} \int_0^{x_e} Nu \, dx = 1.62 \left(\frac{ax_e}{\langle v \rangle D^2} \right)^{-\frac{1}{3}} \quad \text{for} \quad \frac{ax_e}{\langle v \rangle D^2} < 0.05 \qquad (\text{III.46c})$$

The term $ax/\langle v \rangle D^2$ is called the Graetz number (Gz): it has here the same function as the Fourier number for non-stationary conduction. In the entrance region of the pipe (short contact time), the thermal boundary layer is built up and therefore $Nu \sim x^{-1/3}$. At great distances from the entrance, the temperature distributions will not change with x and the Nu number becomes constant.

The above Nu values and heat transfer coefficients are related to a mean fluid temperature $\langle T \rangle$ which is defined as being proportional to the heat flux transported by the liquid. Thus:

$$\langle T \rangle \frac{\pi}{4} D^2 \langle v \rangle = \int_0^{D/2} 2\pi r v_r T_r \, dr \qquad (\text{III.47})$$

This is the temperature we would measure if the flow was collected in a vessel and mixed (mean cup temperature).

Equations (III.46a) and (III.46c) can also be written as:

$$\langle Nu \rangle = 1.62 \, Re^{\frac{1}{3}} \, Pr^{\frac{1}{3}} \left(\frac{x_e}{D} \right)^{-\frac{1}{3}}; \quad \langle Nu \rangle_{min} = 3.66$$

as readers may check for themselves. They can also check that for laminar flow the Nu numbers for pipes of varying diameter and the same length are equal if the volumetric flow rate through the pipes is equal. The ratios of the heat transfer coefficients are then $h_1/h_2 = D_2/D_1$, whereas the total heat flows in both pipes are equal.

The Prandtl number is defined by $Pr = v/a = c_p \eta/\lambda$ and represents the ratio of the rate of momentum and heat transport.

For other types of velocity distributions, boundary conditions and other channel geometries, temperature distributions and heat transfer coefficients have also been calculated theoretically. The results can always be presented by equations such as equations (III.46a,b), but the numeric constants change with the geometry of the flow and the rheology of the fluid. Some theoretical results are presented in

Table III.5. Generally, we can state that the greater the velocity gradient near the wall, the greater the heat transfer coefficient will be. For that reason, under laminar conditions non-Newtonian (pseudoplastic) liquids often show higher heat transfer coefficients than Newtonian fluids. Measurements about temperature distributions and temperature equalization in laminar flow are scarce. Practical experience indicates that, especially in laminar flow, the temperature dependence of the physical properties of the fluid (especially viscosity) has great influence on the rate of heat transport. During heating up of a liquid, the warmer liquid near the wall will have a lower viscosity than the liquid near the centre of the tube. Since the shear stress distribution is fixed (see Section II.1.2), this means that due to the heat transfer process the temperature distribution will be flattened (more uniform). This increases the heat transport.

Table III.5 Local Nu numbers for flow in channels

Geometry	Type of flow	Condition Gz	Boundary $T_w = $ constant Nu =	Condition $\phi_H'' = $ constant Nu =
Circular pipe	plug	$\{$ <0.05	$0.565 \ Gz^{-1/2}$	$0.885 \ Gz^{-1/2}$
		$\l$ >0.1	5.77	8.00
	newtonian	$\{$ <0.05	$1.08 \ \ Gz^{-1/3}$	$1.30 \ Gz^{-1/3}$
		$\l$ >0.1	3.66	4.36
Parallel plates:	plug	$\{$ <0.05	$0.565 \ Gz^{-1/2}$	$0.885 \ Gz^{-1/2}$
two-sided		$\l$ >0.1	4.93	6.00
heat transfer	newtonian	$\{$ <0.05	$0.98 \ \ Gz^{-1/3}$	$1.8 \ \ Gz^{-1/3}$
		$\l$ >0.1	3.76	4.12
Parallel plates;	plug	$\{$ <0.05	$0.565 \ Gz^{-1/2}$	$0.885 \ Gz^{1/2}$
one-sided		$\l$ >0.1	2.47	3.00
heat transfer	newtonian	$\{$ <0.05	$0.78 \ \ Gz^{-1/2}$	$0.94 \ Gz^{-1/3}$
		$\l$ >0.1	2.43	2.69
Square	plug	>0.1	—	6.0
channel	newtonian	>0.1	2.95	3.6

Differences in density caused by temperature differences may cause extra flow (free convection) and also tend to increase the heat transfer. An experimentally checked relationship for the heat transfer during laminar flow of Newtonian liquids is given by Sieder and Tate as:

$$\langle Nu \rangle = 1.86 \ Re^{\frac{1}{3}} \ Pr^{\frac{1}{3}} \left(\frac{x_e}{D} \right)^{-\frac{1}{3}} \left(\frac{\eta}{\eta_w} \right)^{\frac{1}{7}} \tag{III.48}$$

$\langle Nu \rangle_{min} = 3.66$ for heating

where η is the viscosity of the fluid at mean fluid temperature $(\langle T \rangle - T_0)/2$ and η_w the fluid viscosity at wall temperature T_w.

III.3.2 Heat transfer during turbulent flow

The best known empirical correlation for heat transfer during turbulent flow in circular pipes is:

$$\text{Nu} = \frac{hD}{\lambda} = 0.027 \, \text{Re}^{0.8} \, \text{Pr}^{\frac{1}{3}} \left(\frac{\eta}{\eta_w} \right)^{\frac{1}{7}} \tag{III.49}$$

which is valid in the region $2 \times 10^3 < \text{Re} < 10^5$ and $\text{Pr} \geq 0.7$.

We can gain some insight into this type of relationship by using the boundary layer concept (Figure III.11). For turbulent flow in pipes, the total velocity gradient can be thought to be located in a layer of thickness δ_h at the wall (see Section II.2.1). The thickness of this hydrodynamic boundary layer is defined by (equation (II.24)):

$$\tau_w = \eta \frac{\langle v \rangle}{\delta_h} = \frac{f}{2} \rho \langle v \rangle^2$$

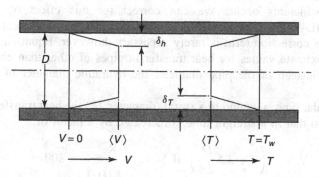

$$V=0 \qquad \langle V \rangle \qquad \qquad \langle T \rangle \qquad T=T_w$$

Figure III.11 Boundary layers in turbulent pipe flow

which gives for the ratio of pipe diameter to hydrodynamic boundary layer (equation (II.25)):

$$\frac{D}{\delta_h} = \frac{f}{2} \, \text{Re}$$

An analogous model can be set up for the temperature distribution by assuming that the total temperature gradient is located in a layer of thickness δ_T. So we can write:

$$h(T_w - \langle T \rangle) = \lambda \frac{T_w - \langle T \rangle}{\delta_T}$$

or:

$$\text{Nu} = \frac{hD}{\lambda} = \frac{D}{\delta_T} = \frac{D}{\delta_h}\frac{\delta_h}{\delta_T} = \frac{f}{2}\text{Re}\frac{\delta_h}{\delta_T} \qquad (\text{III.50})$$

We have already found (Section II.2.1) that the ratio of hydrodynamic boundary layer to thermal boundary layer is given by (equation (II.29)):

$$\frac{\delta_h}{\delta_T} = \text{Pr}^{\frac{1}{3}}$$

Combined with equation (III.50) we obtain:

$$\text{Nu} = \frac{f}{2}\ \text{Re}\ \text{Pr}^{\frac{1}{3}} \qquad (\text{III.51})$$

and using the Blasius equation (II.37) to substitute for the friction factor, we finally obtain for turbulent flow in pipes with smooth walls:

$$\text{Nu} = 0.04\ \text{Re}^{0.75}\ \text{Pr}^{\frac{1}{3}}$$

which is quite similar to equation (III.49) if we add again the Sieder and Tate correction for the influence of the radial viscosity gradients.

During turbulent flow, there is also an entrance region where higher heat transfer coefficients occur. We can correct for this effect by multiplying equation (III.49) by $(1 + 0.7\ D/L)$. Since for most practical heat exchangers $L \gg D$, this correction term is rarely necessary, however. Equation (III.49) will yield approximate values for heat transfer in pipes of other than circular cross-sections if instead of the pipe diameter the hydraulic diameter of the channel is used.

If a circular pipe is wound to a spiral (diameter $2R$) the heat transfer coefficient compared to that in a straight tube is increased by a factor of:

$$\left(1 + 3.5\frac{D}{2R}\right) \quad \text{if } 10 < \frac{\text{Re}}{(2R/D)^{\frac{1}{2}}} < 200 \qquad (\text{III.52})$$

due to the occurrence of systematic eddies.

Using the analogy between mass and heat transfer given in Section IV.3.1, the correlations for mass transfer given in Section IV.3 can also be used for calculations concerning heat transfer.

III.3.3 Partial and total heat transfer coefficients

Heat must often be transferred from one flowing medium to another, which are separated from each other by a wall. We then have (e.g. as shown in Figure III.12) h_1, a wall with thickness d_w and conductivity λ_w, moreover possibly a deposit or dirt layer with d_d and λ_d and finally h_2. The overall heat transfer coefficient U can now be formally defined according to:

$$\phi_H'' = U\{\langle T_1 \rangle - \langle T_2 \rangle\} \tag{III.53}$$

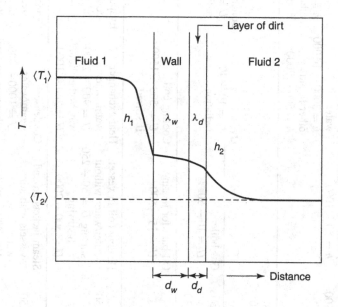

Figure III.12 Heat resistances in series

where $\langle T_1 \rangle$ and $\langle T_2 \rangle$ are the mean temperatures of the flowing liquids. Just as with heat conduction through various solid layers (Figure III.1), here in the one-dimensional case the total resistance $1/U$ equals the sum total of the partial resistances. Thus:

$$\frac{1}{U} = \frac{1}{h_1} + \frac{d_w}{\lambda_w} + \frac{d_d}{\lambda_d} + \frac{1}{h_2} \tag{III.54}$$

When the various surface areas are not of the same size, e.g. in the case of thick-walled cylinders, the varying sizes of the inner and outer surface areas should be taken into account.

The value of U is mainly determined by the greatest resistance in series — by the lowest value of h or λ/d. In metal apparatus, mostly $\lambda_w/d_w \gg h$ (λ steel $\approx$ 45 W/m°C). However, if a dirt layer is present, λ_d/d_d can have a consider-able influence on the overall U (e.g. in water-cooled heat exchangers $\lambda_d/d_d \approx$ $0.35/5 \times 10^{-4} = 600$). The partial heat transfer coefficients increase in the order gas (free convection), gas (forced convection), liquid (free convection), liquid (forced convection), condensing vapour and boiling liquid. Table III.6 gives a

Table III.6 Survey of heat transfer coefficients (W/m²°C)

Heat flow → to: ↓ from:	Gas (free convection) $h = 5 - 15$	Gas (flowing) $h = 10 - 100$	Liquid (free convection) $h = 50 - 1000$	Liquid (flowing) water $h = 3000 - 10000$ other liquids $h = 500 - 2000$	Boiling liquid water $h = 3500 - 60000$ other liquids $h = 1000 - 20000$
Gas (free convection) $h = 5 - 15$	Room/outside air through glass $U = 1 - 2$	Super heaters $U = 3 - 10$			
Gas (flowing) $h = 10 - 100$		Heat exchangers for gases $U = 10 - 30$	Gas boiler $U = 10 - 50$	Oven $U = 10 - 40+$ radiation	Boiler $U = 10 - 40+$ radiation
Liquid (free convection) $h = 50 - 1000$	Radiator central heating $U = 5 - 15$		Oil bath for heating $U = 25 - 500$	Cooling coil $U = 500 - 1500$ if stirred	
Liquid (flowing) water $h = 3000 - 10000$ other liquids $h = 500 - 3000$		Gas coolers $U = 10 - 50$	Heating coil in vessel water/water without stirring $U = 50 - 250$ with stirring $U = 500 - 2000$	Heat exchanger water/water $U = 900 - 2500$ water/other liquids $U = 200 - 1000$	Evaporators of cooling units, brine coolers $U = 300 - 1000$
Condensing vapour water $h = 5000 - 30000$ other liquids $h = 1000 - 4000$	Steam radiators $U = 5 - 20$	Air heaters $U = 10 - 50$	Steam jackets around vessels with stirrers, water $U = 300 - 1000$ other liquids $U = 150 - 500$	Condensers steam/water $U = 1000 - 4000$ other vapour/water $U = 300 - 1000$	Evaporators steam/water $U = 1500 - 6000$ steam/other liquids $U = 300 - 2000$

survey of approximate values for the partial heat transfer coefficient occurring in practice and the consequences for the overall coefficient if two partial coefficients are combined.

In the case of, laminar flow, equation (III.54) yields only approximate results. For an exact solution of this problem, equation (III.45) should be solved applying the correct boundary conditions.

III.3.4 Problems

1. A very long horizontal heat exchanger for warming up liquids consists of internally smooth pipes which are kept at constant temperature; the flow through the pipes is turbulent.

   (a) If the pressure drop is doubled per unit of length of the pipes, by which factor does the heat transfer coefficient increase?

   (b) By which factor should the pipe length be increased in order to attain the same heating up as in the former case?

   Now answer the same questions for a laminar flow if the entrance region is negligibly short with respect to the total pipe length.

   Answer: turbulent flow (a) h 1.39 times as large

   (b) L 1.08 times as large

   laminar flow (a) h constant

   (b) L 1.2 times as large

*2. On the outside of a steel tube (outside diameter 1 in, wall thickness $\frac{1}{32}$ in) steam condenses, whereas there is a turbulent flow of cooling water at a rate v through the tube. From measurements it is found that on variation of v the total heat transfer coefficient U can be represented by:

$$\frac{1}{U} = 0.001 + \frac{0.004}{v^{0.8}}$$

where U is expressed in BTU/ft^2 h °F and based on the external pipe diameter; v is expressed in ft/s. Using the same tube some time later, the same experiments are carried out and it is then found that:

$$\frac{1}{U} = 0.0025 + \frac{0.004}{v^{0.8}}$$

The difference should be ascribed to a deposit layer on the cooling-water side. Determine the heat resistance caused by the condensate layer, tube wall and dirt layer and a relationship for the h on the cooling-water side. Find out whether the latter relationship agrees approximately with the existing empirical correlations for this case. The λ of steel is 26 BTU/ft^2 h °F.

Answer: dirt layer   0.0015 ft^2 h °F/BTU

tube wall   0.0001 ft^2 h °F/BTU

condensate layer   0.0009 ft^2 h °F/BTU

$h = 250\ v^{0.8}$; agrees within 10%

3. For a completely turbulent gas flow (Re $= 10^6$) in a rough pipe ($D = 0.10$ m, relative roughness 10^{-2}), the following relationship is valid between the Nu number and the friction factor:

$$Nu = \frac{f}{2}\ Re$$

Calculate h, if the heat conductivity of the gas is 0.02 W/m °C and of the pipe material 200 W/m °C.

This rough pipe (wall thickness 2 mm) is surrounded by a steam jacket which keeps the temperature of the outside at 120 °C. The inlet temperature of the gas is 20 °C, the outlet temperature 30 °C. What will be the outlet temperature if the gas throughput is doubled, and by which factor is the total amount of heat, which the pipe transfers per unit of time, increased? Discuss the (very simple) answer.

Answer: $h = 950$ W/m^2 °C; ϕ_H = twice as large; $T_{out} = 30$ °C

4. On the outside of a tube, a hydrocarbon is evaporated. The heat necessary for the evaporation is supplied by condensing steam in the tube. During this process a thin layer deposits on the tube wall. Measurements have shown that the amount of deposit is proportional to the total amount of heat transferred. Prove that the heat flux at time t is given by:

$$\phi_H''(t) = \left[\left(\frac{1}{h_s} + \frac{d_1}{\lambda_1} + \frac{1}{h_k} \right)^2 + \frac{2c\Delta T t}{\lambda_2} \right]^{-\frac{1}{2}} \Delta T$$

where:

h_s = the partial heat transfer coefficient on the steam side;
d_1 = thickness of the tube wall;
λ_1 = heat conductivity of the tube wall;
h_k = the partial heat transfer coefficient on the hydrocarbon side;
c = volume of deposit per unit of heat transferred;
λ_2 = heat conductivity of the deposit;
ΔT = the difference between the mean temperatures of the steam and of the hydrocarbon.

5. A heat exchanger must be designed for heating oil from 40 to 60 °C. The heating medium is condensing steam of 100 °C. For this situation we can assume that the heat resistance is entirely on the oil side. The oil flow through the pipe is laminar (Gz > 0.1). For an oil flow of 10 m³/h, 100 pipes with an internal diameter of 2.2 cm and a length of 1 m are found to be necessary.
   (a) If 100 pipes with an internal diameter of 3.3 cm are chosen, what should be the pipe length?
   (b) If 15 m³/h of oil must be heated in the original heat exchanger from 40 to 60 °C, what should be the pipe length?

   Answer: (a) 1 m; (b) 1.5 m

*6. For a flow through a circular pipe (steady-state condition) the velocity distribution and the temperature distribution are measured to be parabolic functions of the radius. Calculate the Nusselt number for this flow.
   Answer: Nu = 6

*7. Show that John took the right steps in the case of the boiling water discussed at the beginning of this section.

Comments on problems

Problem 2

The overall heat transfer coefficient is given by:

$$\frac{1}{U} = \frac{1}{h_c} + \frac{1}{h_w} + \frac{d_s}{\lambda_s} + \frac{d_d}{\lambda_d}$$

For the steel wall, we find $d_s/\lambda_s = 0.0001\text{ft}^2\text{ h °F/BTU}$. The first measurement ($d_d = 0$) yielded:

$$\frac{1}{U} = 0.001 + \frac{0.004}{v^{0.8}} = \frac{1}{h_c} + \frac{1}{h_w} + \frac{d_s}{\lambda_s}$$

Now only h_w is dependent on the fluid velocity. Thus $h_w = 250v^{0.8}\text{BTU/ft}^2\text{h °F}$. Furthermore:

$$\frac{1}{h_c} + \frac{d_s}{\lambda_s} = 0.001$$

i.e.:

$$\frac{1}{h_c} = 0.0009 \text{ ft}^2 \text{ h °F/BTU.}$$

The second measurement analogously yields:

$$\frac{d_d}{\lambda_d} = 0.0015 \text{ ft}^2 \text{ h °F/BTU}$$

We can calculate h_w with equation (III.49) as:

$$h_w = \frac{\lambda}{D} 0.027 \left(\frac{\rho v D}{\eta} \right)^{0.8} \mathrm{Pr}^{\frac{1}{3}} = 4180 v^{0.8} \mathrm{W/m^2\,^\circ C}$$

$$= 283 v^{0.8} \mathrm{BTU/ft^2 h\,^\circ F}$$

Problem 6

The velocity distribution of the flowing fluid is given by

$$v = v_{\max} \left(1 - \frac{r^2}{R^2} \right)$$

We now know, furthermore, that the temperature distribution is parabolic, i.e. $T \sim r^2/R^2$. Defining that $T = T_w$ at $r = R$ and $T = T_c$ at $r = 0$, the temperature distribution is therefore given by:

$$\frac{T - T_c}{T_w - T_c} = \frac{r^2}{R^2}$$

The mean temperature of the flow in the pipe can be calculated from:

$$\langle T \rangle = \frac{\int_0^R 2\pi v T r \, dr}{\int_0^R 2\pi v r \, dr} = \frac{(\frac{1}{6}T_c + \frac{1}{12}T_w) R v_{\max}}{\frac{1}{4} v_{\max} R}$$

to be:

$$\langle T \rangle = \tfrac{2}{3} T_c + \tfrac{1}{3} T_w ; \quad \langle T \rangle - T_w = \tfrac{2}{3}(T_c - T_w)$$

The heat flux through the pipe wall is now given by:

$$\Phi_H'' |_{r=R} = \lambda \left. \frac{dT}{dr} \right|_R = \lambda (T_w - T_c)\frac{2}{R} = h(T_w - \langle T \rangle)$$

from which we find:

$$\mathrm{Nu} = \frac{h 2R}{\lambda} = \frac{4(T_w - T_c)}{T_w - \langle T \rangle} = 6$$

Problem 7 John and the boiling water

John assumed that all heat transfer resistance is caused by the layer of scale. Thus:

$$\phi_H = \frac{\lambda}{d} A \Delta T = 157 \text{ W}$$

$$\phi_m = \frac{\phi_H}{\Delta H} = \frac{\lambda A \Delta T}{d \Delta H} = 6.8 \times 10^{-5} \text{ kg/s}$$

In 4 h, therefore, only about 1 kg of water will evaporate.

III.4 Heat exchangers

> John had just poured out a cup of coffee and filled his pipe
> when his boss asked to see him for a few minutes. 'Hell',
> John thought, 'the coffee will be cold before I come back.'
> Remembering the pages to come, John put milk and sugar
> in his coffee and went to see his boss.

Heat exchangers play an important role in technology, not only for conditioning liquid flows but also for attaining a favourable heat economy. On the basis of the demands made on the heat exchange (the amount of heat ϕ_H which must be transferred per unit of time and the available temperature difference), and of data on the heat resistances, the required exchanger area A of a heat exchanger can be calculated from the formula:

$$\phi_H = UA\Delta T \qquad\qquad (\text{III.55})$$

Usually both U and ΔT are dependent on the place in the heat exchanger, so that the formula for the entire heat exchanger should read:

$$\phi_H = \overline{U} A \overline{\Delta T} \qquad\qquad (\text{III.56})$$

We have already discussed the determination of the overall heat transfer coefficient U for simple pipe flow (Section III.3) and we will return to this subject in later sections (see also Table III.5). Here, we will find out how the mean temperature difference can be determined in a number of simple cases and we will consider the problems which have to be solved when designing heat exchangers.

III.4.1 Determination of mean temperature difference

Let us consider a simple model of a heat exchanger i.e. two concentric tubes (Figure III.13). Here, for example, liquid' gives up heat to liquid", either co-currently or countercurrently. In the former case, T_L' should invariably be $\geq T_L''$. In the latter case, the outlet temperature T_0'' can also be higher than T_L'. In the

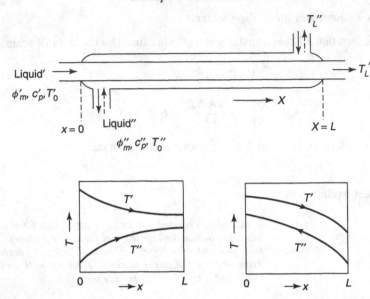

Figure III.13 Heat exchanger and temperature gradients for co- and countercurrent flow

first instance, a heat balance can be set up over the entire heat exchanger, which shows that in the stationary state just as much heat is supplied as removed. If the heat exchange with the surroundings can be neglected and no heat is produced in the apparatus, the heat balance will read:

$$(\phi_m C_p)'(T_0' - T_L') = \pm(\phi_m c_p)''(T_0'' - T_L'') = \phi_H \quad \begin{cases} + \text{ countercurrent} \\ - \text{ cocurrent} \end{cases}$$

(III.57)

This relationship is valid for any heat exchanger which satisfies the above conditions[†] and is indispensable for solving a heat exchanger problem.

In order to calculate the mean value ΔT over the entire apparatus in equation (III.56), we should first know T' and T'' as a function of the place. For the type of heat exchanger in Figure III.13, we make a heat balance over a small part dx:

$$-(\phi_m c_p)' \, dT' = \mp(\phi_m c_p)'' \, dT'' = d\phi_H \qquad (III.58)$$

[†] If heat exchangers are considered in which phase transitions also take place (evaporation or condensation) the heat balance should be given a more general form:

$$\phi_m'(H_0' - H_L') = \pm\phi_m''(H_0'' - H_L'') = \phi_H$$

where H = the enthalpy per unit of mass.

Then, according to equation (III.55), the following applies:

$$d\phi_H = (T' - T'')US\,dx \qquad \text{(III.59)}$$

where $S = $ circumference of the exchange area. After introduction of $\Delta T = T' - T''$, we find from equation (III.57):

$$\Delta T_L - \Delta T_0 = -\phi_H \left[\frac{1}{(\phi_m c_p)'} \mp \frac{1}{(\phi_m c_p)''} \right]$$

and from equation (III.58):

$$dT' - dT'' = d\Delta T = d\phi_H \left[\frac{1}{(\phi_m c_p)'} \mp \frac{1}{(\phi_m c_p)''} \right]$$

Elimination of the term in brackets and of $d\phi_H$ using equation (III.59), finally gives the following differential equation for ΔT:

$$\frac{d\Delta T}{\Delta T} = U \left(\frac{\Delta T_L - \Delta T_0}{\phi_H} \right) S\,dx \qquad \text{(III.60)}$$

For a few special cases the integration can be carried out easily.

(a) If U is independent of x or T (often the case with gases). Then integration of equation (III.60) leads to:

$$\phi_H = USL \frac{\Delta T_L - \Delta T_0}{\ln \Delta T_L / \Delta T_0} = USL(\Delta T)_{\log} \qquad \text{(III.61)}$$

We see that for this case, $\overline{\Delta T}$ in equation (III.56) equals the logarithmic mean of ΔT_0 and ΔT_L, both cocurrently and countercurrently.

Figure III.14 gives a simple graph which can be used for a rough estimation of the logarithmic mean temperature difference. From this figure we find, for example, for $\Delta T_L = 40\,°C$ and $\Delta T_0 = 10\,°C$, that $(\Delta T)_{\log} = 22\,°C$.

(b) If U is dependent on the temperature (hence also on the place). To a first approximation, U can be assumed to be linear in ΔT:

$$U = U_0 + (U_L - U_0)(\Delta T - \Delta T_0)/(\Delta T_L - \Delta T_0)$$

Substitution in equation (III.60) and integration between 0 to L, gives:

$$\phi_H = SL \frac{U_L \Delta T_0 - U_0 \Delta T_L}{\ln(U_L \Delta T_0 / U_0 \Delta T_L)} \qquad \text{(III.62)}$$

So here the product $\overline{U\Delta T}$ is used instead of ΔT in equation (III.56). If the value of U depends on ΔT in a more complicated way, equation (III.61) can be applied to a number of sections of the heat exchanger or the integration can

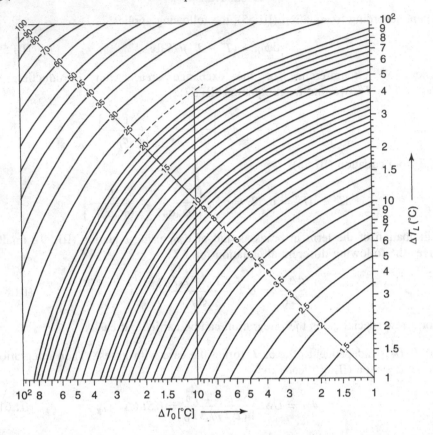

Figure III.14 Logarithmic mean temperature difference as a function of ΔT_L and ΔT_0

be carried out numerically. Cases where this is necessary are, for example, those where combined condensation and after-cooling or pre-heating and evaporation occur. During the transition between the two mechanisms, the value of the transfer coefficient U may change rapidly.

Finally, we have to add that most heat exchangers applied in practice do not belong to the simple 'single-pass' type discussed here but to the 'multiple-pass' principle illustrated in Figure III.15. In addition, for these apparatus a mean temperature difference can be calculated. In practice, however, we prefer to use the logarithmic mean temperature difference of a countercurrent heat exchanger extended by a correction term ψ:

$$\overline{\Delta T} = \psi \frac{\Delta T_L - \Delta T_0}{\ln \dfrac{\Delta T_L}{\Delta T_0}}$$

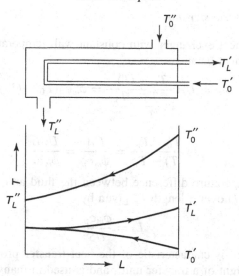

Figure III.15 Principle and temperature gradients of a multiple pass heat exchanger (one shell, two tube passes)

Values for the correction factor ψ can be obtained from engineering handbooks where graphs like Figure III.16 can be found for various situations. These graphs show ψ as a function of the ratio X of the temperature drop of one fluid and the maximum temperature difference, $(T_L' - T_0')/(T_0'' - T_0')$, and of the ratio R of the temperature changes of both streams, $(T_0'' - T_L'')/(T_L' - T_0')$. In order to ensure a high efficiency of a multiple pass heat exchanger, we should try to keep $\psi > 0.75$.

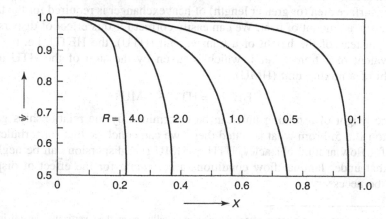

Figure III.16 Correction factor ψ for multiple pass heat exchangers (one shell, even number of tube passes) as a function of R and X

III.4.2 Height of a transfer unit

If we consider a heat exchanger with constant wall temperature T_w, we can set up the following heat balance:

$$\phi_H = UA\frac{T_0 - \langle T \rangle}{\ln\dfrac{T_0 - T_w}{\langle T \rangle - T_w}} = \phi_m c_p(T_0 - \langle T \rangle)$$

or:

$$\ln\frac{T_0 - T_w}{\langle T \rangle - T_w} = \frac{UA}{\phi_m c_p} = \frac{USL}{\phi_m c_p}$$

i.e. the mean temperature difference between the fluid and wall decreases by a factor $1/e$ ($= 0.37$) over a length L_e given by:

$$L_e = \frac{\phi_m c_p}{US} \tag{III.63}$$

This length, which is characteristic of the heat transfer process, is known as 1 HTU, i.e. the 'height of a transfer unit', and is used in many books on chemical engineering.[†] They also introduce the ratio L/L_e which is termed as the number of transfer units (NTU).

We will use the concept of HTUs here to show how we can allow for fluid dispersion when designing a heat exchanger. If fluid dispersion occurs in a heat exchanger, extra heat transport in the direction of flow will be the consequence, and the actual temperature gradients over the length of the exchanger may differ from those shown in the previous section (where dispersion was neglected). This is illustrated by Figure III.17 which shows the temperature gradients over the length of a heat exchanger (as in Figure III.13) for cocurrent plug flow and for dispersed flow of liquid. It will be clear that in the latter case the mean temperature difference between the two fluids is smaller and, consequently, a larger surface area (or greater length) of heat exchanger is required for the transfer of the same amount of heat. We can easily correct for this effect of dispersion by using instead of the height of a transfer unit (HTU), the HETU, i.e. the height equivalent of a transfer unit, which is given by the sum of the HTU and the height of a mixing unit (HMU).

$$\text{HETU} = \text{HTU} + \text{HMU} \tag{III.64}$$

The height of a mixing unit can be determined by the relationships given in Section II.8.5. From what is stated there, we can conclude that for turbulent flow and for flow around obstacles, HTU $\gg$ HMU (i.e. dispersion can be neglected), but that under laminar flow conditions a correction for the effect of dispersion may be necessary.

[†] As heat exchangers are more often placed horizontally rather than vertically, length instead of height would seem logical. However, height is used because of the analogy with mass transfer (see Section IV.4.2).

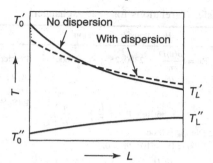

Figure III.17 Temperature gradients with and without dispersion

III.4.3 Design of heat exchangers

With the knowledge now available, it is possible to determine the main dimensions of a heat exchanger. If the overall heat transfer coefficient, all temperatures and the fluid flows are known, we can easily determine the total surface of the exchanger required.

The task remains to determine the number N, diameter D and length L of the pipes as well as the diameter of the shell D_s and the geometric pattern of the pipes in the shell. For the solution of this problem, two momentum balances are available which restrict the applicable velocities, diameters and lengths due to limits put on pressure drops in the pipes and in the shell. Furthermore, we have an economic balance which is used to find the cheapest design for a given problem. Once we have chosen the type of heat exchanger these three balances determine N, L and D, and geometric considerations then further determine shell diameter, etc.

If we realize that, besides the solutions of the above problems, questions on construction material and a great number of constructional details must be solved, we can understand why large engineering firms often have departments specialized in designing heat exchange equipment.

Finally, in Table III.7 some heat transfer correlations are summarized, which are useful when designing heat exchangers. A few of these relationships have already been discussed in Section III.3; others will be treated in following sections.

III.4.4 Problems

1. Through a horizontal smooth pipe (internal diameter 18 mm) water at 90 °C is forced up at a rate of 0.8 m/s. Determine the distance from the entrance at which the water begins to boil if the pressure drop is negligible and the pipe wall over its entire length is kept at 110 °C.

   Answer: $L = 1.5$ m

Table III.7 Heat transfer correlations for heat exchangers

Flow situation	Heat transfer correlation	Remarks
Flow through straight circular pipes Flow along circular pipes without baffles	$\mathrm{Re} = \dfrac{\rho \langle v \rangle D}{\eta} < 2300,\ 25 < \mathrm{Pr} < 4000$ $\langle \mathrm{Nu} \rangle = \dfrac{\langle h \rangle D}{\lambda}$ $= 1.86\ \mathrm{Re}^{\frac{1}{3}}\ \mathrm{Pr}^{\frac{1}{3}} \left(\dfrac{L}{D} \right)^{\frac{1}{3}} \left(\dfrac{\eta}{\eta_w} \right)^{\frac{1}{7}}$ $\langle \mathrm{Nu} \rangle_{\min} = 3.66$ $2300 < \mathrm{Re} < 10^5,\ \mathrm{Pr} > 0.7$ $\langle \mathrm{Nu} \rangle = \dfrac{\langle h \rangle D}{\lambda} = 0.027\ \mathrm{Re}^{0.8}\ \mathrm{Pr}^{\frac{1}{3}} \left(\dfrac{\eta}{\eta_w} \right)^{\frac{1}{7}}$	Physical properties of fluids at $\langle T \rangle = \dfrac{T_{\text{in}} + T_{\text{out}}}{2}$ For correction for spiralled pipe, see equation II.52
Flow along circular pipes with baffles Opening a m^2/m^2	$3 < \mathrm{Re} = \dfrac{\phi_v d\, p}{\langle A \rangle \eta} < 2 \times 10^4,$ $0.5 < \mathrm{Pr} < 500$ $a \approx 0.25$ $\langle \mathrm{Nu} \rangle = \dfrac{\langle h \rangle d}{\lambda} = 0.22\ \mathrm{Re}^{0.62}\ \mathrm{Pr}^{\frac{1}{3}} \left(\dfrac{\eta}{\eta_w} \right)^{\frac{1}{7}}$	n = number of pipes Physical properties at $\dfrac{T_{\text{in}} + T_{\text{out}}}{2}$ Mean area available for flow $\langle A \rangle$ $= \left(\dfrac{S - d}{S} (D - a) \dfrac{\pi p}{4} (D^2 - nd^2) a \right)^{\frac{1}{2}}$
Flow along flat plate (one side)	$\mathrm{Re} = \dfrac{vL}{v} < 3 \times 10^5$ $\langle \mathrm{Nu} \rangle = \dfrac{\langle h \rangle L}{\lambda} = 0.33\ \mathrm{Re}^{\frac{1}{2}}\ \mathrm{Pr}^{\frac{1}{3}}$ $5 \times 10^5 < \mathrm{Re} < 10^7$ $\langle \mathrm{Nu} \rangle = 0.037\ \mathrm{Re}^{0.8}\ \mathrm{Pr}^{\frac{1}{3}}$	See equation (III.65) Physical properties at $\dfrac{T_{\text{in}} + T_{\text{out}}}{2}$
Plate heat exchangers	$\mathrm{Re} = \dfrac{\rho \langle v \rangle 2\delta}{\eta}$ $\langle \mathrm{Nu} \rangle = \dfrac{\langle h \rangle 2\delta}{\lambda} = c\ \mathrm{Re}^a\ \mathrm{Pr}^{\frac{1}{3}} \left(\dfrac{\eta}{\eta_w} \right)^{\frac{1}{7}}$ c and a dependent on type and make of plates	In practice $4 < \delta < 6$ mm For rough estimate (within factor 2) take $c = 0.5$ and $a = 0.57$
Falling film on vertical wall	$\mathrm{Re} = \dfrac{\rho \langle v \rangle 4\delta}{\eta} < 1600$ $\langle h \rangle = 2.05\ \mathrm{Re}^{-\frac{1}{3}} \left(\dfrac{g \rho^2 \lambda^3}{\eta^2} \right)^{\frac{1}{3}}$ $\mathrm{Re} > 3200$ $\langle h \rangle = 0.0087\ \mathrm{Re}^{0.4} \left(\dfrac{g \rho^2 \lambda^3}{\eta^2} \right)^{\frac{1}{3}}\ \mathrm{Pr}^{\frac{1}{3}}$	$\delta_{\text{lam}} = \left(\dfrac{3\eta \phi_m}{W \rho^2 g} \right)^{\frac{1}{3}}$ $\delta_{\text{lam}} = 0.2 \delta_{\text{turb}}\ \mathrm{Re}^{0.2}$ Physical properties at $\dfrac{T_{\text{in}} + T_{\text{out}}}{2}$
Scraped-surface heat exchanger	$\langle h \rangle = 1.13 \psi (\lambda \rho c_p n N)^{\frac{1}{2}}$	n = number of rows of scraper blades N = stirrer speed (1/s) Correction factor $\psi = 1$ for water $0.2 < \psi < 0.6$ for viscous liquids
Boiling	See Figure III.23	

Table III.7 *(continued)*

Flow situation	Heat transfer correlation	Remarks
Natural convection around horizontal pipes	$1000 < \text{Gr Pr} = \dfrac{D^3 g \Delta_\rho}{va} < 10^9$ $\langle \text{Nu} \rangle = \dfrac{\langle h \rangle D}{\lambda} = 0.55 \, (\text{Gr Pr})^{\frac{1}{4}}$	Physical properties σ $\dfrac{T_{\text{wall}} + T_{\text{bulk}}}{2}$
Natural convection on vertical surfaces	$1700 < \text{Gr Pr} = \dfrac{L^3 g \Delta_\rho}{va} < 10^8$ $\langle \text{Nu} \rangle = \dfrac{\langle h \rangle L}{\lambda} = 0.55 \, (\text{Gr Pr})^{\frac{1}{4}}$ $\text{Gr Pr} > 10^8$ $\langle \text{Nu} \rangle = \dfrac{\langle h \rangle L}{\lambda} = 0.13 \, (\text{Gr Pr})^{\frac{1}{3}}$	$\Delta \rho \lvert \rho_{\text{bulk}} - \rho_{\text{bulk}} \rvert$ Influence of combined natural/ forced convection of $\langle a \rangle$ see Section III.6
Condensation on horizontal pipe	$\langle h \rangle = 0.72 \left(\dfrac{g \rho^2 \lambda^3 \Delta H}{D \eta \Delta T} \right)^{\frac{1}{4}}$	Influence of inert gas on condensation (see equation III.79)
Condensation on vertical pipe	$\text{Re} = \dfrac{4 \phi_m}{\pi D \eta} < 1400$ $\langle h \rangle = 0.94 \left(\dfrac{g \rho^2 \lambda^3 \Delta H}{L \eta \Delta T} \right)^{\frac{1}{4}}$ $\text{Re} > 2000$ $\langle h \rangle = 0.0077 \, \text{Re}^{0.4} \left(\dfrac{g \rho^2 \lambda^3}{\eta^2} \right)^{\frac{1}{3}}$	Physical properties of fluids at $T = T_c - \frac{3}{4} \Delta T$ For quick estimation of $\langle h \rangle$ see Figure III.22
Condensation in horizontal pipe	$1000 < \dfrac{\text{Re}}{2} \sqrt{\dfrac{\rho_c}{\rho_v}}$ $= \dfrac{2 \phi_m}{\pi D \eta_c} \sqrt{\dfrac{\rho_c}{\rho_v}} < 5 \times 10^4$ $\dfrac{\langle h \rangle D}{\lambda_c} = 5.03 \left(\dfrac{\text{Re}}{2} \sqrt{\dfrac{\rho_c}{\rho_v}} \right)^{\frac{1}{3}} \text{Pr}_c^{\frac{1}{3}}$ $5 \times 10^4 < \dfrac{\text{Re}}{2} \sqrt{\dfrac{\rho_c}{\rho_v}} < 3 \times 10^5$ $\dfrac{\langle h \rangle D}{\lambda_c} = 0.0265 \left(\dfrac{\text{Re}}{2} \sqrt{\dfrac{\rho_c}{\rho_v}} \right)^{0.8} \text{Pr}_c^{\frac{1}{3}}$	index $c =$ condensate index $v =$ vapour
Condensation in vertical pipe	Use correlation for condensation on vertical pipe If considerable vapour velocity, $\text{Re} = \dfrac{\rho \langle v \rangle D}{\eta} > 10^4$ $\langle h \rangle = 0.00157 \left(\dfrac{\lambda_c \, p_c \, c_{pc} \, p_v v_v}{\eta_c} \right)^{\frac{1}{2}}$	$c_{pc} = c_p$ (condensate)

2. A closed vessel filled with 86 l of water, which is stirred thoroughly, has a temperature of $100\,°\text{C}$, whereas the air temperature is $20\,°\text{C}$. If the total outer surface area of the vessel is $1\,\text{m}^2$ and the mean heat transfer coefficient U over this area is $25\,\text{W/m}^2\,°\text{C}$, how long does it take for the temperature to fall to $50\,°\text{C}$?

Answer: 4 h

3. In a double-tube heat exchanger, benzene is heated from 20 to 60 °C countercurrently with water, which during this process cools down from 88 to 48 °C. During an overhaul, the direction of the water flow is accidentally reversed. Determine the final temperature of the benzene. (The heat transfer coefficient is independent of ΔT.)

   Answer: 52 °C

*4. A turbulent liquid stream at 25 °C flows through a straight pipe, the wall temperature of which is 100 °C. At 3 m distance from the entrance the mean liquid temperature is 50 °C.
   (a) Determine at what distance from the entrance the liquid has a mean temperature of 90 °C.
   (b) To what degree does this distance change if the flow rate is doubled?

   Answer: (a) 15 m; (b) increases by 15%

*5. A gas stream at 10^4 m³/h must be cooled from 200 to 50 °C. Use is made of a bank of steel tubes installed in the open air. The gas flows through the pipes, while water is sprayed on to the bank of tubes and wind flows between the pipes. In this way the outer walls of each tube are kept at a constant temperature of 20 °C. Design the heat exchanger with the smallest surface area (to optimize its costs), which consists of standard pipe lengths ($L = 16$ ft $= 4.8$ m, or multiples thereof) and which shows a pressure drop over the single passes of less than 0.1 bar.

   Data: $a_{gas} = 10^{-5}$ m²/s, $\nu_{gas} = 10^{-5}$ m²/s

   Answer: $L = 4.8$ m, $N = 87$, $D = 0.05$ m.

*6. Show that John drew the right conclusion regarding his cup of coffee discussed at the beginning of this section.

Comments on problems

Problem 4

A heat balance in the stationary state reads:

$$\phi_H = h\pi DL\Delta T_{\log} = \phi_v \rho c_p \Delta T$$

25 °C   50 °C   90 °C

(a) Writing the above equation for two lengths of pipe 1 and 2, we obtain ($h_1 = h_2$, $\phi_{v1} = \phi_{v2}$) after division of the equations:

$$L_2 = L_1 \frac{\Delta T_{\log 1} \Delta T_2}{\Delta T_{\log 2} \Delta T_1} = 14.9 \text{ m}$$

(b) Now the fluid velocity v varies and with it $\phi_v \sim v$ and $h \sim v^{0.8}$. Using the same procedure we find:

$$L_2 = L_1 \frac{\Delta T_{\log 1} \Delta T_2 h_1 \phi_{v2}}{\Delta T_{\log 2} \Delta T_1 h_2 \phi_{v1}} = 14.9 \times 2^{0.2} = 17.1 \text{ m}$$

Problem 5

$$ND^{0.8} L = \frac{\phi_v \rho c_p \Delta T v^{0.8}}{0.027 v^{0.8} \lambda \pi \Delta T_{\log}} = 38.5 (\text{m}^{1.8}).$$

The length should be 4.8 m (why?). If we choose now three practical pipe dimensions, we can, with the help of the Fanning equation (III.18), produce the following table:

D (m)	0.025	0.05	0.10
N	158	87	51
v (m/s)	40.5	16.2	6.95
$4f$	0.019	0.019	0.019
Δp(N/m^2)	3900	310	28.6
A(m^2)	59.5	66	77

We see that any diameter above approximately 0.04 m will satisfy the pressure drop criterion. The smaller this diameter, the smaller the surface area. So, $D = 2$ in($= 0.05$ m), $L = 4.8$ m and $N = 87$ is a good design.

Problem 6 John and the cup of coffee

John wants to keep his coffee as hot as possible. If he added milk and sugar directly, the heat balance:

$$Mc_p(T_0 - T) = M_s c_{ps}(T - T_s) + M_m c_{pm}(T - T_m)$$

shows that the temperature of the mixture is:

$$T = \frac{Mc_p T_0 + M_s c_{ps} T_s + M_m c_{pm} T_m}{Mc_p + M_s c_{ps} + M_m c_{pm}}$$

The further cooling of the coffee is then described by:

$$Mc_p \frac{dT}{dt} = \phi_H = -UA(T - T_{air})$$

which can be integrated to yield:

$$\frac{T - T_{\text{air}}}{T_0 - T_{\text{air}}} = \exp\left(\frac{-UAt}{Mc_p}\right)$$

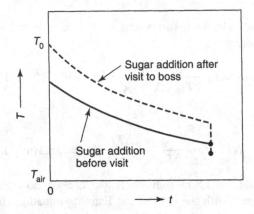

The decrease in temperature is shown in the neighbouring graph. If John had left the coffee without milk and sugar, the temperature difference, $T - T_{\text{air}}$, would be larger. Thus the total system would lose more heat and the final temperature after addition of milk and sugar would be lower.

III.5 Heat transfer by forced convection around obstacles

The body was lying in a draughty corner. The police doctor stated that the man could only have been dead for 1 h and that death was due to a stab in the heart with a pointed weapon of approximately 3 cm diameter. The murder weapon had not yet been found. John noticed that the jacket and shirt of the victim were wet. He estimated the air temperature to be 20°C and the wind velocity in the corner to be 8 m/s. He made a quick calculation and concluded that the man could have been stabbed with an icicle.

III.5.1 Flow along a flat plate

This situation is illustrated in Figure II.37. A thermal boundary layer is present, the thickness (δ_T) of which increases in the x-direction. This causes the local heat transfer coefficients to decrease in the positive x-direction. For the calculation of the heat transfer coefficients we can apply the same technique as used in Section III.3 for the calculation of heat transfer coefficients for turbulent flow in pipes.

We first calculate the thickness of the hydrodynamic boundary layer from equation (II.64):

$$\delta_h = \frac{1}{0.332}\sqrt{\frac{xv}{v}}$$

Using our knowledge about the ratio of the hydrodynamic and the thermal boundary layer thickness (equation (II.29)):

$$\frac{\delta_h}{\delta_T} = \left(\frac{v}{a}\right)^{\frac{1}{3}} = \text{Pr}^{\frac{1}{3}}$$

we can now calculate the Nusselt number as:

$$\text{Nu} = \frac{hx}{\lambda} = \frac{x}{\delta_T} = \frac{x}{\delta_h}\text{Pr}^{\frac{1}{3}} = 0.332\sqrt{\frac{vx}{v}}\text{Pr}^{\frac{1}{3}} \qquad (\text{III.65})$$

This equation is valid for $\text{Re} = vx/v < 3 \times 10^5$. At higher Re values the boundary layer becomes turbulent and is of constant thickness.

III.5.2 Heat transfer to falling films

The derivation of a relationship for heat transfer to falling films follows the lines set out in Section III.3 when discussing laminar and turbulent pipe flows. During laminar film flow the Graetz number is practically always:

$$\text{Gz} = \frac{ax}{\langle v \rangle D^2} > 0.1$$

i.e. the Nusselt number is constant. Under this condition, we find for flow between two parallel plates from an analysis identical to the one applied in Section III.3.1 to circular pipes:

$$\text{Nu} = \frac{h2\delta}{\lambda} = 3.76$$

Consequently, for a laminar film on a flat plate:

$$\text{Nu} = \frac{hD_h}{\lambda} = \frac{h4\delta}{\lambda} = 2 \times 3.76 = 7.52$$

Substituting δ by means of (see problem 2, Section II.1):

$$\delta = \sqrt[3]{\frac{3\eta\phi_v}{\rho g W}} = \sqrt[3]{\frac{3\eta\langle v\rangle\delta}{\rho g}}$$

we find, after some rearrangement:

$$h = \frac{7.52}{4}\sqrt[3]{\frac{4}{3}}\left\{\frac{\rho\langle v\rangle 4\delta}{\eta}\right\}^{-\frac{1}{3}}\left\{\frac{g\rho^2\lambda^3}{\eta^2}\right\}^{\frac{1}{3}}$$

Thus the heat transfer coefficient to a falling film is:

$$h = 2.05 \left\{ \frac{\rho \langle v \rangle 4\delta}{\eta} \right\}^{-\frac{1}{3}} \left\{ \frac{g\rho^2 \lambda^3}{\eta^2} \right\}^{\frac{1}{3}} \qquad \text{(III.66a)}$$

for:

$$\text{Re} = \frac{\rho \langle v \rangle 4\delta}{\eta} = \frac{4\phi_m}{W\eta} < 1600$$

Equation (III.66a) is in agreement with experimental results.

Above Re = 3200, the film flow is turbulent and the heat transfer coefficient can be described by the empirical correlation:

$$h = 0.0087 \left(\frac{\rho \langle v \rangle 4\delta}{\eta} \right)^{0.4} \left(\frac{g\rho^2 \lambda^3}{\eta^2} \right)^{\frac{1}{3}} \left(\frac{v}{a} \right)^{\frac{1}{3}} \qquad \text{(III.66b)}$$

III.5.3 Flow around spheres and cylinders

During flow around cylinders, spheres and similar bodies, a boundary layer is similarly formed which is thicker as x (the distance measured along the surface from the edge) is larger. As has been discussed when treating the flow resistance of bodies (Section II.5.1), the boundary layer becomes detached somewhere behind the greatest thickness of the body; behind it there is the 'dead water' with eddies. The mean heat transfer coefficient $\langle h \rangle$ over the entire surface is mainly determined by what happens at the flow side of the body. The relationships for $\langle h \rangle$ for such bodies therefore have a form which very much resembles equation (III.65). Of the great number of empirical relations for cylinders and spheres, the following are mentioned as a good approximation for gases and liquids (Pr > 0.7).

(a) Long cylinders with circular cross-sections perpendicular to the flow:

$$\frac{\langle h \rangle D_u}{\lambda} = 0.57 \left(\frac{v_r D_u}{v} \right)^{0.50} \left(\frac{v}{a} \right)^{0.33} \quad (1 < \text{Re} < 10^4) \qquad \text{(III.67)}$$

(b) Flow around spheres:

$$\frac{\langle h \rangle D_u}{\lambda} = 2.0 + 0.66 \left(\frac{v_r D_u}{v} \right)^{0.50} \left(\frac{v}{a} \right)^{0.33} \quad (1 < \text{Re} < 10^4) \qquad \text{(III.68a)}$$

In both equations, the physical properties of the fluid should be taken at the average film temperature unless the correction term $(\eta/\eta_w)^{0.14}$ is added. At Re > ca. 10^4, the exponent of Re is slightly higher than 0.5. More detailed information can be found in various handbooks.

At Re = 0, we find for a sphere Nu = 2, the value for pure conductive transport. In the region 0.1 < Re < 1 (compare Section II.4.2), boundary layer theory becomes less adequate than an Oseen-type of approach (a linearization of the viscous terms in the transport of momentum, accounting for velocities downstream which are higher in the neighbourhood of the sphere than Stokes' law predicts). This can be accounted for by using:

$$\frac{hD_u}{\lambda} = 2.0 + \tfrac{1}{2}\text{Pe} \left(\text{Pe} = \frac{v_r D_u}{a} \ll 1 \right) \qquad \text{(III.68b)}$$

At high flow velocities the convective transport is much greater than the conductive transport. For this situation we can estimate the thickness of the hydrodynamic boundary layer on the basis of the penetration of momentum (Section II.1.6) as:

$$\delta_h = \sqrt{\pi v t}$$

Furthermore, realizing that $\delta_h/\delta_T = \text{Pr}^{\frac{1}{3}}$ and that the time t available for creating the boundary layer is $t \approx D/v$ we find:

$$\text{Nu} = \frac{hD}{\lambda} = \frac{D}{\delta_T} \approx 0.56\sqrt{\frac{Dv}{v}}\,\text{Pr}^{\frac{1}{3}}$$

which is in good agreement with equation (III.68a).

The relationships given (equations (III.67) and (III.68a)) are applicable in situations where the heat transport by convection (in the x-direction) is much more important than the conductivity in that direction, which means:

$$\rho c_p v_r \frac{dT}{dx} \gg \lambda \frac{d^2T}{dx^2}$$

i.e. if:

$$\frac{\rho c_p v_r D_u}{\lambda} \gg \frac{d^2T}{d(x/D_u)^2} \left(\frac{dT}{d(x/D_u)} \right)^{-1}$$

The left-hand side of this inequality is a dimensionless constant which is called the Péclet number (Pé = $v_r D_u/a$). The right-hand term is a value which follows from the temperature distribution. This value differs for spheres and cylinders but is of the order of 1. So it could be said that the above relationships hold if Pé $\gg$ 1.

These equations must be applied with care at low flow velocities. In such cases we have to check whether heat transfer by free convection participates significantly (see Section III.6).

III.5.4 Heat transfer in packed beds

The heat transfer between the fluid and the particles in a packed bed can also be described with the relationships for a boundary layer flow (equation (III.65)), provided the hydraulic diameter D_h is taken as the characteristic dimension of the successive short channels of which the bed is built up (see Section II.6.1). The result for a bed of spheres is:

$$\frac{\langle h \rangle D_h}{\lambda} = \frac{2}{3} \frac{\langle h \rangle \varepsilon D_u}{(1 - \varepsilon)\lambda} = \text{constant} \times \left(\frac{2v_0 D_u}{3v(1 - \varepsilon)} \right)^{\frac{1}{2}} \Pr^{\frac{1}{3}} \qquad \text{(III.69)}$$

For the constant for which approximately the value 0.66 is expected, a value of (0.75 ± 0.15) is found in the range $10 < \text{Re}_h = 2v_0 D_u/3v(1 - \varepsilon) < 10^3$. Many authors state that for the mean Nu number of a packed bed ($\varepsilon = 0.40$):

$$\frac{\langle h \rangle D_u}{\lambda} = (1.8 \pm 0.3) \left(\frac{v_0 D_u}{v} \right)^{\frac{1}{2}} \left(\frac{v}{a} \right)^{\frac{1}{3}} \qquad \text{(III.69a)}$$

applies, provided $30 < v_0 D_u/v < 3 \times 10^3$ and $\Pr \geq 1$. The reader may ascertain that this corresponds with the above more fundamental result. So it is again advisable when reading the literature to ascertain which definition of the Reynolds number is used (just as in Section II.6.1 for the pressure drop relationships in fixed beds the choice is: Re_h or $v_0 D_u/v$, of which the first is fundamentally correct).

The heat transfer between the tube wall and the fixed bed (i.e. particles + fluid), which, for example, is important for catalytic reactions in this type of reactor, can still only poorly be described. The problem is that of the total temperature difference between the wall and the centre of the bed, an appreciable part is found in the outer layer of particles (thus at the wall), but that considerable temperature differences also occur over the diameter of the bed. It is therefore obvious to try a model description in which two heat resistances occur: one in the boundary layer at the wall and one in the bed itself. The resistance at the wall $(1/\langle h \rangle)$ can, in principle, be described with the same relationship as that between particles and fluid (equations (III.69) or (III.69a)) because the wall can be considered as a (somewhat large and strange) particle. In practice, this appears to be the case, only the constant is ca. 30% lower than in equations (III.69) or (III.69a). The reason is that the mean dimension of the channels at the wall is slightly larger than in the heart of the bed, where the particles lie closer to each other. The resistance in the bed is usually described with the aid of an effective heat conductivity. Regarding the heat transfer, the bed (particles + fluid) is taken as a solid body with a heat capacity equal to:

$$\rho c_p = (1 - \varepsilon)(\rho c_p)_{\text{particles}} + \varepsilon (\rho c_p)_{\text{fluid}}$$

and an effective λ_{eff}. This heat conductivity is then, of course, dependent on the porosity of the bed, the λ_p of the particles, the λ_f of the fluid and of the flow situation in the bed. A well-known rule of thumb for the effective heat conductivity is:

$$\lambda_{\text{eff}} = \lambda_{\text{eff}}^* + 0.1\rho c_p v_0 D_u \qquad \text{(III.70)}$$

where λ_{eff}^* is the effective conductivity of the bed if no flow occurs. It can be calculated with the help of Figure III.18 in which $\lambda_{\text{eff}}^*/\lambda_f$ has been plotted against λ_p/λ_f for various values of ε.

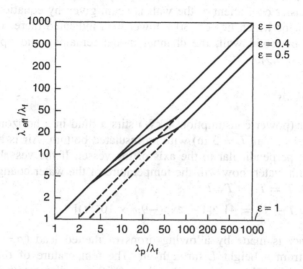

Figure III.18 Effective heat conductivity λ_{eff}^* in a packed bed without flow

At high temperatures, the heat transport from particle to particle is controlled by radiation transfer. This mechanism works parallel to the heat conduction through the fluid (gas). Even then, Figure III.18 can be used provided the conductivity λ_f of the fluid is increased by $4e\sigma \bar{T}^3 D_u$ (the latter term can be understood after reading the Section on radiation, Section III.9).

III.5.5 Heat transfer in fluidized beds

The problem here is that the fluid usually moves through the bed in the form of bubbles (see Section II.6.3). The contact between the fluid and the particles is therefore poorer than expected for a bed in which bubbles are absent (particulate fluidization). Usually the transfer resistance between the bubbles and their surroundings is greater than the transfer resistance between the particles and the fluid that flows around the particles.

In the case of 'particulate fluidization', the situation is analogous to that in the fixed bed. For not too high a porosity ($\varepsilon < 0.75$), when the channel model can still be used, the transfer between the particles and the fluid is described with equation (III.69), whereas at higher porosity the particles 'notice' so little from each other that the transfer is described with the relationship of the single sphere, provided the actual speed in the bed is used as the characteristic speed.

In addition, the heat transfer between a wall and the bed can be described by the considerations discussed for fixed beds, but owing to the mobility of the particles the effective conductivity in the bed is so great that we may assume the temperature in the bed and outside the 'boundary layer' at the wall as being uniform. The heat transfer coefficient at the wall is again given by equation (III.69) with a constant factor which lies ca. 30% lower than indicated there. Because of the presence of the solid wall, the channel model remains valid up to very high porosities ($\varepsilon < 0.9$).

III.5.6 Problems

*1. A stirrer (power consumption 4 kW) stirs a fluid in a horizontal cylindrical vessel ($D = 1$ m, $L = 3$ m) with an insulated bottom. Air is blown at a rate of 9 m/s perpendicular to the axis of the vessel. If the vessel is completely filled with water, how will the temperature of the water change with time if at $t = 0$, $T = T_0 = T_{air}$?

   Answer: $T - T_0 = 41.5[1 - \exp(-9.8 \times 10^{-6}t)]$

*2. Lead shot is made by allowing drops of melted lead ($\rho = 11\,340$ kg/m^3) to fall from a height L through air. The temperature of the air is 20 °C. What should L be to ensure that drops of 2 mm diameter reach the bottom completely solidified? The drops reach the stationary rate of fall very soon after being released. The heat conductivity of lead is so high that in the drops a uniform temperature can be assumed equal to the melting temperature ($T_s = 327$ °C) of lead (melting heat of lead 23.5×10^3 J/kg).

   Answer: 14.8 m

3. Two equally heavy metal spheres of good heat conductivity fall 'stationary' through air. The ratio of the diameters of the spheres is 2 : 1. With the help of dimensional analysis, it can be proved that the product of the stationary rate of fall and the diameter of the one sphere is equal to that of the other sphere ($vD =$ constant). If the initial temperature of the spheres is T_1 and that of the air T_∞, calculate the ratio between the distances the spheres should travel to be cooled to temperature T_2. The product of the density and the specific heat is the same for both spheres.

   Answer: $\dfrac{\text{Distance large sphere}}{\text{Distance small sphere}} = 2$

4. If a thermometer with temperature T_0 is suddenly put in a flowing fluid with a temperature T_f, the mean temperature T of the thermometer will, in the course of time, approach the value T_f. The time constant τ of a thermometer is defined as the time necessary to reach a value of ($T_f - T$), which is equal to ($T_f - T_0$)/e, where e = the base of natural logarithms. If the thermometer

conducts the heat infinitely well, has an external surface area A and a total heat capacity $c(J/°C)$, and if the heat transfer coefficient between the flow and the thermometer is h, then derive that:

$$\tau = \frac{c}{Ah}$$

We now consider a thermometer consisting of a mercury-filled iron cylinder ($D_u = 20$ mm, length $L = 20$ mm, wall thickness $d = 1.5$ mm) under the following conditions:

(a) in air of 20 °C at a flow rate of 10 m/s, perpendicular to the cylinder;
(b) in water of 20 °C at a flow rate of 2 m/s, perpendicular to the cylinder;
(c) in condensing steam with a mean heat transfer coefficient $h = 15 \times 10^3$ W/m² °C.

Calculate τ in all cases, both for a clean thermometer and for a thermometer covered with a 0.5 mm thick dirt layer ($\lambda_{dirt} = 0.6$ W/m °C). Neglect the effect of the ends. The specific heat of Fe is:

$$c_p = 4.6 \times 10^2 \text{ J/kg °C}$$

that of Hg is:

$$c_p = 1.3 \times 10^2 \text{ J/kg °C}$$

Answer: (a) clean: $\tau = 143$ s; dirty: $\tau = 154$ s
(b) clean: $\tau = 1.7$ s; dirty: $\tau = 17$ s
(c) clean: $\tau = 0.85$ s; dirty: $\tau = 16$ s

5. In order to cool spherical particles ($d_p = 3.5$ mm) from 110 to 50 °C, they are dropped in a tower with an upward (countercurrent) air flow ($v = 8.0$ m/s).
   (a) How high has this tower to be in order to obtain the desired cooling if the heat transfer resistance is entirely located in the air?
   (b) Show that the heat transfer resistance in the particles is almost negligible.

   Data: Air: $\eta = 1.8 \times 10^{-5}$ Pa s; $\rho = 1.2$ kg/m³; $c_p = 1.0 \times 10^3$ J/kgK
   $\lambda = 0.026$ W/m K

   Particles: $\rho_p = 1.5 \times 10^3$ kg/m³; $c_{p,p} = 8.0 \times 10^2$ J/kg K;
   $\lambda_p = 2.0$ W/m K

   Answer: (a) $H = 11.3$ m
   (b) $h_p/h(\text{air}) = 16$

6. In order to estimate the heat conductivity of alcohol, two tests are carried out:
   (a) A thermometer filled with alcohol consisting of a thin-walled spherical bulb ($D = 8$ mm), originally at 20 °C is put into flowing water at 70 °C. After 41 s the thermometer indicates 65 °C.

(b) After temperature equalisation in the first test, the thermometer is suspended in stagnant ambient air at 20 °C. After 1110 s it indicates 25 °C.

Estimate the heat conductivity of alcohol, neglecting free convection.

Answer: 0.18 W/m K

*7. Show that John drew the right conclusion in the case of the wet body discussed at the beginning of this paragraph.

Comments on problems

Problem 1

The outside heat transfer coefficient can be calculated with equation (III.67) to be $h = 10.26$ W/m °C. A heat balance over the entire system reads:

$$\phi_H = 4 \times 10^3 = V\rho c_p \frac{dT}{dt} + hA(T - T_0)$$

and after integration between $t = 0$, T_0 and t, T we obtain:

$$T - T_0 = 41.5[1 - \exp(-9.8 \times 10^{-6}t)]$$

Problem 2

The velocity of free fall is found to be (equation (II.69)) $v = 24$ m/s and with equation (III.68) we find for the outside heat transfer coefficient $h = 469$ W/m² °C. The heat balance (steady-state conditions) reads for this situation:

$$\phi_H t = \frac{\pi}{6} d_p^3 \rho \Delta H_s = h\pi d_p^2 \Delta T t$$

from which we find $t = 0.616$ s and $L = 14.8$ m.

Problem 7   John and the body at the corner

The heat transfer coefficient to a cylindrical icicle is given by equation (III.67) and, using $\lambda = 0.025$ W/m °C, $v = 14.2 \times 10^{-6}$ m²/s, $a = 20 \times 10^{-6}$ m²/s, we find:

$$h = 9.5D^{-\frac{1}{2}} \, (\text{W/m}^2 \, ^\circ\text{C})$$

The heat balance per unit length of icicle reads:

$$\phi_H' = -\frac{\pi}{2} D \frac{dD}{dt} \rho \Delta H_m = h\pi D \Delta T$$

Substituting for h the value found above and integrating between $t = 0$, $D = D_0$ and t, $D = 0$, we finally obtain (with $\Delta H_m = 334 \times 10^3$ J/kg) $t = 610$ s, and find that an icicle of 3 cm diameter would melt away in ca. 45 min.

III.6 Heat transfer during natural convection

'Yes', John thought, 'the alibi of the suspected farmer seems to be watertight'. The only way by which he could have returned home in time for the police from the house of the murdered banker on the other side of the lake would have been to cross the lake. The air temperature had been 0°C for weeks and had dropped quite suddenly to −20°C 18 hours ago. An ice layer of 3 cm thickness was necessary on this lake to carry a man. John remembered an old lesson, made a swift calculation and arrested the farmer.

III.6.1 Natural convection

Heat transport by natural convection is of importance in cases where there is no forced flow, e.g. during cooling of solids suspended in quiet air, or heating or cooling of liquids through walls without forced flow, caused by pressure differences or stirring.

Consider, for example, a vertical surface with wall temperature T_w in a stagnant fluid which has a temperature T_∞ at a great distance from the surface (Figure III.19). Near the wall the fluid is heated so that the density there is (usually) lower than at a greater distance from the wall. In the stationary state, a velocity distribution adjusts itself (in Figure III.19, drawn for a height x), which is partly determined by the heat transport from the wall. The heat transport in turn is influenced by the velocity distribution. In this case, therefore, the equation of motion (Navier–Stokes) and the energy equation are coupled.

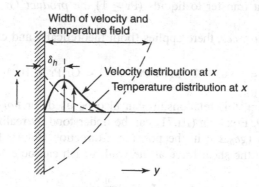

Figure III.19 Free convection along a vertical wall

For laminar natural convection, which in practice occurs most frequently, the velocity and temperature distributions can be calculated for simple geometric situations. As a measure for the heat transfer, once again the local partial heat transfer coefficient h can be introduced. It appears that in the case of natural convection along surfaces in general:

$$\frac{hx}{\lambda} = f\left(\frac{x^3 g}{av}\beta|T_w - T_\infty|\right)$$

applies, where λ, a and v should be taken at the mean boundary layer temperature and the coefficient of expansion β at the ambient temperature T_∞. This relationship shows that h decreases with increasing x because the boundary layer thickness increases with greater x.

For a vertical surface with height L, the mean heat transfer coefficient $\langle h \rangle$ can be represented by:

$$\frac{\langle h \rangle L}{\lambda} = f\left(\frac{L^3 g}{av}\beta|T_w - T_\infty|\right)$$

The quantity $\beta|T_w - T_\infty|$ can also be expressed in densities:

$$\beta|T_w - T_\infty| = \frac{|\rho_\infty - \rho_w|}{\rho_w} = \frac{\Delta\rho}{\rho_w}$$

The dimensionless group which determines the heat transfer can then be written as:

$$\frac{L^3 g}{av}\beta|T_w - T_\infty| = \frac{L^3 g}{av}\frac{\Delta\rho}{\rho_w} = \frac{L^3 g \Delta\rho}{v^2 \rho_w}\frac{v}{a} = \text{Gr Pr}$$

Gr is the dimensionless Grashof number[†] which is often encountered when describing transport phenomena during flow caused by density differences (although in heat transfer to liquids (Pr $\geq$ 1) the *product* Gr Pr almost always occurs).

For *vertical surfaces*, there applies (both theoretically and experimentally):

$$\frac{\langle h \rangle L}{\lambda} = \langle \text{Nu} \rangle = 0.55(\text{Gr Pr})^{\frac{1}{4}}\,(10^3 < \text{Gr Pr} < 10^8) \tag{III.71}$$

By approximation, this relationship can also be used for *horizontal pipes*, if L is replaced by D. Equation (III.71) can be understood by realizing that the force per surface area $(g\delta_h\Delta\rho)$ in the positive x-direction (Figure III.19) must be in equilibrium with the shear force at the wall, which can be estimated to be:

$$\tau_w \approx \eta\frac{v_{max}}{\delta_h}$$

[†] Gr can be regarded as a combination of Re $= vL/v$ and the Froude number $v^2\rho/Lg\Delta\rho$ where the velocity no longer occurs:

$$\text{Gr} = \text{Re}^2/\text{Fr}$$

Thus:

$$g\delta_h \Delta\rho = \eta \frac{v_{max}}{\delta_h} \tag{III.72}$$

The thickness δ_h of the hydrodynamic boundary layer can now be estimated with the help of the penetration theory as:

$$\delta_h = \sqrt{\pi v t} = \sqrt{\frac{\pi v x}{v_{max}}} \tag{III.72a}$$

Substituting v_{max} using equation (III.72), we find for the thickness of the hydrodynamic boundary layer δ_h:

$$\delta_h = \left(\frac{\pi v^2 x \rho}{\Delta\rho g}\right)^{\frac{1}{4}}$$

Remembering that the ratio of the thermal and hydrodynamic boundary layer thickness (see Section II.2.1) is given by (equation (II.29)):

$$\frac{\delta_T}{\delta_h} = \left(\frac{v}{a}\right)^{-\frac{1}{3}} = Pr^{-\frac{1}{3}}$$

we find for the Nu number:

$$Nu = \frac{x}{\delta_T} = \frac{x}{\delta_h} Pr^{\frac{1}{3}} = \left(\frac{x^3 \Delta\rho g}{\pi v^2 \rho}\right)^{\frac{1}{4}} Pr^{\frac{1}{3}} = 0.75 \ Gr^{\frac{1}{4}} \ Pr^{\frac{1}{3}} \tag{III.73}$$

Hence we see, that with the simple model chosen here, we obtain a relationship quite similar to the exact solution given by equation (III.71).

If Gr Pr becomes $> 10^8$, the flow along the surface is turbulent and h no longer depends on x, and by approximation, $\langle h \rangle$ no longer depends on L. It is clear that in this case in a relationship of the type of equation (III.71), the exponent of Gr Pr must be equal to $\frac{1}{3}$. For this case has been found:

$$\langle Nu \rangle = 0.13(Gr \ Pr)^{\frac{1}{3}} \tag{III.74}$$

A similar situation occurs on cooling a horizontal plate. Above the plate thermally generated eddies occur. If the dimension of the plate L is great with respect to the size of the eddies (condition: Gr Pr $> 10^7$) $\langle h \rangle$ is no longer dependent on L. In this case:

$$\langle Nu \rangle = 0.17(Gr \ Pr)^{\frac{1}{3}} \tag{III.75}$$

During heat transfer by forced convection, in principle free convection is always involved, because there are temperature differences in the medium. The combination of both effects is very complicated and as yet little is known about it.

The effect of free convection on heat transfer is negligible compared to that of forced convection if the liquid flow velocities are as follows:

$$v_{\text{free convection}} < v_{\text{forced convection}}$$

From equations (III.72) and (III.72a), the maximum fluid velocity can be calculated approximately and we find:

$$v_{\text{free convection}} = \sqrt{\frac{\pi g x \Delta \rho}{\rho}} < v_{\text{forced convection}}$$

and, by multiplying both sides of this equation by x/v, we finally obtain:

$$\sqrt{\frac{\pi g x^3 \Delta \rho}{\rho v^2}} < \frac{x v_{\text{forced convection}}}{v}$$

or:

$$\sqrt{\pi \text{Gr}} < \text{Re} \qquad\qquad\qquad\qquad (\text{III.76})$$

So we see that free convection heat transfer effects can be neglected if $\text{Re} > \sqrt{\text{Gr}}$, but that convective heat transport is most important if $\text{Re} \ll \sqrt{\text{Gr}}$. In the region $\text{Re} \approx \sqrt{\text{Gr}}$, we can add the fluid velocities due to free and forced convection and use the sum to calculate Nu as a function of the Re number.

III.6.2 Problems

1. Water at 50 °C is passed through a non-insulated horizontal pipe in air of 20 °C. Through the wall an amount of heat of 100 W is lost. The heat loss is completely controlled by free convection. Later on, water at 80 °C is passed through the pipe. Calculate the heat loss.

   Answer: 0.24 kW

2. Liquid condenses in an air-cooled pipe of 10 cm diameter and 2 m length. Calculate the ratio of the condensation capacities if the pipe is put in the horizontal and vertical positions respectively and if:

   (a) $\dfrac{g\Delta\rho}{v\rho_w a} = 3 \times 10^6$

   and

   (b) $\dfrac{g\Delta\rho}{v\rho_w a} = 3 \times 10^8$

   Answer: (a) horizontal:vertical $= 2.2 : 1$

   (b) horizontal:vertical $= 2.7 : 1$

*3. A metal thermometer well (see Figure III.20) (outside diameter D, wall thickness δ) is inserted over a length L into a gas pipe through which a warm

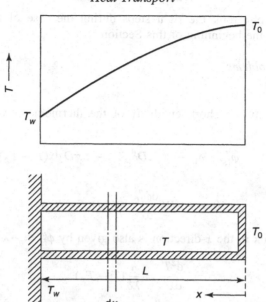

Figure III.20 Thermometer well

gas with temperature T_g flows. The temperature T_w of the well near the wall is lower than T_g, so that by conduction through the metal (λ_m) heat flows through the well to the wall. In the stationary state, this heat is supplied by the gas (heat transfer cocfficient h). Show that for the temperature of the sleeve the following applies:

$$\frac{d^2 \Delta T}{dx^2} - \frac{h}{\lambda \delta} \Delta T = 0, \ \Delta T = T - T_g$$

$$x = 0, \ \frac{d \Delta T}{dx} = 0$$

$$x = L, \ \Delta T_w = T_w - T_g$$

The solution of this equation is:

$$\Delta T = \Delta T_w \frac{\cosh px}{\cosh pL}, \ \text{where} \ p = \sqrt{\frac{h}{\lambda \delta}}$$

There is no heat flow through the plane $x = 0$. Calculate T_0, if $T_g = 100\,°C$, $T_w = 60\,°C$, $D = 20$ mm, $\delta = 1$ mm, $L = 0.15$ m, $\lambda = 100$ W/m °C and $h = 25$ W/m² °C.

Answer: $T_0 = 92.6\,°C$

*4. Show that John took the right steps during the case of the dead banker described at the beginning of this Section.

Comments on problems

Problem 3

A heat balance over a short length dx of the thermometer well reads in the stationary state:

$$0 = \phi_H'' \pi D\delta|_x - \phi_H'' \pi D\delta|_{x+dx} - h\pi D\,dx(T - T_g)$$

or:

$$0 = -\pi D\delta \frac{d\phi_H''}{dx} - h\pi D(T - T_g)$$

Now the heat flux in the x-direction is also given by $\phi_H'' = -\lambda\,dT/dx$. Thus:

$$\frac{d^2T}{dx^2} = \frac{h}{\lambda\delta}(T - T_g)$$

We simplify this equation by writing it in the following form:

$$\frac{d^2\Delta T}{dx^2} = \frac{h\Delta T}{\lambda\delta} = p^2\Delta T; \quad \text{where } p = \sqrt{\frac{h}{\lambda\delta}}$$

This differential equation must now be integrated between the boundary conditions:

$$x = L, T = T_w, \Delta T = T_w - T_g$$

$$x = 0, \quad \phi_H'' = 0, \quad \frac{d\Delta T}{dx} = 0$$

Multiplying throughout with $d\Delta T/dx$ yields:

$$2\frac{d\Delta T}{dx}\frac{d^2\Delta T}{dx^2} = \frac{d}{dx}\left(\frac{d\Delta T}{dx}\right)^2 = 2\frac{d\Delta T}{dx}p^2\Delta T = \frac{d}{dx}(p\Delta T)^2$$

or:

$$\frac{d\Delta T}{dx} = \pm p\Delta T$$

and thus:

$$\Delta T = A\,e^{px} + B\,e^{-px}$$

Using the boundary conditions stated, the values of the constants A and B can be found to be:

$$A = B \quad \text{and} \quad T_w - T_g = A\,e^{pL} + B\,e^{-pL}$$

So we find, finally, for the temperature distribution along the thermometer well:

$$\frac{T - T_g}{T_w - T_g} = \frac{e^{px} + e^{-px}}{e^{pL} + e^{-pL}} = \frac{\cosh px}{\cosh pL} \quad \text{with} \quad p = \sqrt{\frac{h}{\lambda \delta}}$$

The temperature measured at $x = 0$ is thus given by:

$$\frac{T_0 - T_g}{T_w - T_g} = \frac{2}{e^{pL} + e^{-pL}}$$

and with the data given we finally find for the temperature measured $T_0 = 92.65\,°C$ instead of the actual gas temperature $T_g = 100\,°C$.

This example shows that considerable mistakes can be made in temperature measurement if thermometer wells are applied. The deviations increase with decreasing p and thus with decreasing heat transfer coefficients (gases, free convection) and increasing heat conductivity and wall thickness of the well material.

Problem 4 John on the ice

The figure shows the situation during freezing up of the lake. Assuming $T_c \approx T_w$ (we will check later!) we can calculate the heat transfer coefficient due to free convection by equation (III.75) to be $\langle h \rangle = 6.8 \text{ W/m}^2\,°C$ (using $\lambda = 0.025$ W/m °C, $v = 10^{-5}$ m^2/s, $\rho_c = 1.29$ kg/m^3, $\rho_a = 1.385$ kg/m^3, $a = 17.3 \times 10^{-6}$ m^2/s).

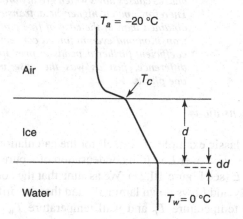

For setting up a heat balance over a thin layer dd, we can neglect the specific heat of the ice formed because at most:

$$\Delta T_{\text{ice}} c_{\text{pice}} \approx 4.2 \times 10^4 \ll \Delta H_{\text{ice}} = 3.35 \times 10^5 \text{ J/kg}$$

Thus our heat balance becomes:

$$\phi_H'' = h(T_c - T_a) = \frac{\lambda}{d}(T_w - T_c) = -\Delta H \rho_{\text{ice}} \frac{dd}{dt}$$

Solving for the ice surface temperature T_c we find:

$$T_c = \frac{\left(\dfrac{\lambda}{d} T_w + h T_a\right)}{h + \dfrac{\lambda}{d}}$$

(for conditions here, $T_c = -2.1\,°C$; thus, indeed, $T_c \approx T_w$). Substituting T_c in the heat balance leads to:

$$\left(1 + \frac{hd}{\lambda}\right) \frac{\mathrm{d}d}{\mathrm{d}t} = -h \frac{T_w - T_a}{\Delta H \rho}$$

and, after integration between $t = 0$, $d = 0$ and t, d, we find that it takes roughly 16 h to form an ice layer of 3 cm thickness. Thus, the farmer could have been back in time, as John rightly concluded.

III.7 Heat transfer during condensation and boiling

> *Condensation and boiling heat transfer can be regarded as special cases of convective heat transfer. The heat transfer is attended by a change of state (vapour → liquid, or vice versa) and the great difference in density between the two phases causes flows which greatly promote the heat transfer. Therefore, much higher heat transfer coefficients can be obtained than in the case of free convection without phase transition and even in forced convection. The heat transfer coefficient further depends in principle on the temperature difference, just as was the case in free convection, in one phase.*

III.7.1 Film condensation

We shall use the classic example of Nusselt for the calculation of the heat transfer coefficient $\langle h \rangle$ averaged over L in film condensation of a pure vapour on a vertical plane with height L (see Figure III.21). We assume that the condensate film covers the surface entirely and flows down laminarly and that the difference ΔT between the condensation temperature T_c and wall temperature T_w is constant over the entire height. When δ is the thickness of the liquid film the local heat transfer coefficient is $h = \lambda/\delta$ (λ = conductivity of condensate). In order to calculate $\langle h \rangle$ over the height L, we must know δ as a function of x.

When on a surface element of height $\mathrm{d}x$ and width W, an amount $\mathrm{d}\phi_v$ condenses, the condensate flow ϕ_v increases over $\mathrm{d}x$ by an amount $\mathrm{d}\phi_v$, which is equal to $\mathrm{d}(\langle v \rangle W \delta)$ (mass balance). Furthermore, on the condensate surface an amount of heat $\Delta H_v \rho \, \mathrm{d}\phi_v$ is released which in the case of laminar flow should

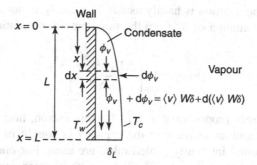

Figure III.21 Film condensation

be transported to the wall by conduction only; so the energy balance will be:

$$\Delta H_v \rho \, d\phi_v = \frac{\lambda}{\delta} \Delta T W \, dx$$

Now it is assumed that for each value of x the relationship between ϕ_v and δ is given by:

$$\delta = \sqrt[3]{\frac{3\eta \langle v \rangle \delta}{\rho g}}$$

(see Section II.1.7, problem 2), although strictly speaking this equation applies only to laminar vertical films of constant thickness. So:

$$\psi_v = \rho g \delta^3 W / 3\eta, \quad \text{and} \quad d\phi_v = \frac{\rho g \delta^2 W}{\eta} \, d\delta$$

Substitution into the energy balance gives:

$$\Delta T \lambda \, dx = \frac{\Delta H_v \rho^2 g \delta^3}{\eta} \, d\delta$$

After integration over x between 0 and L and over δ between 0 and δ_L we get:

$$\lambda \Delta T L = \frac{\Delta H_v \rho^2 g \delta_L^4}{4\eta}$$

The average heat transfer coefficient $\langle h \rangle$ is defined by:

$$\langle h \rangle \, \Delta T W L = \Delta H_v \rho \phi_v |_{x=L} = \frac{\Delta H_v \rho^2 g W \delta_L^3}{3\eta}$$

From the last two equations we find:

$$\langle h \rangle = \frac{4}{3} \frac{\lambda}{\delta_L}$$

but in this form the solution is hardly usable because δ_L is not an easily measurable constant. Elimination of δ_L from the three previous equations gives:

$$\langle h \rangle = 0.94 \left[\frac{\Delta H_v \rho^2 \lambda^3 g}{L \eta \Delta T} \right]^{\frac{1}{4}\dagger} \tag{III.77}$$

So $\langle h \rangle$ is inversely proportional to $\sqrt[4]{L}$; for this reason, horizontal condensers are often applied and an increase in the capacity is sought in W rather than in L. Usually, horizontal internally cooled tubes are used. For one horizontal tube, a formula analogous to equation (III.77) but with a coefficient of 0.72 can be derived if instead of L the external pipe diameter D_u is used. This relationship can be extended to the case where n horizontal pipes are placed under each other. The general equation for this is then:

$$\langle h \rangle = 0.72 \left[\frac{\Delta H_v \rho^2 \lambda^3 g}{n D_u \eta \Delta T} \right]^{\frac{1}{4}} \tag{III.78}$$

The values of $\langle h \rangle$ found in practice for laminar condensate films are on an average 20% higher than follows from theory. Figure III.22 gives a nomograph from which partial heat transfer coefficients for condensation can be rapidly found. The validity is limited to those cases where the condensate film still flows laminarly $(4\langle v \rangle \delta_L / \nu < 10^3)$.

The presence of a non-condensable gas in the vapour greatly reduces the rate of condensation. The vapour must now diffuse through the gas which has accumulated at the condenser surface. The partial vapour pressure at the condenser surface is then considerably lower than the total pressure. The presence of relatively small amounts of gas can therefore decrease the temperature at which the vapour condenses considerably, so that a smaller temperature difference ΔT is available for the heat transport through the condensate film. If $0.01 < p_{inert}/p_{total} < 0.4$, we can correct for this effect by the following empirical rule:

$$\frac{h_{mixture}}{h_{pure\ vapour}} \approx 0.1 \sqrt{\frac{p_{total}}{p_{inert}}} \tag{III.79}$$

Equation (III.79) shows that the presence of, for example, 20% air in a condensing vapour will decrease the heat transfer coefficient by a factor of 4. If $p_{inert}/p_{total} > 0.4$, a heat transfer coefficient should be calculated, assuming no condensation but just cooling of the gas mixture.

---

† The constants ρ, η and λ should be taken at a mean temperature T_f of the condensate film which equals $T_f = T_c - \frac{3}{4}\Delta T$.

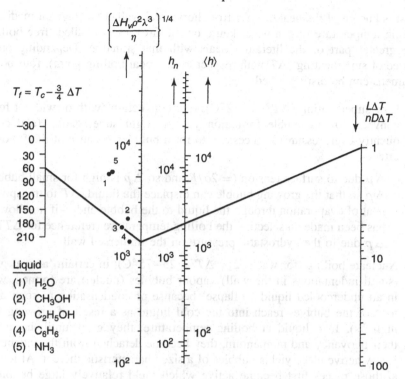

Figure III.22 Nomograph for the calculation of heat transfer coefficients of condensing liquids if $4\langle v \rangle \delta_L / v = 4\phi_m / W\eta < 1000$

III.7.2 Dropwise condensation

If the cooled surface is poorly wettable by the condensate, so-called dropwise condensation occurs. Parts of the surface are then almost dry and have a very high local heat transfer coefficient. As soon as some condensate has formed, a drop collects which after having grown sufficiently then rolls down and sweeps part of the surface clean. The average heat transfer coefficients can be up to ten times higher than for film condensation.

III.7.3 Boiling

The boiling phenomenon is complex. There are eight principal situations, as shown by the following table of possibilities, (a), (b) and (c):

(a) *Wall*   (b) *Liquid*   (c) *Medium*
    horizontal   at boiling temperature   flowing
    vertical   undercooled   stagnant

Most is known of the heat transfer from horizontal walls to a stagnant medium at boiling temperature (e.g. a water kettle on the fire), the so-called 'free boiling'. The greater part of the literature deals with this situation. Depending on the degree of superheating (ΔT with respect to the local boiling point), four boiling regimens can be distinguished:

(1) Low superheating ($\Delta T < $ ca. 2 °C), free convection (with or without forced convection), no bubble formation yet. A slight superheating (and consequently overpressure) is necessary to form bubbles because of the following effects:

   Δp due to surface tension ($= 2\sigma/R$ and so Δp is high for small bubbles), Δp so that the growing bubble can displace the liquid, ΔT to transport the heat of evaporation through the liquid to the bubble and — if no allowance has been made on selecting the boiling temperature (reference for ΔT) — a Δp due to the hydrostatic pressure on the immersed wall.

(2) Nucleate boiling (for water: $2 < \Delta T < 25$–75 °C): in certain 'active' places (small indentations in the wall) vapour bubbles (nuclei) are formed, which in an undercooled liquid 'collapse' because of condensation as soon as the tops of the bubbles reach into the cold liquid as a result of growing (very high hs). In a liquid at boiling temperature, they grow until, because of their buoyancy and momentum, they become detached from the indentation. Every active place yields bubbles of a size characteristic thereof. At low ΔT s, those places first become active which yield relatively large bubbles. If superheating is increased, the frequency at which nuclei are formed in these places becomes higher until the maximum rate of discharge of the bubbles formed has been attained for the relative active place. Next, smaller indentations which produce smaller bubbles are producing at a higher frequency and bring about more 'agitation' in the interface, so that h increases rather strongly with ΔT (approximately proportional to ΔT^3). It appears from the foregoing that in this regime the surface conditions of the wall (nature and size of the irregularities, wettability) play an important role. It is therefore difficult to carry out reproducible experiments, particularly because when a wall has been exposed to boiling once its surface changes (aging).

(3) Leidenfrost regime: the bubbles of neighbouring active places coalesce and form a closed vapour layer. As soon as this takes place the heat transfer coefficient decreases considerably with respect to the values which were obtained in regime (2), because the vapour layer has a fair heat resistance. If instead of a given wall temperature, a given thermal load is applied (electric heating element, nuclear reactor), the temperature of the wall rises considerably on transition from regime (2) to (3), as a result of which the heat transfer coefficient decreases further until finally the heating element melts. In these cases, one should therefore stay at a safe distance from the critical ΔT where the Leidenfrost phenomenon occurs.

(4) A stable vapour layer has formed from which bubbles with a dimension characteristic of the liquid are released (for stagnant, clean water, ca. 15 mm); in the vapour layer the heat transfer takes place by conduction and particularly by radiation (ΔT for water $>500\,°C$).

III.7.4 Heat transfer in evaporators

Evaporators are used for concentrating solutions, if necessary combined with crystallization of the dissolved substances. There are a great many versions, from 'boiling pans' to pipe evaporators, in which the solution flows through banks of, for example, vertical pipes which are usually heated by condensing steam.

During the evaporation in vertical pipes, a combination of heat transfer mechanisms occurs. In the lower part of the pipe the incoming liquid is heated to the boiling point (heat transfer by forced flow). In the upper part the evaporation begins. After a short boiling height, so much vapour has been produced that in the upper part of the pipe mainly a vapour flow occurs which entrains the liquid which has not yet evaporated. In this range, the heat transfer is determined by the flow conditions of the two-phase mixture rather than by the boiling phenomenon.

In certain types of pipe evaporator the liquid is pumped into the pipes (forced circulation). However, there are also many evaporators with natural circulation (thermosiphon action), the rate of circulation in these depending on the vapour production and on the heat transfer. The latter is influenced by the rate of circulation.

Pipe evaporator design is largely based on experience supplemented by experimental data which have been obtained with model set-ups. For dilute aqueous solutions, the h values usually lie between 2000 and 5000 $W/m^2\,°C$ and for hydrocarbons in the range from 500 to 1000 $W/m^2\,°C$. Figure III.23 gives a survey of experimentally determined heat transfer coefficients during boiling as a function of heat flux. This figure can be used for quick estimations of boiling heat transfer coefficients. In order to avoid the 'Leidenfrost' regime just discussed, it is advisable to limit the vapour velocity at the evaporator surface to roughly 0.2 to 0.3 m^3/m^2 s.

III.7.5 Problems

*1. A chemical reaction is carried out in ethanol solution under reflux. The reflux condenser consists of a straight inner glass tube ($\lambda_{glass} = 1$ $W/m\,°C$) of 12 mm diameter and 2 mm wall thickness and an outer pipe of 30 mm inner diameter and 40 cm length.

   (a) If 15 1/min cooling water at 15 °C are available, calculate the amount of heating energy that can be supplied to the reaction mixture so that (just) no vapour leaves the condenser.

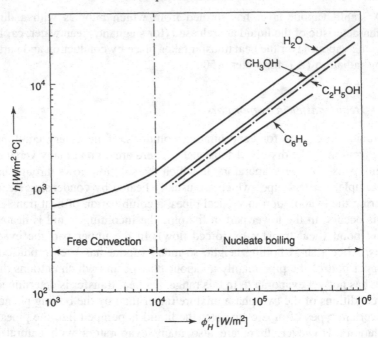

Figure III.23 Heat transfer coefficients during boiling of liquids

(b) What would be the amount if the chemists could be persuaded to use stainless steel equipment?

Answer: (a) 315 W; (b) 700 W

2. A condenser consists of a number of copper pipes ($D_i = 22.8$ mm, wall thickness $d = 1.25$ mm). On the outside of the pipes, 9080 kg/h Freon 12 condenses at $32.2\,°C$. The heat of condensation (137.5 kJ/kg) is removed by cooling water in the pipes. The flow rate of the water is 1.22 m/s, the inlet and outlet temperatures are 15.5 and 17.8 °C, respectively, and the total heat transfer coefficient of Freon–water related to the external pipe surface is 1010 W/m² °C. Calculate the number of pipes n the condenser consists of and the length L of these pipes.

Answer: $n = 72$ pipes; $L = 3.9$ m

3. A packet of 6 × 6 horizontal tubes of 30 mm outer diameter and 1.5 m length is surrounded by saturated steam at 180 °C.
   (a) If the surface of the tubes is kept at 120 °C, calculate the heat flux through the wall.
   (b) What would it be if the pipes were in a vertical position?

Answer: (a) 300 000 W/m²; (b) 236 000 W/m²

Comments on problems

Problem 1

The maximum amount of heat that can be removed is given by $\phi_H = UA\Delta T$. Now ΔT and A are known. Thus, after determining the overall heat transfer co-efficient, U, ϕ_H can be calculated.

For the various partial heat transfer coefficients, we find $h_{water} = 2000$ W/m^2 °C, $h_{glass} = 500$ W/m^2 °C, $h_{condensate} = 950$ W/m^2 °C (from Figure III.22). Thus, $U = 286$ W/m^2 °C and $\phi_H = 315$ W. If the inner pipe is made from stainless steel $h_{steel} = 13\,000$ W/m^2 °C and $U = 630$ W/m^2 °C, so $\phi_H = 700$ W.

III.8 Heat transfer in stirred vessels

Heat transfer in stirred vessels is strongly dependent on the flow pattern produced by the stirrer. As we have already seen in Section II.7.1 (Figure II.46), a propeller-type stirrer creates two systematic eddies which take up nearly half the vessel, the flow in which is parallel to the wall and to the shaft of the stirrer. A paddle-type stirrer mainly produces an eddy perpendicular to the stirrer shaft, whereas a turbine-type stirrer produces two eddies, one above and one below the stirrer.

For the flow along the wall of the vessel this means that a propeller or turbine stirrer produces a flow in the axial direction, whereas a paddle-type stirrer induces flow in the tangential direction. The axial flow along the wall is caused by a jet of liquid which comes from the stirrer and hits the wall at a well-defined point. From this point a boundary layer is built up along the wall. The thickness of this boundary layer increases with distance from the stirrer. During tangential flow a boundary layer along the wall of the vessel is formed also, but this layer has a constant thickness.

It is understandable that it is not possible to produce a simple heat transfer model on the basis of this complex hydrodynamic picture, especially if we realize that the rate at which the boundary layer during axial flow along the wall is formed is dependent on the energy of the eddy and thus on the Re number of the stirrer, nd^2/v. For relatively low Re numbers, the momentum transport from the eddy to the wall boundary layer is small and we can make a rough guess of the heat transfer coefficient at the wall on the basis of momentum penetration:

$$h \sim \frac{\lambda}{\delta_T} = \frac{\lambda}{\delta_h}\frac{\delta_h}{\delta_T} = \frac{\lambda}{\delta_h}\Pr^{\frac{1}{3}} \simeq \frac{\lambda}{\sqrt{\pi v x/v_w}}\Pr^{\frac{1}{3}}$$

where x is the distance along the wall from the point of contact with the liquid jet and v_w the mean velocity of the flow along the wall (which is proportional to nd).

This model predicts that:

$$\frac{hD}{\lambda} \sim \left(\frac{nd^2}{\nu}\right)^{\frac{1}{2}} \mathrm{Pr}^{\frac{1}{3}} \left(\frac{D}{d}\right)^{\frac{1}{2}} \left(\frac{D}{x}\right)^{\frac{1}{2}}$$

or:

$$\mathrm{Nu} \sim \mathrm{Re}^{\frac{1}{2}} \, \mathrm{Pr}^{\frac{1}{3}} \left(\frac{D}{d}\right)^{\frac{1}{2}} \left(\frac{D}{x}\right)^{\frac{1}{2}}$$

where the diameter of the stirrer d and the diameter of the vessel D are introduced because h-is independent of D.

For relatively high Re numbers, the momentum transport between the core of the eddy and the boundary layer determines the boundary layer thickness, which is practically uniform (analogous to the boundary layer thickness during turbulent pipe flow) if the flow situation in the vessel is completely turbulent. This results in (see Section III.3.2):

$$\mathrm{Nu} \sim \mathrm{Re}^{0.8} \, \mathrm{Pr}^{\frac{1}{3}} \left(\frac{D}{d}\right)^{0.8}$$

For turbine-type stirrers which induce mainly axial flow, the experimentally found exponent of the Re number varies between 0.5 (low Re, $\mathrm{Re} < 10^4$) and 0.7 to 0.8 (very high Re numbers, $\mathrm{Re} > 5 \times 10^5$). A rough description of the dependency of the Nu value on the Re number in the region of practical interest ($10^3 < \mathrm{Re} < 10^6$) leads to Re exponents of 0.65 to 0.70.

The dependency of Nu on $\mathrm{Pr}^{\frac{1}{3}}$ as predicted by both models (for high and low Re numbers) has been experimentally substantiated by a number of authors. The influence of the parameters D/d and D/x has been studied by only a few scientists and no agreement has been reached. Both theories predict too strong a dependency of Nu on D/d or on D/x. This is due to the fact that the ratio v_w/nd (which we assumed to be constant) is a function of both d/D and x/D. For turbine stirrers, for example, we might prove that $v_w/nd \sim d/D$, because the width of the jet of liquid leaving the stirrer remains constant until it reaches the wall (see Section II.7.1); it is easily proven that in this case:

$$\mathrm{Nu} \sim \mathrm{Re}^{\frac{1}{2}} \, \mathrm{Pr}^{\frac{1}{3}} \left(\frac{D}{x}\right)^{\frac{1}{2}}$$

Only a few results of measurements of local heat transfer coefficients (see Figure III.25) are available because measurement of these data is complicated and, anyway, for design purposes a knowledge of heat transfer coefficients averaged over the height of the vessel is sufficient.

For flow creating a systematic eddy perpendicular to the axis, as is the case with paddle-type stirrers, we practically always find:

$$\mathrm{Nu} \sim \mathrm{Re}^{\frac{2}{3}} \, \mathrm{Pr}^{\frac{1}{3}}$$

which is another reason why for turbine- and propeller-type stirrers (which always produce tangential flow) the mean Re exponent is found to be ca. 0.65 for $10^3 < \mathrm{Re} < 10^6$.

In the foregoing we have considered heat transfer to the wall of a stirred vessel. The presence of a coil makes no principal difference but complicates the picture because the flow of liquid is more complicated. The presence of a coil will decrease wall heat transfer coefficients (by up to 25% for normal configurations) because of decreased flow velocity along the wall. Coil heat transfer data are correlated with a coil-Nusselt number and are generally found to be:

$$\mathrm{Nu} = \frac{hd_c}{\lambda} \sim \mathrm{Re}^{\frac{2}{3}} \, \mathrm{Pr}^{\frac{1}{3}}$$

Because the combination of vessel and stirrer is not completely standardized — which is partly due to the fact that a stirrer is often not selected for heat transfer performance only but also for other tasks (e.g. suspending, mixing, etc.) — the chemical engineer designing a stirred vessel has to search the literature in order to find heat transfer data which can be applied for solving his problem. In doing so, the chemical engineer will have to answer the following questions:

Which type of liquid flow is required?
Which Re range is required?
Are the literature data found applicable to the specific geometric configuration of the design?
Are local or average heat transfer data required?

In the following, a few experimental correlations for heat transfer to a 'standard turbine stirred vessel' are given (see Figure III.24).

The average heat transfer coefficient to the wall of a standard vessel according to Figure III.24 is:

$$\langle \mathrm{Nu} \rangle = \frac{\langle h \rangle D}{\lambda} = 0.75 \left(\frac{nd^2}{\nu} \right)^{\frac{2}{3}} \mathrm{Pr}^{\frac{1}{3}} \left(\frac{\eta_w}{\eta} \right)^{0.14} \qquad \text{(III.80)}$$

for $10^3 < \mathrm{Re} < 10^5$. For slightly different configurations, the relationship:

$$\langle \mathrm{Nu} \rangle = \frac{\langle h \rangle D}{\lambda} = 1.01 \left(\frac{nd^2}{\nu} \right)^{\frac{2}{3}} \mathrm{Pr}^{\frac{1}{3}} \left(\frac{\eta_w}{\eta} \right)^{0.14} \left(\frac{d}{D} \right)^{0.13} \left(\frac{L}{D} \right)^{0.12} \qquad \text{(III.80a)}$$

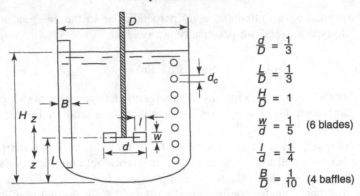

$$\frac{d}{D} = \frac{1}{3}$$

$$\frac{L}{D} = \frac{1}{3}$$

$$\frac{H}{D} = 1$$

$$\frac{W}{d} = \frac{1}{5} \quad \text{(6 blades)}$$

$$\frac{l}{d} = \frac{1}{4}$$

$$\frac{B}{D} = \frac{1}{10} \quad \text{(4 baffles)}$$

Figure III.24 Standard configuration of stirred vessel

can be used $(0.17 < d/D < 0.75; 0.1 < L/D < 0.7)$. For the standard configuration, equation (III.80a) yields equation (III.80) with a constant of 0.77.

Local heat transfer data to the wall can be calculated from the relationship (standard configuration with $z =$ vertical distance from stirrer):

$$\text{Nu} = \frac{hD}{\lambda} = 0.17 \ \text{Re}^{\frac{2}{3}} \ \text{Pr}^{\frac{1}{3}} \left(\frac{\eta}{\eta_w}\right)^{0.14} \left(\frac{D}{z}\right)^{0.8} \quad \text{if } \frac{D}{z} < 15$$

or:

$$\text{Nu} = 1.5 \ \text{Re}^{\frac{2}{3}} \ \text{Pr}^{\frac{1}{3}} \left(\frac{\eta}{\eta_w}\right)^{0.14} \quad \text{if } \frac{D}{z} > 15$$

if $10^3 < \text{Re} < 10^6$.

Figure III.25 illustrates the dependence of the ratio of $h_{\text{local}}/\langle h \rangle$ on the height of the vessel. The average heat transfer values calculated from the local values by integration over the vessel height are in agreement with the values found from equation (III.80), i.e. the surface area under the line $h/\langle h \rangle = f(D/z)$ in Figure III.25 is unity.

The average heat transfer coefficient to coils in a vessel with the configuration of Figure III.24 has been found to be:

$$\langle \text{Nu} \rangle = \frac{\langle h \rangle d_c}{\lambda} = 0.17 \ \text{Re}^{0.67} \ \text{Pr}^{0.37} \left(\frac{d}{D}\right)^{0.1} \left(\frac{d_c}{D}\right)^{0.5} \tag{III.81}$$

for

$$0.25 < \frac{d}{D} < 0.58; \quad 0.03 < \frac{d_c}{D} < 0.15 \text{ and } 400 < \text{Re} < 2 \times 10^5$$

A much-applied scaling-up rule for geometrically similar vessels is to keep the stirrer energy input per unit of volume of the vessel (which is $\sim n^3 D^2$) constant.

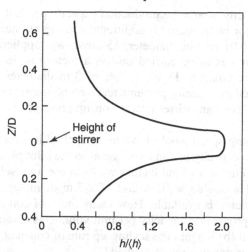

Figure III.25 Local heat transfer coefficients in a standard vessel

In this case, for double-walled vessels, the amount of heat transferred per unit of time and of volume (which is $\sim hD^2/D^3$) is proportional to $D^{-4/3}$ (for low Re values) to $D^{-10/9}$ (for average Re values). For geometrically similar vessels, therefore, the heat transport decreases with increasing size of the vessel if the scaling-up rule of constant power per unit of volume is applied. During scaling-up, steps have therefore to be taken to increase the heat flow (extra surface, larger temperature difference), or higher power consumption of the stirrer has to be accepted.

If vessels with heating or cooling coils are used and the heat transfer surface per unit of volume is kept constant, the heat transferred per unit of time and volume is proportional to $D^{-1/3}$ to $D^{-1/9}$ if the above scaling-up rule is applied. Under these circumstances, only a slight correction of stirrer speed (and thus of power consumption) is required to keep the heat transport during scaling-up constant.

III.8.1 Problem

1. Two water-like liquids which react instantaneously with each other are pumped through a continuously stirred tank reactor. Pilot plant experiments with a double-walled tank (diameter of vessel = height of double wall = D = 0.4 m; stirring speed n = 360 rpm) according to Figure III.24 showed that the production capacity of the reactor is determined by the rate of removal of the heat of reaction. It appeared that the heat transfer coefficient was independent of the cooling water flow rate.

(a) By what factor would the production capacity per unit volume of the pilot plant reactor be increased if additional cooling by means of a coil (0.3 m diameter, 0.01 m tube diameter, 15 turns) was applied?

(b) The reaction is to be carried out on a factory scale in a double-walled reactor, according to Figure III.24, of 3 m diameter. By what factor is the production capacity per unit reactor volume increased if the scaling-up rule of constant stirrer power consumption per unit reactor volume is applied?

(c) What stirring speed would have to be applied in situation (b) if the same production capacity per unit volume as in the pilot plant reactor is desired?

(d) The chief engineer of the factory wants to use a vessel with a cooling coil instead of a double wall. A coil of 2.5 m diameter, made from pipe of 5 cm diameter, is available. How many loops of coil are needed to assure the same production capacity per unit volume as obtained with the reactor in situation (b) if again the scaling-up rule of constant power consumption per unit reactor volume is applied?

(e) What is the ratio of the cooling surfaces of cases (b) and (d)?

Answer: (a) $2.13 \times$; (b) $0.11 \times$; (c) 2700 rpm;
$\qquad$ (d) ca. 15 turns; (e) $A_{wall}/A_{coil} = 1.6$

III.9 Heat transport by radiation

> *'Heat rays' are electromagnetic waves $(0.5 < \lambda < 10 \mu m)$[†] which are radiated by a body owing to thermal agitation of the constituent molecules. Between two surfaces radiation exchange may take place if the interspace is more or less diathermanous for the entire or part of the wavelength range. There is, then, heat transport in both directions and, if the temperatures of the surfaces differ, the difference between the opposite heat flows is not equal to zero.*

By means of thermodynamics it was derived that the heat flow which is emitted per unit of surface, ϕ_z'', by a so-called black surface with temperature T is given by the Stefan–Boltzmann law:

$$\phi_z'' = \sigma T^4 \qquad (III.82)$$

where the radiation constant $\sigma = 5.67 \times 10^{-8}$ W/m^2 K^4 and T the absolute temperature of the surface.

For a black body the intensity distribution over the wavelength range is given by the Planck radiation law. The wavelength at which this intensity is maximal

---

[†] Other examples of electromagnetic waves are radio waves $(\lambda > 10^4 \ \mu m)$, visible light $(0.4 < \lambda < 0.8 \ \mu m)$, X-rays $(\lambda \approx 10^{-4} \ \mu m)$ and γ-rays $(\lambda \approx 10^{-6} \ \mu m)$.

is inversely proportional to T (Wien's displacement law):

$$\lambda_{max}T = 2880 \text{ (micron K)} \tag{III.83}$$

At room temperature, $\lambda_{max} = 10$ μm (far infrared) and at 6000 K $\lambda_{max} = 0.5$ μm (yellow).

A completely black surface emits the maximal amount of radiation relevant to its temperature; it has the emission coefficient $e = 1$. According to Kirchhoff's law (see equation (III.84)), the absorption coefficient for the radiation on the surface is then also $a = 1$. A body which absorbs all radiation and does not reflect any is indeed black.

In general, technical surfaces have an emission and an absorption coefficient smaller than 1; the value depends on the material and the roughness of the surface, the temperature and the wavelength of the radiation. If in a certain wavelength range preferential absorption occurs, then just as in the case of visible light the surface is 'coloured'.

In engineering, mean absorption coefficients over the entire wavelength range will often suffice. In other words, such surfaces are considered as 'grey', i.e. between the extremes 'black' and 'white'. Of most non-metallic surfaces $a > 0.8$, so they are almost black for the infrared radiation. Metals reflect the heat rays only satisfactorily if polished and clean; the absorption coefficient a lies then between 0.05 and 0.20. Oxidation and contamination increase a. Figure III.26 shows absorption coefficients for a number of different materials.

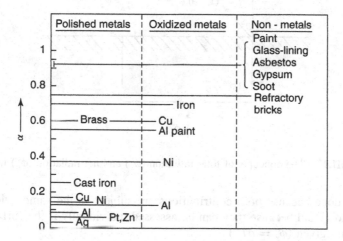

Figure III.26 The absorption coefficient for heat radiation

The amount of heat which is transported between two or more surfaces per unit of time depends not only on the temperature but also on:

the spatial configuration (how far can the surfaces 'see' each other?);
the absorption and emission capacity of these surfaces;
the possible absorption by the intermediate medium (water vapour and carbon dioxide, for example, have strong absorption bands in the infrared);
the emission of the intermediate medium (luminous flames).

Only the first two will be treated further here. For calculating the net transfer by radiation a good balance of the radiation transport has to be drawn up. This can best be done by using Poljak's scheme of calculations for a closed space. He considered that an efficient protocol for calculating radiation transfer is to start with the total radiation from each surface (ϕ''), being the sum of its own emission ($e\phi_z''$, given by its temperature) and its reflection of the as yet unknown foreign irradiation, $(1 - a)\ \phi_i''$ (compare Figure III.27):

$$\phi'' = e\phi_z'' + (1 - a)\phi_i'' \tag{III.84}$$

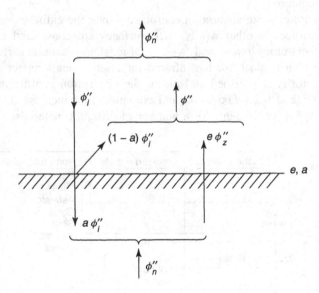

Figure III.27 The concepts of total radiation (ϕ'') and net radiation (ϕ_n'') in W/m^2

This is done because both contributions are subject to the same "view"-factors (see below) and because they can be assessed individually if the surface temperatures are given ($\phi_z'' = \sigma T^4$).

From equation (III.84), two important facts can be derived:

(a) If a wall is enclosed by black bodies, then in the case of thermodynamic equilibrium:

$$\phi_i'' = \phi'' = \phi_z'' = \sigma T^4$$

As a consequence, $e = a$, i.e. the emission coefficient numerically equals the absorption coefficient (Kirchhoff's law).

(b) From the conclusion drawn under (a) it follows that in the case of equilibrium in a closed system the total energy flux of radiation through **each** plane equals

$$\phi_z'' = \sigma T^4.$$

For all N surfaces the $(N-1)$ foreign irradiations have to be assessed as the weighted sum of the $(N-1)$ total radiations from the other surfaces. Each weight-factor is the so-called "view"-factor mentioned above, giving the fraction F_{jk} of the total radiation of the wall j ($A_j \phi_j''$, with A_j its surface area) reaching the wall k. From this assessment, the net radiation transport, ϕ_n'', from a surface may be obtained (compare again Figure III.27):

$$\phi_n'' = \phi'' - \phi_i'' \qquad \text{(III.85)}$$

under the condition that the law of energy conservation is respected:

$$\sum_{j=1}^{N} \phi_{nj}'' A_j = 0$$

Poljak now simplified this often tedious 'bookkeeping' by rewriting equations (III. 84) and (III. 85), using Kirchhoff's law ($e = a$), as follows:

$$\phi'' = \phi_z'' + (1-a)(\phi_i'' - \phi_z'') = \phi_z'' - \frac{1-a}{a}\phi_n'' \qquad \text{(III.86)}$$

which relates the total radiation flux from a wall to its temperature (ϕ_z''), its net transport (ϕ_n'') and its own absorption coefficient.

Equations (III.85) and (III.86) constitute together a smooth procedure for treating radiation problems, as will now be shown for the net radiation transport between two bodies of which the one (1) is completely enclosed by the other (2). Hence:

$$\phi_{n1}'' A_1 + \phi_{n2}'' A_2 = 0 \text{ (law of energy conservation)}$$

Using the two equations in succession and considering that all irradiation on body 1 comes from body 2, the analysis results in one statement:

$$\phi_{n1}'' = \phi_1'' - \phi_2'' = \phi_{z1}'' - \frac{1}{a_1}\phi_n'' - \phi_{z2}'' - \frac{A_1}{A_2}\frac{1-a_2}{a_2}\phi_{n1}''$$

or

$$\left(1 + \frac{1-a_1}{a_1} + \frac{A_1}{A_2}\frac{1-a_2}{a_2}\right)\phi_{n1}'' = \phi_{z1}'' - \phi_{z2}'' = \sigma(T_1^4 - T_2^4) \qquad \text{(III.87)}$$

This is an important relationship which is often used for calculating radiation transport.

If the temperature difference, $\Delta T = T_1 - T_2$, is small compared to the absolute temperatures, T_1 and T_2, $(T_1^4 - T_2^4)$ can be approximated by $4\overline{T}^3 \Delta T$ ($\overline{T}$ being the mean of T_1 and T_2). In these cases, a heat transfer coefficient can be defined as:

$$h_r = 4\sigma\overline{T}^3 \left(1 + \frac{1 - a_1}{a_1} + \frac{A_1}{A_2}\frac{1 - a_2}{a_2} \right)^{-1} \tag{III.88}$$

which at room temperature is of the order of 5 W/m^{2}K and is, e.g. already of the order of the heat transfer coefficients found for free convection. At higher temperatures, h_r increases considerably and thus has nearly always to be taken into account.

For more complex geometries, Poljak's scheme works just as efficiently, especially in computational work. Handbooks (e.g. the VDI Wärmeatlas[†]) need to be used in order to assess the 'view'-factors F_{jk}. They are mostly calculated by cumbersome integrations (summations over small surface elements), which take into account that the radiation intensity depends on the angle ϕ between the direction of the radiation and the normal to the surface (approximately proportional to $\cos\phi$ according to Lambert's law). However, for a number of geometrically simple cases, such 'view'-factors can be found in a relatively simple way. Check that for infinite parallel walls (1 and 2): $F_{12} = F_{21} = 1$. In general, it can be proven that for two walls with surface areas A_1 and A_2, then $A_1 F_{12} = A_2 F_{21}$, which is readily seen by realizing that in equilibrium, $A_1 F_{12}\phi_z'' = A_2 F_{21}\phi_z''$. If wall 2 encloses wall 1, as in the example treated above, $F_{12} = 1$ and thus $F_{21} = A_1/A_2$. This is implicit in equation (III. 87).

III.9.1 Problems

1. A long rod (diameter 1.1 cm, emission coefficient 0.8) with a temperature of 327 °C is suspended in quiet air at 27 °C. Calculate the contribution of free convection and radiation to the total heat flow. Assume the air to be an ideal gas with a density of 1.0 kg/m^3 at 27 °C.

   Answer: horizontal rod: 46% free convection
   54% radiation
   vertical rod: 29% free convection
   71% radiation

*2. A small ceramic sphere (diameter 1 cm), e.g. of a thermocouple holder, is situated in the middle of a long tube with a diameter of 20 cm. Air at 150 °C

[†] *VDI-Wärmeatlas*, 8th edition, Springer-Verlag, 1997

is passed through the tube at a mean velocity of 20 m/s. The temperature of the tube wall is 100 °C. Calculate the temperature of the sphere.

Answer: 147 °C

3. Check that if between two large flat walls with equal emission coefficients a third equal wall is placed, then the net heat transport by radiation is halved. An example of this type of radiation shield is the layers of aluminium foil in the walls of refrigerators and cold store rooms.

4. The maximum intensity of sunlight occurs at a wavelength of approximately 0.5×10^{-6} m. Calculate the surface temperature of the sun.

*5. Consider an enclosure consisting of a black heat source A_1, a black heat sink A_2, and a refractory surface A_R, from which there is no net radiant heat flux. Prove that the net radiation flux from A_1 is given by;

$$\phi_{n1}'' = \left(F_{12} + \frac{F_{1R}F_{R2}}{1 - F_{RR}} \right) \sigma(T_1^4 - T_2^4)$$

6. Consider the same situation as in problem 5, but in this case the surfaces A_1 and A_2 are now grey. Give an expression for the net radiation flux from A_1. How is this relation related to the relationship (III.87) for the net radiation transport between A_1 and A_2 when A_1 completely encloses A_2.

Answer:
$$\left[\left(F_{12} + \frac{F_{1R}F_{R2}}{1 - F_{RR}} \right)^{-1} + \left(\frac{1 - a_1}{a_1} \right) + \frac{A_1}{A_2} \left(\frac{1 - a_2}{a_2} \right) \right] \phi_{n1}''$$
$$= \sigma(T_1^4 - T_2^4)$$

Comments on problems

Problem 2

In steady-state conditions, heat transport by radiation from the sphere must equal convective transport to the sphere. Thus:

$$\langle h \rangle (T_g - T_s) = h_r(T_s - T_w)$$

Now Re = 7000, thus $\langle h \rangle = 183$ W/m² °C (equation (III.68a). Assuming $T_s \approx T_g$, $\bar{T} = (T_g + T_s)/2 = 125$ °C ≈ 400 K. Thus, employing equation (III.88) we find ($a_1 = 0.9$, $A_1/A_2 \ll 1$):

$$h_r = 13 \text{ W/m}^2 \text{ °C}$$

and thus $T_s = 147$ °C.

Problem 5

Start with the following balances:

$$\phi''_{n1} = \phi''_1 - F_{11}\phi''_1 - F_{12}\phi''_2 - F_{1R}\phi''_R$$

$$\phi''_{n2} = \phi''_2 - F_{21}\phi''_1 - F_{22}\phi''_2 - F_{2R}\phi''_R$$

$$0 = \phi''_R - F_{R1}\phi''_1 - F_{R2}\phi''_2 - F_{RR}\phi''_R$$

Derive from these an expression for ϕ''_{n1} in terms of $(\phi''_1 - \phi''_2)$ and continue with relationship (III.86).

CHAPTER IV

Mass Transport

In this chapter, the transport of matter will be treated. The starting point of our analysis is the law of conservation of matter, which reads for a substance A (see Section I.1):

$$\frac{d}{dt}(V\rho_A) = \phi_{v,in}\rho_{A,in} - \phi_{v,out}\rho_{A,out} + V r_A \qquad (IV.1)$$

The following pages are no more than an elaboration of this law.

Apart from agreement in the description of the transport of matter and heat transfer, there are distinct differences which can be divided into non-fundamental and fundamental differences:

Non-fundamental differences:

(a) The interface between two media between which matter is transferred (e.g. between two liquids or a gas and a liquid) is usually mobile, contrary to the interface between two media between which heat is transferred (usually a solid wall).

(b) The order of magnitude of the diffusion coefficients $\mathbb{D}$ of liquid and solid substances is much smaller than, for example, the thermal diffusivity or the kinematic viscosity (for gases at 20 °C and 1 atm, $\mathbb{D} \approx 10^{-5}$ m^2/s; for liquids at 20 °C, $\mathbb{D} \approx 10^{-9}$ m^2/s; and for solid substances, $\mathbb{D}$ varies widely between 10^{-10} and 10^{-15} m^2/s). This means for liquids and solid substances that in the first place the penetration of matter in a given time takes place over (much) smaller distances than the penetration of heat or momentum; in the second place, already with convective currents of low velocities, which frequently occur in the case of mobile interfaces, the convective transport exceeds the diffusion transport in the direction of flow.

Fundamental differences:

(a) If the net flux of matter by diffusion is not equal to zero, then the diffusion process will nearly always induce a drift velocity, which may not be negligible. This will be shown in Section IV.1.1 for a simple case.

(b) In Section I.2, a proportional relationship between mass flux and concentration gradient is defined according to Fick's law. From a thermodynamic point of view, chemical potential instead of concentration has to be used.

As a consequence, the diffusion coefficient may be a strong function of concentration, particularly in non-ideal systems.

(c) In heat transfer, the equilibrium condition at a phase boundary is simple and thermodynamically sound, i.e. the temperatures on both sides of the interface are equal. In mass transfer, this would also be the case if chemical potentials are used.

   If instead we use concentrations, the equilibrium condition at a phase boundary becomes less simple because concentrations, in contrast to chemical potentials, are unequal on both sides of the interface (see Section IV.1.2).

(d) In mass transfer, there can be other driving forces than concentration gradients, e.g. gradients in electrical potential (e.g. during the transport of ions) and pressure gradients (e.g. during transport through membranes). These will not be dealt with in this present book.

(e) In systems with more than two components, the "Fickian" approach cannot be applied if the system cannot be considered as pseudo-binary, i.e. transport of one component at low concentrations in a mixture of other components.

The problems arising from the fundamental differences between heat and mass transfer can be avoided by using the so-called Maxwell–Stefan equations. The reader who is interested in this approach will find a good introduction to this subject in J.A. Wesselingh and R. Krishna, *Mass Transfer*, Ellis Horwood, 1990.

From the foregoing, it appears that the application of Fick's law may lead to over simplifications and even to serious errors in systems which are not ideal, consist of more than two components, and contain component(s) to be transported at high concentrations.

Due to the use of concentrations instead of chemical potentials, the following dilema arises. We formulated a law of conservation of mass because mass is the most characteristic property of matter. Accordingly, we defined the diffusion, coefficient in Section I.2 (equation (I.13)), as the proportionality constant between the mass flux and the gradient in the mass concentration:

$$\phi''_{mA,x} = -\mathbb{D}_A \frac{d\rho_A}{dx} \quad \text{(Fick)} \qquad \text{(IV.2a)}$$

In addition, production rates in chemical reactions are functions of the molar concentrations, which we shall indicate by c_A, etc. ($kmol/m^3$). So we can imagine that for many calculations it offers advantages to define the diffusion coefficient as the ratio between the mole flux ($kmol/m^2$ s) and the gradient in the molar concentration:

$$\phi''_{\text{mol } A,x} = -\mathbb{D}_A \frac{dc_A}{dx} \quad \text{(Fick)} \qquad \text{(IV.2b)}$$

Both thus defined diffusion coefficients are in principle different (although in practice they are often almost equal) and in applying them we should realize on which concentration the description is based.

The diffusion coefficient has been defined in such a way that the net transport through a fixed plane (in Figure 1.6 the plane n = constant) is zero. If we restrict ourselves to binary systems (with components A and B) for practical purposes two cases can be distinguished in which the net transport is zero:

(a) For every unit of mass A which diffuses through a fixed plane, a unit of mass B moves in the opposite direction by diffusion ('barocentric' system).
(b) For every mole A, a mole B moves in the other direction through the plane.
sub (a) Net mass flow = 0, ρ = constant.

For the diffusion transport of substance A we can write:

$$\phi''_{mA,x} = -\mathbb{D}_{AB}\frac{d\rho_A}{dx}$$

whereas for the transport of substance B the following applies:

$$\phi''_{mB,x} = -\mathbb{D}_{BA}\frac{d\rho_B}{dx}$$

where $\phi''_{mA,x}$ is the mass flux of A (kg/m^2 s) in the x-direction for a net mass flow = 0; ρ_A is the mass concentration of A (kg/m^3).
Because the conditions have been such that:

$$\phi''_{mA,x} + \phi''_{mB,x} = 0$$

whereas:

$$\rho_A + \rho_B = \rho = \text{constant} \quad \text{or} \quad d\rho_A + d\rho_B = 0$$

it follows from these relationships that $\mathbb{D}_{AB} = \mathbb{D}_{BA}$. In other words, the definitions have been determined in such a way that the diffusion coefficients of A in B and of B in A have the same numerical value. These definitions are used for describing the diffusion transport in solid substances and liquids: here, in the case of not too high concentrations of, for example, A in B, the density may be assumed to be constant.
sub (b) Net mole flux = 0, the total concentration c = constant (kmol/m^3).
Now we can write for the mole flux (kmol/m^{2}s) in the x-direction:

$$\phi''_{\text{mol A},x} = -\mathbb{D}_{AB}\frac{dc_A}{dx}$$

and:

$$\phi''_{\text{mol B},x} = -\mathbb{D}_{BA}\frac{dc_B}{dx}$$

where c_A and c_B are molar concentrations (kmol/m^3); for ideal gases and constant pressure and temperature, their sum total is constant and equals c. These definitions are very suitable for describing the diffusion transport in gases. In the case of equimolar diffusion:

$$\phi''_{\text{mol A},x} + \phi''_{\text{mol B},x} = 0$$

Also, here, the diffusivities have been defined in such a way that:

$$\mathbb{D}_{AB} = \mathbb{D}_{BA}$$

According to the two definitions the dimension of $\mathbb{D}$ is $(\text{length})^2/\text{time}$, so that $\mathbb{D}$ in the SI system is expressed in (m^2/s). This was likewise the case for the two other transport coefficients, i.e. v (kinematic viscosity) and a (thermal diffusivity). However, in the case of diffusion it is advisable to distinguish between two cases, i.e. $\rho = \text{constant}$ and $c = \text{constant}$. For these cases, $\mathbb{D}$ has to be defined using, respectively, the gradient of the mass or of the molar concentration.

IV.1 Stationary diffusion and mass transfer

John noticed that, before he had left his office, the lawyer who had disappeared had prepared a pot of coffee. A cup on his desk was still half full and from the scale on the sides, caused by evaporation of the water, he could see that the cup had been filled to 1 cm below the rim. He estimated the cup to be 8 cm high, remembered the pages to come and concluded that the lawyer must have left his office approximately 3 weeks before.

IV.1.1 Stationary diffusion

During most practical cases of diffusion, neither the net mass flux nor the net mole flux discussed in the foregoing section will be zero. A practical example in which the net mole flux is almost zero is distillation in which, for every mole of the light component which passes into the gas phase, approximately one mole of the heavy component enters the liquid. However, on absorption and extraction of one component from a mixture the net transfer is invariably unequal to zero. In addition, during a chemical reaction, e.g. at a catalyst surface, the mole fluxes of the reactants to be fed and the mole fluxes of the products to be discharged are usually unequal (e.g. in a dimerization).

When describing the mass transfer in these cases, allowance must be made for the entrainment of substances in the flow caused by the mass transfer as such. For a binary mixture moving with a velocity v, relative to a plane $x = \text{constant}$, the steady-state diffusion flux of component A relative to this velocity is:

$$(v_A - v)c_A = -\mathbb{D}\frac{dc_A}{dx}$$

The rate of molar transport relative to the plane $x = $ constant is then, since $vc = \phi''_{\text{mol A}} + \phi''_{\text{mol B}}$:

$$\phi''_{\text{mol A},x} = -\mathbb{D}\frac{dc_A}{dx} + (\phi''_{\text{mol A},x} + \phi''_{\text{mol B},x})\frac{c_A}{c} \qquad \text{(IV.3a)}$$

whereas the mass transport is given by:

$$\phi''_{mA,x} = -\mathbb{D}\frac{d\rho_A}{dx} + (\phi''_{mA,x} + \phi''_{mB,x})\frac{\rho_A}{\rho} \qquad \text{(IV.3b)}$$

The first right-hand term in these equations represents the molar/mass flux resulting from diffusion, whereas the second term is a measure of the molar/mass flux resulting from total (drift) flow. Equations (IV.3a) and (IV.3b) can only be solved if the relationship between the molar fluxes of A and B is known. We can distinguish two extreme situations:

(a) $\phi''_A + \phi''_B = 0$, i.e. equimolar diffusion;
(b) $\phi''_B = 0$, i.e. diffusion of A through stagnant B.

In practice, situations lying between these extremes will often occur, but since the differences between cases (a) and (b) are often small, it is nearly always possible to produce a good estimate of the molar flux by making a choice between these cases. In the following, we will analyze the two extreme cases stated in greater detail.

(a) *Equimolar diffusion*, $\phi''_A + \phi''_B = 0$
Equation (IV.3) can now be simplified to:

$$\phi''_{\text{mol A},x} = -\mathbb{D}\frac{dc_A}{dx} \qquad \text{(IV.4)}$$

If we consider one-dimensional stationary diffusion in the x-direction, the mass balance (IV.1) yields (there is no production of A):

$$\phi''_{\text{mol A},x,\text{in}} = \phi''_{\text{mol A},x,\text{out}} = \text{constant}$$

or $\phi''_{\text{mol A},x} = $ constant for every plane $x = $ constant. Solution of this equation yields, with the boundary conditions, $c_A = c_{A1}$ at $x = x_1$ and $c_A = c_{A2}$ at $x = x_2$, the concentration distribution:

$$\frac{c_A - c_{A1}}{c_{A2} - c_{A1}} = \frac{x - x_1}{x_2 - x_1} \qquad \text{(IV.5)}$$

We see that the concentration distribution (Figure IV.1) is linear.

The mole flux of A is found after substitution of dc_A/dx in equation (IV.4) by the value calculated from equation (IV.5) to be:

$$\phi''_{\text{mol A},x} = -\mathbb{D}\frac{c_{A2} - c_{A1}}{x_2 - x_1} \qquad \text{(IV.6)}$$

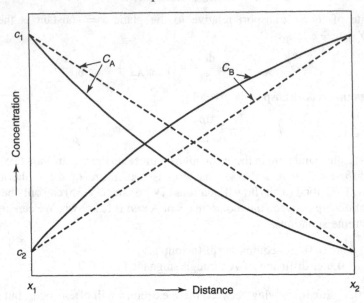

Figure IV.1   Concentration profile during equimolar diffusion (broken line) and during diffusion through a stagnant fluid (solid line)

So, during equimolar diffusion the mole flux (and analogously also the mass flux) is directly proportional to the concentration gradient $\Delta c / \Delta x$. We notice that mass transport by equimolar diffusion is analogous to the transport of heat by conduction. For one-dimensional heat conduction, we had found in Section III.1.1 (equation (III.2a)):

$$\phi_H'' = -\lambda \frac{T_2 - T_1}{x_2 - x_1} = -a \frac{\rho c_p T_1 - \rho c_p T_2}{x_2 - x_1}$$

This equation is identical to equation (IV.6) if we replace the molar diffusivity $\mathbb{D}$ by the thermal diffusivity a and the molar concentrations c (mol/m^3) by the 'heat concentrations' $\rho c_p T$ (W/m^3). We can therefore also adapt the relationships found for stationary heat conduction around cylinders and spheres to diffusion. By using equations (III.6) and (III.8), we find for equimolar diffusion between coaxial cylinders with diameters D_1 and D_2 and concentrations c_{A1} and c_{A2}:

$$\phi_{\text{mol A},r}' = -\mathbb{D} \frac{2\pi(c_{A1} - c_{A2})}{\ln D_2/D_1} \tag{IV.7}$$

and for equimolar diffusion between concentric spheres (diameters D_1 and D_2, concentrations c_{A1} and c_{A2}):

$$\phi_{\text{mol A}} = -\mathbb{D} \frac{2\pi D_1 D_2}{D_2 - D_1}(c_{A1} - c_{A2}) \tag{IV.8}$$

(b) *Diffusion through a stagnant body, $\phi''_{mol\,B} = 0$*

This process can be observed, for example, during selective absorption or extraction of material A from a mixture of A and B. For this situation, equation (IV.3a) can be simplified to:

$$\phi''_{mol\,A,x} = -\mathbb{D}\frac{c}{c - c_A}\frac{dc_A}{dx} \qquad (IV.9)$$

Application of the mass balance (equation (IV.1)) yields, again, that under stationary conditions and no production of A, the molar flux of A through all planes $x = $ constant must be the same. So:

$$\phi''_{mol\,A,x} = -\mathbb{D}\frac{c}{c - c_A}\frac{dc_A}{dx} = \text{constant}$$

The above equation can be integrated and, applying the boundary conditions x_1, c_{A1} and x_2, c_{A2}, we find for the concentration distribution:

$$\frac{c - c_A}{c - c_{A1}} = \left(\frac{c - c_{A2}}{c - c_{A1}}\right)^{(x-x_1)/(x_2-x_1)} \qquad (IV.10)$$

This concentration distribution is also shown in Figure IV.1 for materials A (e.g. solvent) and B (e.g. solute). With the aid of equation (IV.10) we find from equation (IV.9) for the molar flux for diffusion through a stagnant body:

$$\phi''_{mol\,A,x} = \frac{-\mathbb{D}c}{x_2 - x_1}\ln\frac{c - c_{A2}}{c - c_{A1}} \qquad (IV.11)$$

If c_A is at all places much smaller than the total concentration c, the logarithmic factor in this relationship can be approximated by:

$$\ln\frac{c - c_{A2}}{c - c_{A1}} \approx -\frac{c_{A2} - c_{A1}}{c}$$

and we find:

$$\phi''_{mol\,A,x} = -\frac{\mathbb{D}(c_{A2} - c_{A1})}{x_2 - x_1} \qquad (IV.12)$$

i.e. the same result as for equimolar diffusion.

Mass transport through a stagnant fluid is a factor of

$$f_D = \frac{c}{c_{A1} - c_{A2}}\ln\frac{c - c_{A2}}{c - c_{A1}} = 1 + \frac{1}{2}\frac{c_{A1} + c_{A2}}{2} + \ldots$$

larger than during equimolar diffusion. The correction factor f_D is named after Stefan; if this factor is applied to equations (IV.7) and (IV.8), these equations will also correctly describe, for the geometries considered, diffusion through a stagnant fluid.

IV.1.2 Mass transfer coefficients

Analogous to the introduction of the heat transfer coefficient h in Section III.1.3, which was defined by:

$$h = \frac{\phi''_H}{T_w - \langle T \rangle} = -\lambda \frac{\left. \frac{dT}{dx} \right|_{x=0}}{T_w - \langle T \rangle}$$

we can now introduce a mass transfer coefficient k, which is defined as the ratio of the mass flux and the concentration gradient (which constitutes the driving force) in one phase:

$$k_A = \frac{\phi''_{mol,A}}{c_{A0} - \langle c_A \rangle} = -\mathbb{D}_A \frac{\left. \frac{dc_A}{dx} \right|_{x=0}}{c_{A_i} - \langle c_A \rangle} \tag{IV.13}$$

Here, $\phi''_{mol\ A}$ is the mass flux of A into phase B and $c_{A_i} - \langle c_A \rangle$ is the difference between the concentration at the interface and the average concentration of A in phase B (mean cup concentration). For diffusion from a sphere into an infinite stagnant medium, we find, combining equations (IV.13) and (IV.8) (the latter for the case $D_2 \longrightarrow \infty$):

$$k_A = 2\frac{\mathbb{D}}{D} \quad \text{or} \quad \frac{k_A D}{\mathbb{D}} = 2$$

The dimensionless number $k_A D/\mathbb{D}$ is called the Sherwood number, Sh, which plays the same role in mass transfer as the Nusselt number in heat transfer. For diffusion from a sphere into a stagnant infinite fluid, Sh = 2, analogous to Nu = 2 for heat conduction. The dimension of the mass transfer coefficient is the same as that of velocity, i.e. distance/unit of time.[†]

If a certain component is exchanged between two mobile phases (e.g. gas absorption, extraction, distillation, etc.) we encounter an overall mass transfer coefficient K, which is composed of the partial transfer coefficients in the different phases. Contrary to heat transfer (Section III.2.2), however, in the case of mass transfer it is not possible to calculate the overall coefficient by summing up the partial resistances $(1/k)$ to the total resistance $(1/K)$. This is due to the fact that the concentrations in both phases are generally not equal if the two phases are in equilibrium.

[†] In mass transfer with gases, the partial pressure instead of the molar concentration is often used for calculations. A mass transfer coefficient is then defined as:

$$\phi''_{mol\ A} = k_g(p_{A,w} - p_{A,f})$$

since for ideal gases, $c = p/RT$, $k_g = k_A/RT$.

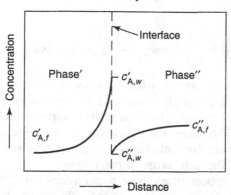

Figure IV.2 Concentration profiles near the interface of two phases

Figure IV.2 shows an interface between two fluid phases through which a stationary mass flux ϕ'' of a certain component occurs; the mole concentration distributions drawn on either side of this interface are representative of the mass exchange between two turbulent flows. According to the definition of the mass transfer coefficient k, we can write:

$$\phi''_{\text{mol A}} = k'(c'_{A,w} - c'_{A,f}) = k''(c''_{A,f} - c''_{A,w}) \qquad \text{(IV.14)}$$

It may now be assumed that there is equilibrium between the two phases at the interface, so that $c'_{A,w}$ and $c''_{A,w}$ are related according to the equilibrium relationship for this system.

In principle, with this equilibrium, relationship $c'_{A,w}$ and $c''_{A,w}$ can be eliminated from the previous relationships, so that a relationship between $\phi''_{\text{mol A}}$, k', k'', $c'_{A,f}$ and $c''_{A,f}$ remains. An analytically simple relationship is only obtained if the equilibrium can be represented by means of a relationship of the form:

$$c'_{A,w} = mc''_{A,w}$$

where m represents the distribution coefficient. At low concentrations, the distribution of a component over two phases approximately satisfies such a relationship.[†]

Elimination of $c'_{A,w}$ and $c''_{A,w}$ with this distribution law gives:

$$\phi''_{\text{mol A}} = \left(\frac{1}{k''} + \frac{1}{mk'}\right)^{-1}\left(c''_{A,f} - \frac{c'_{A,f}}{m}\right) = K''\left(c''_{A,f} - \frac{c'_{A,f}}{m}\right) \qquad \text{(IV.15a)}$$

[†] Henry's law for gas–liquid and Nernst's distribution law for liquid–liquid systems: the relationship between the distribution coefficient m and the Henry coefficient He is $m = \text{He}/RTc_l$, where He is the ratio of the partial pressure in the gas phase and the mole fraction in the liquid phase.

or also:

$$\phi''_{\text{mol A}} = \left(\frac{m}{k''} + \frac{1}{k'}\right)^{-1} (mc''_{A,f} - c'_{A,f}) = K'(mc''_{A,f} - c'_{A,f}) \qquad \text{(IV.15b)}$$

It appears that the total mass transfer resistance ($1/K''$, if related to the phase'' and $1/K'$ is related to the phase') is composed additively of the partial resistances $1/k'$ and $1/k''$, but that at the same time the relationship of the equilibrium concentrations is involved. The reader may check that the partial resistance of the phase in which the equilibrium concentration is lowest is (relatively) the most important if k' and k'' are of the same order of magnitude.

Furthermore, it appears from these relationships for K that the concentration difference, or 'the driving force', with the help of which k'' and k' have been defined, represents the deviation from the equilibrium between the two phases. If we write instead of $c'_{A,f}/m$, $c''^*_{A,f}$ (i.e. the concentration phase'' would have if it was in equilibrium with $c'_{A,f}$), and instead of $mc''_{A,f}$, $c'^*_{A,f}$, we get the more general definition for k' and k'':

$$\phi''_{\text{mol A}} = K'(c'^*_{A,f} - c'_{A,f}) = K''(c''_{A,f} - c''^*_{A,f})$$

When in a certain apparatus (mass exchanger), a mass flow $\phi_{\text{mol A}}$ of a certain component must be transferred from one medium to another, the required exchange area A follows from an equation of the form:

$$\phi_{\text{mol A}} = A\langle K(c_A - c_A^*)\rangle$$

where K and c are related to one of the two phases. If K does not vary too much over the apparatus, in the case of cocurrent or countercurrent flow of both phases, for $< c_A - c_A^* >$ the logarithmic mean between the driving forces at the inlet and outlet of the apparatus can be taken (see also Section IV.4).

IV.1.3 General approach for the calculation of concentration distributions

The partial differential equation which forms the basis for calculating concentration distributions is obtained if the law of conservation of matter is applied to a small (e.g. Cartesian) volume element of the medium. If the flow which may come from the diffusion itself is neglected, so that Fick's law (equation (IV.2)) can be used for a fixed surface element in the medium, the derivation is completely analogous to that of the general differential equation for heat transport (see Section III.1.4).

The resulting equations which we can now derive are for a component A:

$$\frac{\partial \rho_A}{\partial t} + v_x \frac{\partial \rho_A}{\partial x} + v_y \frac{\partial \rho_A}{\partial y} + v_z \frac{\partial \rho_A}{\partial z} = \mathbb{D}_A \left\{ \frac{\partial^2 \rho_A}{\partial x^2} + \frac{\partial^2 \rho_A}{\partial y^2} + \frac{\partial^2 \rho_A}{\partial z^2} \right\} + r_A$$

$$\text{(IV.16a)}$$

where r_A = production of A in kg/m^3 s (barocentric system) and:

$$\frac{\partial c_A}{\partial t} + v_x \frac{\partial c_A}{\partial x} + v_y \frac{\partial c_A}{\partial y} + v_z \frac{\partial c_A}{\partial z} = \mathbb{D}_A \left\{ \frac{\partial^2 c_A}{\partial x^2} + \frac{\partial^2 c_A}{\partial y^2} + \frac{\partial^2 c_A}{\partial z^2} \right\} + R_A$$

(IV.16b)

where R_A = production of A in kmol/m^3 s. Strictly speaking, these two equations only apply at constant $\mathbb{D}_A$ and if $\rho_A \ll \rho$ or $c_A \ll c$. In this case, they can also be used if more than one component is present in a low concentration in the medium.

The diffusion equation for low concentrations with constant $\mathbb{D}$ and without chemical reaction (also called Fick's second law) corresponds entirely with Fourier's differential equation (equation (III.2)). The solution of a great many diffusion problems can therefore be found in books on heat conductivity (see, for example, Carslaw and Jaeger[†]).

The distribution of the neutron concentration in a nuclear reactor is in principle also described with an equation such as (IV.16), where r_A, the neutron production by nuclear fission, is proportional to the neutron concentration.

The production terms r_A and R_A usually depend on the concentration in the mixture in a way which is determined by the chemical kinetics of the reaction. For simple kinetics and for simple geometric situations, solutions of equation (IV.16) are known, especially for the stationary state. We shall return to this subject in Section IV.5.

IV.1.4 Film model

Analogous to the boundary layer theory in heat transfer (see Section III.3.2), Lewis and Whitman introduced the concept of a boundary layer to mass transfer. They assumed the mass transfer resistance to be located in a stagnant boundary layer of thickness δ_c (see Figure IV.3) through which mass transport has to take place. The mass balance (equation (IV.16b)) reads for this case (stationary state):

$$0 = \mathbb{D}_A \frac{d^2 c_A}{dx^2}$$

(IV.17)

At the interface ($x = 0$), the equilibrium concentration c_{A0} is present, whereas on the other side of the film ($x = \delta_c$), the concentration equals the bulk concentration c_{A1}. Solving equation (IV.17) with these boundary conditions yields the concentration distribution:

$$\frac{c_A - c_{A1}}{c_{A0} - c_{A1}} = 1 - \frac{x}{\delta_c}$$

(IV.18)

[†] M.S. Carslaw, J.C. Jaeger, *Conduction of heat in solids*, 2nd edition, Oxford, 1959.

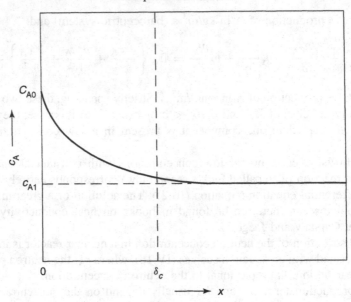

Figure IV.3 Film model concentration distribution

With the aid of this concentration distribution we can now calculate the mass transfer coefficient, for which we find:

$$k = -\mathbb{D}_A \frac{\left.\dfrac{dc_A}{dx}\right|_{x=0}}{c_{A0} - \langle c_A \rangle} = \frac{\mathbb{D}_A}{\delta_c} \tag{IV.19}$$

Thus, the Sherwood number is given by:

$$\text{Sh} = \frac{kD}{\mathbb{D}_A} = \frac{D}{\delta_c} \tag{IV.20}$$

and we see the formal analogy between the Sh and the Nu numbers ($\text{Nu} = D/\delta_T$) and the dimensionless expression $\frac{1}{2}f\text{Re}(= D/\delta_h)$, which all represent the ratio between geometrical scale and the thickness of the boundary layer for the transport process concerned. We may not conclude from equation (IV.20) that the mass transfer coefficient is linearly dependent on the diffusivity because δ_c is also dependent on $\mathbb{D}$, as will be discussed later.

The thickness of δ_c of the mass transfer boundary layer cannot be determined directly but can be calculated if the mass transfer coefficient is known. So, the film model does not actually help us to predict k values. The advantage is its simplicity and the possibility for estimating the order of magnitude of the film thickness over which the mass transfer process occurs. A further advantage of this model is the fact that the influence of a chemical reaction on the mass transfer process can easily be studied. We will therefore use the film model extensively in Section IV.5 when treating mass transfer with chemical reaction.

IV.1.5 Problems

*1. The vapour pressure of naphthalene at room temperature is approximately 0.05 mmHg. If a mothball (diameter 1 cm) consisting of this material is suspended in still air, calculate the initial rate of vaporization. How long does it take before the diameter of the mothball is reduced to half its initial value ($\mathbb{D}$ of naphthalene in air $= 0.7 \times 10^{-5}$ m^2/s, $\rho = 1150$ kg/m^3)?

    Answer: $\phi_m = 1.54 \times 10^{-10}$ kg/s; $t = 51$ d

2. If in a gas phase a reactant A diffuses to a catalyst surface for dimerization, the mole flux of A is given by:

$$\phi''_{mol\ A,x} = -\mathbb{D}_A \frac{2c}{2c - c_A} \frac{dc_A}{dx}$$

    Outline the proof for this.

3. If in a desiccator, a wet plate is at a distance of $L = 0.1$ m from a layer of a drying agent, calculate the drying time if:

    total concentration $c = 4.5 \times 10^{-2}$ kmol/m^3;
    equilibrium vapour concentration of water under the prevailing conditions $c = 1.1 \times 10^{-3}$ kmol/m^3;
    diffusion coefficient of water vapour $\mathbb{D} = 2.5 \times 10^{-5}$ m^2/s;
    surface area of the plate $= 12 \times 10^{-4}$ m^2;
    amount of water to be evaporated $= 1.2 \times 10^{-3}$ kg.

    The desiccator can be considered to be isothermal and convective flow can be neglected.

    Answer: 55 h

4. In a closed gas burette, two equal volumes of a pure gas and a pure liquid are brought into contact with each other under atmospheric pressure and at room temperature. After shaking vigorously for some time, the pressure appears to have become two-thirds atmospheric, whereas the temperature has not changed. Calculate the distribution coefficient of the gas in the liquid.

    Answer: $m = $ concentration in liquid/concentration in gas phase) $= 0.5$

*5. Small bubbles (original diameter d_0) of a pure gas are brought into a liquid in which they ascend very slowly. During ascending, the bubbles disappear by absorption (Sh $= 2$). It appears that the time of solution t and the original bubble diameter satisfy the relation $d_0^2 = $ constant $\times t$ where the constant has the value 25×10^{-9} m^2/s. Give a theoretical explanation for this relationship and calculate from the constant the diffusion coefficient of the gas in the liquid, if the solubility of the gas is given by $m = 0.5$ (concentration in the gas phase divided by concentration in the liquid phase). Assume that the pressure, and consequently the concentration in the bubbles remains constant.

Answer: $\mathbb{D} = 25/16 \times 10^{-9}$ m^2/s

*6. A bottle filled with diethyl ether is connected with the surrounding air ($T = 15\,°C$) by a long vertical tube ($L = 50$ cm, $D = 2$ cm). The diffusion in the tube is determinative of the rate of evaporation of the ether; 16.5×10^{-9} kg ether evaporates per second.
   (a) How much ether (ϕ_m) would evaporate per second if the tube were twice as long and twice as wide?
   (b) What is the diffusion coefficient of ether in air?
   (c) Make a rough estimate of the time one should wait after filling the bottle before the diffusion coefficient can be measured with this set-up.
   The vapour pressure of ethyl ether at $15\,°C$ is 4.85×10^4 N/m^2.

   Answer: (a) 3.3×10^{-8} kg/s
            (b) 1.28×10^{-5} m^2/s
            (c) ca. 120 min

7. Air at $20\,°C$ flows with a velocity of 9 m/s through a smooth circular pipe of 12 cm inner diameter.
   (a) How large is the Kolmogorow length scale in this flow?
   (b) Why is it quite likely that mass transfer from particles or droplets of 10 μm diameter to this air flow can be described by $Sh \cong 2$?

   Answer: (a) $\lambda_k = 8.7 \times 10^{-5}$ m

8. Somebody measures the diffusion coefficient of water in oil by pouring a layer of oil on to a layer of water in a glass vessel ($D_u = 6.8$ cm). The vessel is put in a thermostat ($21\,°C$) and the space above the oil is kept dry with phosphorus pentoxide (P_2O_5). The weight decrease of the vessel is measured as a function of time. The results obtained are as follows:

Layer thickness (cm)	Weight loss glass vessel (g/day)
0.00052	3.14
0.13	0.0116
0.37	0.0042
0.63	0.0032
1.35	0.0046
2.26	0.0053

Calculate the diffusion coefficient of water in this oil and explain the results. The solubility of water in the oil at 20 °C is 0.004 wt%; ρ of oil $= 880$ kg/m^3.

Answer: $\mathbb{D} = 1.4 \times 10^{-9}$ m^2/s

*9. Show that John drew the right conclusion in the case of the missing lawyer discussed at the beginning of this section.

Comments on problems

Problem 1

Since the air is stagnant, Sh $= 2$ is valid (Sections IV.1.2 and IV.1.4.). We can assume the partial pressure of naphthalene to be equal to zero at a great distance from the mothball and to be equal to the saturation pressure p^* at the surface of the ball (no resistance at the interface). So, the molar flow rate is given by:

$$\phi_{mol} = kAc^* = 2\frac{\mathbb{D}}{D}\pi D^2 \frac{p}{RT}$$

which leads to $\phi_{mol} = 1.20 \times 10^{-12}$ kmol/s and, with the molar weight of naphthalene being $M = 128$ kg/kmol:

$$\phi_m = M\phi_{mol} = 1.54 \times 10^{-10} \text{ kg/s}$$

A mass balance of the mothball shows:

$$\phi_m = -\rho\frac{dV}{dt} = -4\pi\rho D^2\frac{dD}{dt} = M\phi_{mol}$$

Substituting for ϕ_{mol} the expression found above, and integrating between $t = 0$, D_0; t, D, we obtain:

$$t = \frac{\rho(D_0^2 - D^2)RT}{8M\mathbb{D}p^*}$$

Problem 5

The concentration of the pure gas A in the bubbles is $c_{A,g} = 1$ m^3/m^3 and we can assume the concentration in the bulk of the liquid to be $c_{A,L} = 0$. We further assume the concentration at the interface $c_{A,i}$ to equal the equilibrium concentration $c_A^*(= c_{A,g}/m)$ (no resistance of interface, pure gas, therefore no gas phase resistance). The volumetric flow rate of A from the bubble is then:

$$\phi_v = Akc_A^* = -\frac{dV}{dt}$$

and, using Sh $= 2$ to substitute for k, we find after integration that:

$$\mathbb{D} = \frac{\text{constant } m}{8\mathbb{D}c_{A,g}} = \frac{25}{16} \times 10^{-9} \text{ m}^2/\text{s}$$

Problem 6

We can assume diffusion of ether vapour through stagnant air and that equation (IV.11) can be applied. The partial vapour pressure of ether at the liquid surface is the equilibrium pressure p^* (no transport resistance at the interface); at the end of the tube the partial pressure is zero. In order to roughly estimate the time before the vapour starts to leave the tube we assume a sharp interface of ether vapour/air which moves at a velocity dL/dt. A material balance then shows:

$$\phi_{mol} = c^* A \frac{dL}{dt}$$

and after combining this with equation (IV.11) and integrating between $t = 0$, $L = 0$ and t, L, we find $t = 121$ min.

Problem 9 John and the case of the disappeared lawyer

The situation is illustrated by the drawing below. At the top of the cup, the water vapour pressure is zero (assuming dry air in the office); at the coffee surface, the equilibrium water vapour pressure p^* is present (~ 20 mmHg at room temperature). Water vapour diffuses through stagnant air and equation (IV.11) yields the molar flow as a function of the distance L from the top of the cup (total concentration $c = p/RT = 10^5/8.2 \times 10^3 \times 298 = 0.042$ kmol/m^3).

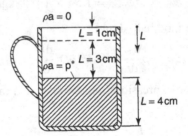

A mass balance, on the other hand, gives for the mass flow of water vapour:

$$\phi_m = \rho \frac{dV}{dt} = \rho A \frac{dL}{dt}$$

Combining both equations, we find (M = molar weight of water):

$$\phi_{mol} = \frac{A \mathbb{D} p}{LRT} \ln \frac{p - 0}{p - p^*} = \frac{\phi_m}{M} = \frac{\rho A}{M} \frac{dL}{dt}$$

Integrating with the boundary conditions $t = 0$, $L = L_1 = 1$ cm and t, $L_2 = 4$ cm, we find, with $\mathbb{D} = 2.5 \times 10^{-5}$ m^2/s, $t = 533$ h and, once again, John is right.

IV.2 Non-stationary diffusion

The skin diver, busily filling the cylinders first with oxygen and then with nitrogen at the desired pressure, told John

*that the search for the drowned man had been delayed by
the fact that no bottles with pressurized air were available.
He had found, however, cylinders with pure oxygen and
nitrogen and he would start the search in a few minutes.
John thought: the diver's cylinders are standing upright and
each has a diameter of 20 cm and a length of 60 cm. In order
to avoid another casualty, he suggested that the diver fill the
cylinders when they were in the horizontal position and wait
half an hour before using them.*

Because of the analogy between the microbalances for mass transfer
(equation (IV.16)) and that for the heat transport (equation (III.4)) the non-
stationary penetration of matter into a medium can be treated easily and quickly.
If in a rigid medium the concentration is c_{A0} and at time $t = 0$ the interfacial
concentration is brought to c_{Ai}, the mass transfer is given as a function of the
time with the solutions in Section III.2 if the following substitutions are made:

$$a = \lambda/\rho c_p \longrightarrow \mathbb{D}_A$$

$$(\rho c_p T) \longrightarrow c_A$$

$$(\rho c_p T_0) \longrightarrow c_{A0}$$

$$(\rho c_p T_1) \longrightarrow c_{Ai}$$

$$\phi_H'' \longrightarrow \phi_{mol\,A}''$$

For short times (Fo $= \mathbb{D}_A t/D^2 < 0.05$), the penetration theory applies and we
find for the mole flux:

$$\phi_{mol\,A}'' = (c_{Ai} - c_{A0})\sqrt{\frac{\mathbb{D}_A}{\pi t}} \tag{IV.21}$$

(analogous to equation (III.24)). For long times (Fo $= \mathbb{D}_A t/D^2 > 0.1$), Sh $=$
constant applies and Figures III.8 and III.9 — after the correct substitutions — can
be used. (It is customary to indicate the characteristic number for non-stationary
heat (and for mass penetration) as the Fourier number, Fo.) We see that in the
case of short times (equation (IV.21)), the mass transfer coefficient k at any time
t is given by:

$$k = \sqrt{\frac{\mathbb{D}_A}{\pi t}} \tag{IV.22}$$

If the total mass transfer operation lasts a time t_e, the average mass transfer
coefficient is found as:

$$\langle k \rangle = \frac{1}{t_e}\int_0^{t_e}\sqrt{\frac{\mathbb{D}_A}{\pi t}}\,dt = 2\sqrt{\frac{\mathbb{D}_A}{\pi t_e}} \tag{IV.23}$$

Since the penetration depth for mass transfer ($= \sqrt{\pi \mathbb{D} t}$) is usually very small, because of the low values for the diffusion coefficient and the short contact times characteristic of mass transfer (usually $\ll 1$ s), the penetration theory is even more frequently used for the description of mass transfer than for heat transfer.

Particularly in cases where the interface (because it flows) is refreshed, the penetration theory is usable. For ascending bubbles in a liquid the maximal contact time of the surface elements is given by $t_e = d/v$ ($d =$ bubble diameter, $v =$ rate of ascension, the 'jacket' of the bubble is renewed every time it has moved a diameter). For a CO_2 bubble which ascends in water at a rate of 0.5 m/s and which has a diameter of 10^{-2} m, $t_e = 2 \times 10^{-2}$ s. Furthermore, $\mathbb{D}_{CO_2} = 2 \times 10^{-9}$ m^2/s (20 °C, 1 atm), so in this case the mean mass transfer coefficient in the liquid phase $\langle k \rangle = 2\sqrt{\mathbb{D}/\pi t_e} = 3.6 \times 10^{-4}$ m/s (mass transfer coefficients for aqueous solutions under normal conditions are invariably of this order of magnitude).

Based on the idea that in many apparatuses for liquid–liquid or gas–liquid contact the interface is refreshed because there is flow, it has been proposed to introduce an age distribution for surface elements in the apparatuses, and on this basis to calculate mean mass transfer coefficients $\langle k \rangle$ with the penetration theory. If $\psi(t)\,dt$ is the fraction of the surface with ages between t and $t + dt$, according to this theory:

$$\langle k \rangle = \int_0^\infty \psi(t)\sqrt{\frac{\mathbb{D}}{\pi t}}\,dt \qquad (IV.24)$$

Usually the age distribution $\psi(t)$ is unknown. In the literature, two theories about $\psi(t)$ can be found. Higbie assumed that the probability of finding a surface element of age t for all times between $t = 0$ and $t = t_e$ (the maximum possible age) is equally great, so:

$$\psi(t)\,dt = \frac{dt}{t_e} \quad \text{for} \quad 0 \le t \le t_e$$

and:

$$\psi(t) = 0 \quad \text{for} \quad t > t_e$$

This is a supposition which is correct for, e.g. the absorption of a gas in jets of liquid, in laminar liquid films and in swarms of falling, rigid liquid droplets.

In these cases, the mean mass transfer coefficient is:

$$\langle k \rangle = 2\sqrt{\frac{\mathbb{D}}{\pi t_e}}$$

as we had already found with equation (IV.23). Danckwerts supposed that every surface element, independent of its previous history, has at any moment a chance

$s\,dt$ to disappear in the subsequent period of time dt. In this case:

$$\psi(t) = s\,e^{-st}$$

This can be realized by writing this relationship as follows:

$$-d\psi(t) = \psi(t)s\,dt$$

which in words means: 'The chance that the surface element disappears in the time between t and $t + dt$ equals the chance that it is still there at time t multiplied by $s\,dt$'. Using this supposition, we find with equation (IV.24):

$$\langle k \rangle = \sqrt{\mathbb{D}s} \qquad \text{(IV.25)}$$

The constant s is called the surface renewal frequency (unit, s^{-1}). The practical use of Danckwerts' theory is restricted to those cases where a priori something can be said about the renewal frequency; such cases are rare, unfortunately. This theory can be applied, for example, for calculating the mass transfer coefficients for liquids flowing over a packing material. If the liquid has, at any contact place between two packings, a chance p to be completely mixed, the frequency s is then given by the ratio of p and the residence time on one piece of packing.

The same is valid for a dispersion of small, separate droplets which have a certain chance (independent of their previous history) to leave the dispersion. The droplets must stay in the continuous phase for a short time only, so that they are not completely extracted. For small droplets staying in the continuous phase for a long time, $k = 2\mathbb{D}/d_p$ describes the mass transfer coefficient in the continuous phase.

IV.2.1 Problems

1. A jet of water (diameter 2 mm, temperature 20 °C) with uniform velocity distribution ($v = 5$ m/s) falls vertically through practically pure CO_2 gas under atmospheric pressure. How much CO_2 is absorbed per unit of time by the first 10 cm of the jet? The solubility of CO_2 in water at 20 °C and 1 atm CO_2 pressure is 1.73 kg CO_2/m^3. The diffusion coefficient $\mathbb{D} = 1.7 \times 10^{-9}$ m^2/s.

   Answer: 3.6×10^{-7} kg/s

2. A wall is coated on one side with a 0.2 mm thick layer of paint consisting of a very volatile component and a heavier, non-volatile component. The paint is allowed to dry, i.e. the volatile component is caused to evaporate.
   (a) Calculate with a simple physical model how long it takes before the drying front has reached the interface paint wall.
   (b) How long does it take before 99% of the volatile component has evaporated?

The diffusion coefficient of the volatile component in the paint is $\mathbb{D} = 2.2 \times 10^{-11} \text{ m}^2/\text{s}$.

Answer: (a) 10 min
        (b) 4 h

*3. A layer of water and a layer of toluene are brought together at time $t = 0$. Both layers contain 10 kg/m^3 of an iodine compound. The ratio of the equilibrium concentrations of this iodine compound in toluene and in water is 10:1; the ratio of the diffusion coefficients of the iodine compound in water and in toluene is 4. In what direction is the iodine compound transported? Calculate for relatively short times the concentration at the interface for both phases, assuming that the transport takes place by diffusion only. Sketch the concentration distribution in both phases.

Answer: $c_{iw} = 2.5 \text{ kg/m}^3$; $c_{it} = 25 \text{ kg/m}^3$

*4. A short laminar jet of pure water (temperature 20 °C) falls through pure SO$_2$. The gas has a temperature of 20 °C and a pressure of 1 atm. With what theory can the rate of absorption of the gas be described?
   (a) Calculate the surface temperature of the water jet if under the prevailing conditions the solubility of SO$_2$ in water is 1.54 kmol/m^3, the heat of solution is 6.7 kcal/mole and the Lewis number (Le $= a/\mathbb{D}$) is 90. In this calculation neglect the heat transport to the gas phase.
   (b) How much higher will the rate of absorption be if the jet falls twice as fast?
   (c) What is the influence of the jet diameter on the rate of absorption? (In practice it appears that the jet contracts because it is accelerated by gravity. What will be the influence of the contraction on the rate of absorption considering the answer to the previous question?)

Answer: (a) $T = 21.09\,°\text{C}$
        (b) $\sqrt{2}$ higher
        (c) no influence if flow rate is constant

5. The discontinuous phase of a continuously operated emulsion reactor shows the residence time distribution of an ideal mixer (63% leaves the reactor within 5 s). Give an estimate of the lower limit of the partial mass transfer coefficient in the continuous phase. Can the partial mass transfer coefficient in the discontinous phase also be estimated ($\mathbb{D} > 0.5 \times 10^{-9} \text{ m}^2/\text{s}$)?

Answer: $k \geq 10^{-5}$ m/s; yes, $\geq 10^{-5}$ m/s if $d_p > 2 \times 10^{-4}$ m

6. For large bubbles with a free-flowing interface (model: a half-sphere with diameter D) the rate of ascension is:

$$v_r = \tfrac{2}{3}\sqrt{gD}$$

Prove that the mass transfer coefficient for the free-flowing interface of these bubbles is given by:

$$\langle k \rangle = 1.6\sqrt{\frac{\mathbb{D}v_r}{D}}$$

7. A water droplet with a diameter of 2 mm falls 4 m through surrounding air (20 °C). Originally the droplet contains no oxygen. What is its mean oxygen concentration after 4 m? If it contains originally 5 mg O_2/l, what would have been its mean oxygen concentration after 4 m? The distribution coefficient of O_2 in H_2O is 0.033 (20 °C).

   Answer: 0.85 mg O_2/l; 5.34 mg O_2/l

*8. The liquid hold-up H_l in a packed column at very low gas velocities (below loading point) can be estimated from the relationships for a free-falling liquid film to be:

$$H_l = a\left(\frac{3v_0\nu}{ga}\right)^{1/3} \quad (\mathrm{m^3/m^3})$$

   where $v_0 =$ the superficial liquid velocity (m³/m² s). Outline the proof for this.
   Thus the mean residence time of the liquid is given by:

$$\tau_l = H_l\frac{L}{v_0}$$

   Estimate the mass transfer coefficient in a column packed with 1 in Raschig rings ($a = 200$ m²/m³) at a liquid load of $v_0 = 10^{-3}$ m³/m² s of water.

   Answer: $k = 4.2 \times 10^{-5}$ m/s

9. A drug has to be dosed slowly into the body of the patient; to this end, an aqueous solution of the drug is put into spherical capsules. In the patient's body the drug slowly diffuses through the wall of the capsule. The concentration of the drug outside the capsule is negligible, while the wall of the capsule may be considered as a flat membrane.
   Which percentage of the drug has diffused out of the capsule after 12 h?

   Data: Internal diameter of the capsule: $d_i = 4$ mm
   Thickness of the wall of the capsule: $d_w = 0.3$ mm
   Diffusion coefficient of the drug in water: $\mathbb{D} = 10^{-9}$ m²/s
   Diffusion coefficient of the drug in the wall of the capsule:
   $\mathbb{D}_w = 10^{-11}$ m²/s

   Answer: 88%

*10. Show that John gave the right answer to the diver in the case of the drowned man discussed at the beginning of this section.

Comments on problems

Problem 3

Since c_t/c_w at equilibrium $= 10$, the iodine compound is transported into the toluene phase. For short times we can apply the penetration theory and we find:

$$\phi''_m = (c_w - c_{iw})\sqrt{\frac{D_w}{\pi t}} = (c_{it} - c_t)\sqrt{\frac{D_t}{\pi t}}$$

Assuming equilibrium at the interface, i.e. $c_{it}/c_{iw} = 10$, we find with the given ratio of diffusion coefficients:

$$c_{it} = 25 \text{ kg/m}^3; \quad c_{iw} = 2.5 \text{ kg/m}^3$$

The penetration depth is then given by:

$$\delta = \sqrt{\pi D t}$$

and since $D_t = \frac{1}{4}D_w$, we find:

$$\delta_t = \frac{1}{2}\delta_w$$

which enables us to draw the concentration distribution curves in both phases.

Problem 4

Applying the penetration theory, the mass flux of SO_2 into the jet at any time is given by:

$$\phi''_{SO_2} = c_i\sqrt{\frac{D}{\pi t}}$$

If we assume that:

  all heat of solution is produced at the surface of the jet;
  there is no heat loss into the gas phase;
  the temperature in the centre of the jet remains constant

we can describe the transport of heat into the jet by the penetration theory:

$$\phi''_H = \rho c_p \Delta T \sqrt{\frac{a}{\pi t}}$$

Furthermore, the heat to be transported is related to the mass flux of SO_2 into the jet by:

$$\phi''_H = \phi''_{SO_2} \times \Delta H_s$$

and we find:

$$\Delta T = \frac{c_i \Delta H_s}{\rho c_p}\sqrt{\frac{D}{a}} = 1.09\,°C$$

The total mass flow into the jet is given by (using $t = L/v$):

$$\langle \phi_{SO_2} \rangle = 2c_i \pi \, dL \sqrt{\frac{Dv}{\pi L}} = 4c_i \sqrt{DL\phi_v}$$

So, if the jet velocity is doubled, ϕ_v is doubled and ϕ_{SO_2} increases by a factor of $\sqrt{2}$. We further see that, at constant ϕ_v, the diameter has no influence on the mass flow rate.

Problem 8

For the thickness of a falling laminar film we found in Section II.1:

$$d = \left(\frac{3\nu \langle v \rangle d}{g} \right)^{1/3}$$

where:

$$\langle v \rangle d = \frac{\phi_v}{\text{width}} = \frac{\phi_v L}{A} = \frac{\phi_v L}{aV} = \frac{v_0}{a}$$

Since $H_l = ad$, we find:

$$H_l = a \left(\frac{3v_0 \nu}{ga} \right)^{1/3}$$

If we assume the probability that the liquid is completely mixed after each packing particle to be $p = 1$, we find for the frequency of surface renewal:

$$s = \frac{1}{\tau_p} = \frac{L}{d_p \tau_L} = \frac{v_0}{H_l d_p} = \frac{1}{ad_p} \left\{ \frac{v_0^2 ga}{3\nu} \right\}^{\frac{1}{3}} = 1.71 \text{ s}^{-1}$$

and with Dankwaerts' theory we find:

$$k = \sqrt{Ds} \simeq \sqrt{10^{-9} \times 1.77} = 4.1 \times 10^{-5} \text{ m/s}$$

Problem 10   John and the case of the drowned man

The contents of the bottles will be practically homogeneous if:

$$Fo = \frac{Dt}{4L^2} \gtrsim 0.2$$

(see Figure III.8). Estimating the diffusion coefficient of O_2 in N_2 at 2×10^{-5} m^2/s, John thus finds $t = 4$ h for the bottles filled while standing upright. If the bottles are filled when they are in the horizontal position, the distance L over which concentration equalization has to occur is $\frac{1}{3}$ and so the time needed is $\frac{1}{9}$ of 4 h.

IV.3 Mass transfer with forced convection

> *John was busy doing work he hated, namely cleaning his*
> *pipes. He realized that the rate of growth of the tar layer*
> *in the stem was completely determined by mass transfer.*
> *Contemplating that the air flow in the pipe was laminar,*
> *he concluded that he could decrease the growth rate of*
> *the layer by a factor of 0.79 if he halved the rate of air*
> *flow. Wondering whether this measure would influence his*
> *smoking pleasure, he started to fill the pipe from the humidor*
> *on his desk.*

IV.3.1 Analogy with heat transfer

For mass transfer between a wall and a convective flow, the same analysis can
be followed as for heat transfer between a wall and a forced flow (Section III.3)
provided mass transfer is considered at low concentrations and without chemical
reaction. This can be understood from the fact that with the following 'transpo-
sitions', the entire description of the mass transfer (without chemical reaction) is
also valid for the description of the heat transfer.

Heat transfer	Mass transfer (component A)
$\rho c_p T$	c_A or ρ_A
$\lambda / \rho c_p = a$	$\mathbb{D}_A$
ϕ_H''	$\phi_{mol\ A}''$ or ϕ_{mA}''
$h / \rho c_p$	k_A

Thus, if for a certain geometric case of heat transfer with forced convection it is
concluded that, for example:

$$\text{Nu} = f_1(\text{Re, Pr, Gr, Fo, Gz, } \ldots)$$

it follows from the above that, for the same case with the same type of boundary
conditions, the following applies:

$$\text{Sh} = f_2(\text{Re, Sc, Gr, Fo, Gz, } \ldots)$$

We have already come across the Sherwood number, $\text{Sh} = kD/\mathbb{D}$, in
Section IV.1. The Schmidt number, $\text{Sc} = v/\mathbb{D}$, performs in mass transfer the
role which is played by the Pr number in heat transfer; it is a measure for the
ratio between the thicknesses of the hydrodynamic and concentration boundary
layers. For gas mixtures, Sc is about 1, while for liquids Sc is considerably higher
($10^2 - 10^3$ for normal liquids), because $\mathbb{D}$ is so much lower than v. However,

for gases Sc hardly depends on the temperature, while Sc decreases strongly for liquids if the temperature increases.

Before it can be concluded that the results for heat transfer (i.e. the function f_1) can be transposed to a reliable relationship for mass transfer (function $f_1 = f_2$), it should be ascertained whether the mass transfer process does indeed proceed entirely analogously to the (assumed to be known) heat transfer process. This means that the functions f_1 and f_2 are only equal to each other if the Re range used for mass and heat transfers is equal, and if the Sc range in which the relationship has to be applied equals the Pr range in which the heat transfer relationship applies. For gases where Sc $\approx$ Pr, this 'transposition' can be carried out frequently; the relationship for heat transfer for liquids cannot always be transposed to usable relationships for mass transfer because, for liquids, Sc $\gg$ Pr. On the basis of the analysis for heat transfer between a wall and a turbulent flow (Section III.3.2), the reader can now establish for himself that in the range $2 \times 10^3 < $ Re $ < 10^5$, Sc > 0.7, a usable relationship for mass transfer between a wall and a turbulent flow is given by:

$$Sh = 0.027 \, Re^{0.8} \, Sc^{0.33} \qquad \text{(IV.26)}$$

With these analogy relationships, mass transfer values, for example, can be calculated from theoretical or experimental data for heat transfer and vice versa. Now, particularly if Re is not too low, the local or the mean transfer coefficients over a surface can be calculated with good approximation from formulae such as:

$$Nu = C \, Re^m \, Pr^n$$
$$Sh = C \, Re^m \, Sc^n$$

The exponent m varies from ca. 0.33 (for the laminar entrance region in a pipe) via 0.5 (in the case of flow around spheres) to ca. 0.8 (for turbulent flow in pipes), but in all these cases (except with liquid metals) the exponent n of Pr (and hence also of Sc) is about equal to 0.33. Chilton and Colburn have used this for representing the analogy between heat and mass transfers in the following form. They defined a heat transfer value:

$$j_H = Nu \, Re^{-1} \, Pr^{-\frac{1}{3}}$$

and a mass transfer value:

$$j_D = Sh \, Re^{-1} \, Sc^{-\frac{1}{3}}$$

Using these definitions, the analogy for geometrically similar cases and not too low values of Re becomes:

$$j_H = j_D = C \, Re^{m-1} \qquad \text{(IV.27)}$$

For turbulent flow through pipes and along flat plates, this analogy can be extended to:

$$j_H = j_D = \tfrac{1}{2} f \qquad\qquad (IV.28)$$

where f is the so-called Fanning friction factor (see Section II.2, equation (II.22)). This analogy with momentum transfer, however, applies only in those cases where f is caused by wall friction. The flow resistance of a body caused by eddies, which can be an important part of the overall flow resistance, is not attended by analogous effects in the heat or mass transfers. So for these cases equation (IV.28) does not apply, while equation (IV.27) does.

The analogy between heat and mass transfers also applies to free convection. The density difference $\Delta\rho$ which occurs in the Grashof number (see Section III.6) might originate from concentration differences as well as from temperature differences. If there are differences both in concentration and in temperature (combined heat and mass transfers), the influence of both on Gr can be taken into account.

IV.3.2 *Mass transfer during laminar flow*

The mass transfer process can be calculated accurately for a number of simple flow situations, with the results then being collected in a database. Although a certain problem is hardly ever likely to be completely equal to a case from this database, it is possible by proper estimation to approach the actual problem with a database case. We can identify the 'chemical engineer' by this 'estimation'.

It should be borne in mind that three types of data for the velocity field near the boundary layer are of decisive importance for the mass transfer. These are, in order of importance:

(1) the velocity of the boundary layer and that of the fluid (v_i and v_x);
(2) the velocity gradient perpendicular to the boundary layer;
(3) the velocity perpendicular to the interface near the surface of the mass transfer boundary layer (v_y).

We shall discuss here the five cases which are represented in Figure IV.4. These examples contain the characteristics for the flow field just mentioned. The transfer process is described with the following mass balance (see equation (IV.16b) and the boundary conditions:

$$0 = \mathbb{D}\frac{\partial^2 c}{\partial y^2} - v_x \frac{\partial c}{\partial x} - v_y \frac{\partial c}{\partial y} \qquad\qquad (IV.29)$$

with $c = c_0$ for $y = 0$ and for all x (at the interface the equilibrium concentration prevails), $c = 0$ if $x = 0$ for all y (the inflowing liquid has the concentration

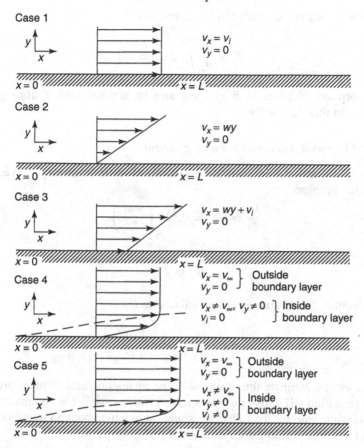

Figure IV.4 Five examples of laminar flow situations

zero), and $c = 0$ if $y = \infty$ for all x (no accumulation of substance very far from the interface).

Case 1. Uniform velocity parallel to the interface

Here $v_x = v_i$ and $v_y = 0$, and the boundary layer also moves at a velocity v_i. This case has been discussed by Higbie. By moving with the flow, it becomes a case of non-stationary diffusion (penetration theory with $t = x/v_i$). The solution is:

$$k = \sqrt{\frac{D}{\pi t}} = \sqrt{\frac{D v_i}{\pi x}}$$

or:

$$\frac{kx}{D} = \mathrm{Sh}_x = \frac{1}{\sqrt{\pi}} \left(\frac{v_i x}{v}\right)^{1/2} \left(\frac{v}{D}\right)^{1/2} = 0.565\, \mathrm{Re}_x^{1/2} \mathrm{Sc}^{1/2} \qquad (\text{IV.30})$$

The mean mass transfer coefficient is given by:

$$\langle k \rangle = \frac{1}{L} \int_0^L k \, dx = 2k_L$$

The absorption of gases in liquid jets and in laminar falling films proceeds according to this description.

Case 2. Flow with a constant velocity gradient

Here $v_x = wy$ and $v_y = 0$, and the boundary layer is stationary. Lévêque[†] has solved this problem:

$$\frac{kx}{D} = Sh_x = 0.539 \left(\frac{wx^2}{D} \right)^{1/3} \tag{IV.31}$$

and:

$$\langle k \rangle = \tfrac{3}{2} k_L$$

For a laminar, Newtonian pipe flow (diameter D):

$$w = \frac{dv_x}{dy} \bigg|_{y=0} = \frac{8 \langle v_x \rangle}{D}$$

The reader can compare this result with the analogous heat transfer problem of Graetz (equation (III.46b)). For mass transfer, this case is not so important but it is a borderline case for the following situation in which both an interface velocity (v_i) and a velocity gradient ($w = dv_x/dy|_{y=0}$) play a role.

Case 3. Interface velocity and velocity gradient

Here, $v_x = v_i + wy$ and $v_y = 0$. The complete solution of this case is known, but in practice, approximations of this solution are sufficient. These approximations apply to w or v_i being sufficiently small. The characteristic dimensionless value is $w^2 Dx/v_i^3$. If this value is lower than 1, the penetration theory (case 1) gives (to within 15%) the correct mass transfer coefficient. If this value is higher than 1, the Lévêque theory (case 2) gives the correct result to within 15%.

In general, w is of the order of magnitude of v_i/L. The dimensionless group mentioned can then be written as $D/v_i L$ (Péclet number). If we consider bubbles and droplets with a diameter of at least 1 cm and a velocity of at least 0.1 m/s, both for gases ($D \approx 10^{-5}$ m^2/s) and for liquids ($D \approx 10^{-9}$ m^2/s), this value

---

[†] *Ann. des Mines*, 1928, **13**, 201, 305, 381.

is lower than 1. So in these cases the penetration theory can be used. For gas bubbles with small diameters and velocities the penetration theory is no longer applicable.

Case 4. Stationary interface and a velocity $v_x = v_\infty$ sufficiently far from the interface

For the velocity and shear stress distributions along the interface, the following applies, according to Schlichting's boundary layer theory:

$$\tau_w = 0.332 \left(\frac{v_x x}{\nu}\right)^{-\frac{1}{2}} \frac{1}{2} \rho v_x^2$$

The following also applies:

$$\tau_w \equiv \eta \frac{dv_x}{dy}\bigg|_{y=0} = \eta \frac{v_x}{\delta_h}$$

where δ_h is the thickness of the hydrodynamic boundary layer. Furthermore, the following relationship holds:

$$\mathrm{Sh}_x = \frac{kx}{\mathbb{D}} = \frac{x}{\delta_c}$$

If the mass transfer boundary layer δ_c is much thinner than the hydrodynamic boundary layer δ_h, then the following relationship for the two boundary layer thicknesses applies:

$$\frac{\delta_h}{\delta_c} = \mathrm{Sc}^{1/3}$$

from which follows:

$$\mathrm{Sh}_x = 0.332 \, \mathrm{Re}_x^{1/2} \, \mathrm{Sc}^{1/3} \tag{IV.32}$$

and:

$$\langle k \rangle = 2k_L$$

Now $\delta_c \ll \delta_h$ only applies when $\mathrm{Sc} \gg 1$. However, for $\mathrm{Sc} = 1$, equation (IV.32) also yields reasonably accurate results.

For separate solid particles, e.g. in a fluidized bed or in a not too dense swarm of rigid liquid droplets, equation (IV.32) describes the mass transfer satisfactorily if $\mathrm{Sc} \gg 1$ has been satisfied (L is in this case the particle diameter d_p). In the case of particles in a suspension or rigid droplets in an emulsion, equation (IV.32) can also be used. The problem in these cases is that the relative velocity v_x of the particle with respect to the fluid is difficult to determine. This velocity is mainly determined by the type of stirrer, the number of revolutions, the diameter of the stirrer (D) and the diameter of the particle d_p (see also Section II.6.2).

Case 5. Moving interface and a velocity far from the interface

This problem lies between case 4 ($v_i = 0$) and case 1 ($v_i = v_\infty$). The complete solution is known.[†] In this case, no simple expression is found for the mass transfer coefficient.

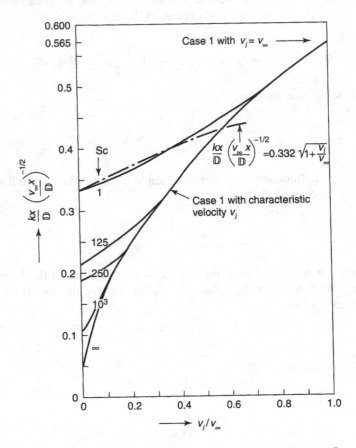

Figure IV.5   The mass transfer coefficient in the flow situation for case 5

The result of this analysis can be seen in Figure IV.5. This figure shows that for Sc > 200 the penetration theory can be applied if the interface velocity v_i is higher than 10% of the velocity of the bulk v_∞ (case 1 at velocity v_i). From this figure it also follows that for Sc = 1 (e.g. with a gas flow through a column

---

[†] *Appl. Sci. Res.*, 1961, **10A**, 241.

with packing over which liquid drips), the boundary layer theory (case 4) can be applied to predict the mass transfer in the gas phase, if we use as the characteristic velocity the sum total of the effective gas rate and the surface rate of the film $(v_\infty + v_i)$. This result has been confirmed experimentally.

IV.3.3 Mass transfer during turbulent flow

If the laminar boundary layer exceeds a certain thickness, the boundary layer flow becomes turbulent (approximately for $v_x \delta_h / v > 5 \times 10^4$). The eddies then penetrate into the area where the mass transfer takes place. The (statistical) velocity components in the y-direction contribute, together with the diffusion, to the transport perpendicular to the interface. Quite near the interface where the eddies hardly penetrate (laminar sublayer), transport takes place exclusively as a result of diffusion. At larger distances from the interface, the (statistical) transport prevails by entrainment. In between lies an area where both processes are equally important.

Since the transport by turbulent motion is statistical, just as for the diffusion, this transport is described by means of an eddy mass diffusivity, E, which depends on y. Equation (IV.30) then becomes:

$$\phi''_{mol,y} = -(\mathbb{D} + E)\frac{dc_A}{dy} \qquad \text{(IV.33)}$$

Since the transport of matter and momentum proceeds analogously by the eddies, the momentum flux can be described with the same diffusivity E:

$$\tau = -\rho(v + E)\frac{dv_x}{dy} \qquad \text{(IV.34)}$$

The next steps are to measure $v_x(y)$, to assume τ in all layers near to the interface to be constant and equal to the shear stress on the wall, to determine $E(y)$ using equation (IV.34), and then to use this result for calculating k using equation (IV.33). To this end, equation (IV.33) is written as follows:

$$\text{Sh}_x = \frac{kx}{\mathbb{D}} = \frac{x\phi''_{mol,y}}{\mathbb{D}\int_0^{c_0} dc_A} = x\left[\int_0^\infty \frac{dy}{\left(1 + \dfrac{E}{\mathbb{D}}\right)}\right]^{-1} \qquad \text{(IV.35)}$$

A drawback to this method, of course, is that particularly close to the wall where the resistance to the mass transfer is found, $v_x(y)$ is difficult to measure. As a consequence, the $E(y)$ values derived from the velocity distribution are inaccurate, which causes errors in the estimation of Sh_x by this route.

Most data are known about steady-state turbulent flow along a solid flat wall. The number of suggestions for $E(y)$ is large. All of these suggestions can be summarized as follows:

$$\frac{k}{v_\infty} = St = \frac{f/2}{1 + g(Sc)\sqrt{f/2}}$$ (IV.36)

where:

St (Stanton number) = Sh/Re Sc;
f = Fanning friction factor for the solid wall = $2\tau_w/\rho v_x^2$;
$g(Sc)$ = function of Sc.
In the case where Sc = 1, $g = 0$ applies and $2k = f v_\infty$ (Reynolds analogy).
Furthermore, g is the following function of Sc:

Sc	1	5	10	100	500	1000	2000
g	0	27	52	330 ($\pm$10%)	1000 ($\pm$15%)	1800 ($\pm$20%)	2800 ($\pm$30%)

The greater error in the function g for higher values of Sc is mainly caused by the various approximations of $E(y)$ near the wall. Deissler, for example, finds $g = 9.0$ Sc$^{3/4}$ (Sc > 200); Reichardt finds, for the same Sc range, $g = 8.8$ Sc$^{4/5}$, while Lin finds $g = 18$ Sc$^{2/3}$.

These results can be used in the description of mass transfer between a pipe wall and a turbulent flow (e.g. the free space in a packed bed can be considered as a system of irregular channels). The friction factor f in equation (IV.36) should then exclusively describe the friction resistance and not the form resistance. This factor should also be related to a Reynolds number in which the hydraulic diameter of the channels occurs as the characteristic length.

Using the analogy between heat and mass transfer given in Section IV.3.1, the correlations for heat transfer can also be used for mass transfer.

IV.3.4 Problems

1. An air freshener is suspended vertically in still air. Calculate the thickness which evaporates each day.

   Data: molecular weight: 200   vapour pressure: 10^{-3} atm
   $\nu = 1.5 \times 10^{-5}$ m^2/s $L = 25$ cm
   $\mathbb{D} = 0.5 \times 10^{-5}$ m^2/s $\rho = 1500$ kg/m^3

   Answer: 0.1 mm/day

2. The inner wall of a tube has, in the course of time, been caked with a solid layer of brine of 2 mm thickness. The heat transfer coefficient h at the wall is 1000 W/m^2°C. Calculate the time necessary to dissolve the salt if the tube is

flushed under the same conditions with distilled water. ($\rho_{salt} = 2500$ kg/m^3; the Lewis number under these conditions is Le $= a/\mathbb{D} = 100$ and the solubility of the salt is 300 kg/m^3.)

Answer: 25 min

3. A metal foil has to be etched. To this end, it is drawn through an etching liquid which flows at the same speed (1 m/s) with the foil. The etching process is fully determined by mass transfer. By what factor does the velocity of the etching process change if the foil is stopped and the liquid continues to flow (Sc $= 10^3$)?

Answer: 0.2 times

*4. The Obersalzberg, a mountain in the German/Austrian Alps near Salzburg, consists of relatively pure sea salt. The salt is mined by drilling and digging out chambers of ca. 60×60 m and 1 m height some hundreds of metres deep into the mountain. These chambers are filled with water. Fresh water is added and a 90% saturated salt solution is withdrawn continuously. Part of the mines can be visited and the guide will tell you that the height of these chambers increases at a rate of ca. 1 cm/day. What speed of solution of the salt would you predict? ($\rho_{salt} = 2165$ kg/m^3, solubility $= 300$ kg/m^3, $\rho_{saturated\ solution} = 1200$ kg/m^3, $\mathbb{D}_{NaCl} = 1.5 \times 10^{-9}$ m^2/s and $\nu = 10^{-6}$ m^2/s).

Answer: 1.48 cm/day

*5. In a diluted emulsion, the rate of circulation of the surface of the droplets is one quarter of the relative velocity between the two phases. After a small amount of surface-active substance has been added, the droplets have become rigid. By what factor has the external mass transfer coefficient changed after the addition (Sc $= 10^3$)?

Answer: 0.37 times

6. In a cylindrical vessel (diameter 2 m), a 5 mm thick crust of potassium permagenate (KMnO) crystals has deposited on the side wall in the course of a crystallization process. The crystals are removed by filling the vessel with a diluted KMnO$_4$ solution ($c_0 = 3$ kg/m^3) at 20 °C and by stirring vigorously. It is known that under the same conditions the heat transfer coefficient of the liquid to the wall is 1100 W/m^2 °C. After what time has the entire crust dissolved? The solubility of KMnO$_4$ at 20 °C is 63.2 kg/m^3, the density of crystallized KMnO$_4$ is 2700 kg/m^3, and the diffusion coefficient of KMnO$_4$ in water is 10^{-9} m^2/s. Let all other substance properties be equal to those of water at 20 °C.

Answer: 518 min

7. A liquid flows turbulently along a rough wall. The temperature of the wall rises by 5 °C, thus changing the Sc number from 10^3 to 500. How does the mass transfer coefficient change?

Answer: 1.8 times

*8. A small cooking salt crystal is dissolved at 20 °C in a large amount of clean water which is stirred thoroughly. At what value does the difference between the water and crystal temperatures adjust itself?

Data: $\mathbb{D}_{NaCl} = 1.5 \times 10^{-9}$ m²/s
$\Delta H_s = 9.3 \times 10^4$ J/kg
solubility = 300 kg/m³

Answer: 0.3 °C

9. A plate of a water-soluble solid material is put vertically in a tank filled with water, whereupon the plate gradually dissolves in the water
   (a) By which mechanism does the plate dissolve?
   (b) Derive the following equation which relates the mass concentration of dissolved material (ρ_A) and the time elapsed (t):

$$(\rho_A^* - \rho_A)^{-1/3} - \rho_A^{*-1/3} = 0.086 \frac{LB}{V} \left(\frac{\mathbb{D}_{A}^2 g}{\nu \rho} \right)^{1/3} t$$

in which ρ_A^* is the solubility of the plate material, L the length of the plate, B the width of the plate, V the water volume in the tank and ρ the density of the aqueous solution. Mass transfer from the edges of the plate may be neglected.
   (c) Calculate the time required to dissolve the entire plate.

Data: $\rho_A^* = 15$ kg/m³; $L = B = 0.30$ m;
            $\mathbb{D}_A = 1.2 \times 10^{-9}$ m²/s; $\nu = 1.0 \times 10^{-6}$ m²/s;
            $V = 0.18$ m³; $\rho = 1.0 \times 10^{-3}$ kg/m³

Initial thickness of the plate: $d = 1.0 \times 10^{-2}$ m
Density of the plate material: $\rho_s = 2.0 \times 10^3$ kg/m³

Answer: (c) 48 h

*10. Show that John drew the right conclusion in the case of his pipe, discussed at the beginning of this section.

Comments on problems

Problem 4

We assume the bulk of the salt solution to have a uniform concentration of 90% of saturation, so $c_\infty = 270$ kg/m³. At the ceiling, a thin film of liquid is completely saturated, $c_i = 300$ kg/m³. Because of the difference in density, free convection occurs, which enhances mass transfer.

By analogy with equation (III.75) we can write:

$$\langle Sh \rangle = \frac{\langle k \rangle L}{\mathbb{D}} = 0.17(Gr\ Sc)^{1/3} = 0.17 \left(\frac{L^3 g \Delta \rho}{\nu^2 \rho_i} \frac{\nu}{\mathbb{D}} \right)^{\frac{1}{3}}$$

Estimating ρ_∞:

$$\rho_\infty = 1000 + 0.9 \times 200 = 1180 \text{ kg/m}^3$$

we find:

$$\langle k \rangle = 1.23 \times 10^{-5} \text{ m/s}$$

A material balance shows that:

$$\phi_m'' = \langle k \rangle (c_i - c_\infty) = -\rho \frac{dH}{dt}$$

and we find:

$$-\frac{dH}{dt} = 1.48 \text{ cm/d}$$

The salt contains small amounts of gypsum and iron oxide which do not dissolve but settle at the bottom of the chamber. Apparently, these insoluble impurities slightly lower the speed of dissolution.

Problem 5

The drawings show the flow situations around the stagnant and the mobile droplet. These flow situations were discussed in Section IV.3.2 (Figure IV.4, cases 4 and 5). So for the stagnant droplet, equation (IV.32) is valid:

$$Sh = 0.332 \left(\frac{v_\infty d_p}{\nu} \right)^{1/2} Sc^{1/3}$$

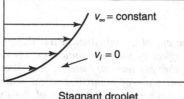

Stagnant droplet

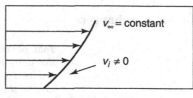

Mobile droplet

whereas for the mobile droplet, because $Sc > 200$, the penetration theory (equation (IV.30)) can be applied:

$$Sh = 0.565 \left(\frac{v_0 d_p}{\nu} \right)^{1/2} Sc^{1/2}$$

Consequently:

$$\frac{k_{\text{stagnant}}}{k_{\text{mobile}}} = \frac{0.332}{0.565} \left(\frac{v_{\infty}}{v_i}\right)^{1/2} \text{Sc}^{-1/6} = 0.37$$

because $v_i = \frac{1}{4} v_{\infty}$.

Problem 8

We assume that all heat of solution produced at the crystal–liquid interface is transported into the liquid with uniform bulk temperature T_{∞}. A simple heat balance shows:

$$\phi_H'' = h(T_i - T_{\infty}) = \phi_m'' \Delta H_s = k c_i \Delta H_s$$

Now, because of the analogue transport mechanism of heat and mass:

$$\frac{\text{Nu}}{\text{Sh}} = \left(\frac{\text{Pr}}{\text{Sc}}\right)^{1/3} \quad \text{or} \quad \frac{k}{h} = \frac{\mathbb{D}}{\lambda} \left(\frac{a}{\mathbb{D}}\right)^{1/3}$$

so:

$$T_i - T_{\infty} = \Delta T = \frac{k}{h} c_i \Delta H_s = c_i \Delta H_s \frac{\mathbb{D}}{\lambda} \left(\frac{a}{\mathbb{D}}\right)^{1/3} = 0.32\,°C$$

Problem 10   John and the case of the pipe

In the stem of John's pipe there is laminar flow with flow velocity zero at the boundary (case 1 of Section IV.3.2). For this case, the mass transfer coefficient is proportional to the third root of the velocity gradient, $w^{1/3}$ (Lévêque, equation (IV.31)). Since for laminar pipe flow, $w = 8\langle v_x\rangle/D$, the mass transfer coefficient is proportional to $\langle v_x\rangle^{1/3}$ and the rate of accumulation of tar is also proportional to $\langle v_x\rangle^{1/3}$. If, therefore, the volumetric flow rate is halved, the rate of accumulation is decreased to $(\frac{1}{2})^{1/3} = 0.79$ of its initial value.

IV.4 Mass exchangers

> *John put down his cup of tea with distaste and thought about the detrimental effect of water pollution on his drinking habits. He knew that his secretary used a coal filter to decrease the concentration of impurities in the water by a factor of 10. Remembering that because of the long dry period the level of impurities had increased by a factor of 3, he advised his secretary to increase the length of the coal column by 48%.*

The calculation of mass exchangers in which no chemical reactions occur proceeds analogously to the calculation of heat exchangers. Mass transfer usually takes place in apparatus in which the two exchanging phases are brought into close contact with each other. To this end, the one phase is finely dispersed in the other so that a large exchange area is formed.

For selecting the correct type of mass transfer apparatus and for determining the main dimensions, the following data are of importance:

(a) the thermodynamic equilibrium between the phases in question under the prevailing conditions (temperature and pressure);
(b) the rate of mass transport in the chosen apparatus as a function of the operational variables (e.g. flow rate, stirring intensity, etc.).

IV.4.1 Thermodynamic equilibrium

During heat transport under equilibrium conditions, the temperatures in all phases are equal, i.e. $T_a = T_b$, $\Delta T = 0$. During mass transport under equilibrium conditions the chemical potentials of a given substance in both phases are equal. These chemical potentials are seldom known and therefore the more practical concept of a distribution coefficient (see also Section IV.1.2) was introduced. We will use here exclusively the simplest correlation possible, $c_A = mc_B$. This simplifies the calculations considerably but it is not essential for the approach chosen.

For a large number of cases the m value, and consequently the slope (gradient) of the equilibrium line, is constant at constant temperature. In such cases there are sufficient thermodynamic data for calculating apparatus for physical absorption if the value of m, which is sometimes also called the Bunsen absorption coefficient, is known as a function of the temperature. Table IV.1 gives a survey of distribution coefficients of some gases in water.

Table IV.1 Distribution coefficient m for various gases in water at 20 °C; concentration in water divided by concentration in gas phase at equilibrium

Gas	m
Air	0.019
Ammonia	735
Carbon dioxide	0.92
Chlorine	2.41
Hydrogen	0.019
Hydrogen chloride	462
Nitrogen	0.016
Oxygen	0.033
Sulphur dioxide	41.4
Hydrogen sulphide	2.6

Table IV.1 shows that solubilities may differ by a factor of 10^4 and that a small difference in chemical structure of the component to be absorbed may have a great influence on the solubility.

As soon as the thermodynamics are known, some considerations regarding the attainable equilibrium should be made prior to selecting the type of mass exchanger. The thermodynamic equilibrium between the phases under the conditions prevailing in the apparatus indicates what could be achieved in the case of very intense contact between the phases at very long contact times. To this end, we take as an example the case where a well-mixed gas flow ϕ_g with inlet concentration c_0 comes to equilibrium by absorption, with a likewise well-mixed liquid flow ϕ_L with inlet concentration zero.

From a mass balance (mass flow in = mass flow out), one derives that the concentration in the emerging liquid is given by:

$$c = c_0 \left(1 + \frac{m\phi_{vl}}{\phi_{vg}} \right)^{-1} \tag{IV.37}$$

To ensure the highest possible concentration in the liquid, e.g. the oxygen concentration in the manufacturing of drinking water, the ratio $m\phi_{vl}/\phi_{vg}$, known as the extraction factor, should be as small as possible. As a consequence, the ratio of the volume flows ϕ_{vl}/ϕ_{vg} to be chosen for a sound design depends on the distribution coefficient of the gas to be absorbed. It is left to the reader to analyze desorption or scrubbing in a similar way. The limitations of this kind of reasoning are, of course; that in absorption the feed flow is not too highly exhausted, while in desorption the scrub flow is not too highly saturated.

IV.4.2 Choice of the apparatus

The type of apparatus with which the mass transfer process has to be carried out is chosen on the basis of a number of semi-quantitative considerations. If one is not good at selecting the right type of apparatus, further development of the apparatus does not help much. The final result will then be the best apparatus of a poor type.

As a first step in the selection procedure, the degree of exhaustion is considered. If $m\phi_L/\phi_g$ is high, the g phase contains ultimately only little of the substance to be transferred, which can be the aim of the process. If, on the other hand, the degree of exhaustion is low, the L phase becomes almost saturated under the conditions of the ingoing gas, which can also be the aim of the process.

Since when choosing physical absorption apparatus it has first to be decided whether the liquid or the gas has to be distributed, it is clear on the basis of the above considerations that readily soluble gases can only be transferred efficiently at high values of ϕ_{vg}/ϕ_{vL}, so that the liquid will be distributed. For some types of gas–liquid contact apparatus, approximate values are given in Table IV.2, with

Table IV.2 Some approximate values for gas–liquid contact apparatus. The flow rate is $v_{0L} = 2 \times 10^{-2}$ m^3/m^{2}s

Type of contactor	Maximum gas flow rate v_{0g} (m^3/m^{2}s)	Maximum interfacial area a (m^2/m^3)	Liquid fraction (m^3/m^3)	Power consumption per kg liquid (W/kg)
Bubble column	0.20	300	>0.6	2
Distillation stage (with bubble caps)	0.3	75	>0.6	2.5
Stirred tank reactor	0.15	1000	>0.5	6
Sieve plate	1	150	>0.15	1
Falling film	2	100	<0.05	1
Packed column (two phases)	1	250	<0.05	1.5

which the semi-quantitative considerations, on choosing the apparatus, can be supported.

If on the basis of thermodynamic considerations the preference of some types of apparatus has been expressed, some other consideration may play a part, such as the pressure drop over the apparatus, the heat transfer, corrosion and the question of whether a countercurrent way of contacting the phases is possible.

IV.4.3 Size determination of the mass exchanger

If on the basis of the complex of qualitative and quantitative considerations discussed in Section IV.4.2 the type of apparatus has been determined, we have to know the mass transfer coefficient and the specific interface as a function of the operational conditions of the apparatus, so that this apparatus can be designed.

In the case of physical gas absorption the resistance to mass transfer is mostly entirely on the liquid side and the total transfer coefficient related to the liquid phase equals the partial transfer coefficient in this phase. The values of the mass transfer coefficients can be calculated from the data in Section IV.4.2 and appear to be of the order of 10^{-4} m/s for all practical cases of gas absorption. The same relationships can also be used for calculating mass transfer coefficients between two liquid phases if their flow behaviour is known.

The area of the interface between the two phases is more difficult to estimate and Table IV.2 only gives rough indications. For gas–liquid contacting in a packed column, the surface of the packing material can be taken as the actual interface (accurate within ±20%). The size of the gas–liquid interface in a bubble column or in a stirred vessel is strongly dependent on power input, gas flow rate, gas hold-up and properties of the liquid. Reliable correlations for the estimation of the size of the interface over a wide range of operating conditions are still lacking. If the volume fraction of gas $(1-\varepsilon)$ and the average diameter of the gas bubbles is known, the specific interface can be calculated from:

$$a = \frac{6(1-\varepsilon)}{d_p} \tag{IV.38}$$

This relationship can also be used for calculating the size of the specific interface for dispersions of liquids. Having determined the size of the mass transfer interface, we can now find the necessary dimensions of the mass exchanger from a mass balance (equation (IV.1)).

As a first example, we take the case in which two phases 1 and 2 with flow rates ϕ_{v1} and ϕ_{v2}, respectively, are contacted in countercurrent plug flow in a column. For this case we can set up the following mass balance (see also Figure IV.6) for phase 1 (K_1 = total mass transfer coefficient based on phase 1) over a differential height dx:

$$-\phi_{v1}\mathrm{d}c_1 = K_1 a(c_1 - mc_2)A\,\mathrm{d}x$$

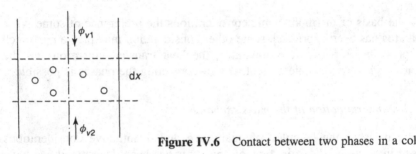

Figure IV.6 Contact between two phases in a column

Integrating over the total apparatus we obtain:

$$-\int_{c_1(x=0)}^{c_1(x=L)} \frac{\mathrm{d}c_1}{c_1 - mc_2} = \int_0^L \frac{K_1 a}{\phi_{v1}} A\,\mathrm{d}x = \int_0^L \frac{\mathrm{d}x}{(HTU)_1} = \frac{L}{(HTU)_1} \tag{IV.39}$$

The terms of equation (IV.39) are dimensionless. The left-hand term is called the number of transfer units (NTU). The right-hand term is a quotient of two lengths: the length of the column (L) and the height of a transfer unit (HTU). The HTU is a measure of the efficiency of the column (small HTU → good transfer). The HTU can be defined as that length of the column for which the integral of the left-hand term of equation (IV.39) has the value of 1.

The HTU as used in equation (IV.39) is called an 'overall' HTU because the overall driving force $(c_1 - mc_2)$ has been used (here related to the '2' phase). If we had only considered one phase (e.g. the '1' phase) with driving force $= c_1 - c_{1i}$ (where c_{1i} is the concentration at the interface of phases 1 and 2) we would have obtained a partial height of a transfer unit $(htu)_1$.

It can easily be derived from equations (IV.15) and (IV.39) that:

$$(HTU)_1 = (htu)_1 + E(htu)_2 \qquad (IV.40)$$

where $E(= m\phi_1/\phi_2)$ is the extraction factor. So the HTU is dependent on the extraction factor.

In fact, equation (IV.39) is the same as the design equation for a plug flow heat exchanger obtained in Section III.4. This will now be shown. Integration of equation (IV.39) gives:

$$\frac{c_{10} - c_{1L}}{\Delta c_0 - \Delta c_L} \ln \frac{\Delta c_0}{\Delta c_L} = \frac{L}{HTU} \qquad (IV.41)$$

where

$$\Delta c_0 = c_{10} - mc_{20}$$

$$\Delta c_L = c_{1L} - mc_{2L}$$

If we set up a material balance over the entire apparatus we obtain:

$$0 = \phi_{v1} c_{10} - \phi_{v1} c_{1L} + \phi_m$$

where ϕ_m is the mass flow from phase 1 to phase 2. Substituting the concentration difference, $c_{10} - c_{1L}$, with the help of equation (IV.41) we find for the mass flow rate, if both phases are in countercurrent plug flow:

$$\phi_m = \phi_{v1} \frac{L}{HTU} \frac{\Delta c_0 - \Delta c_L}{\ln \dfrac{\Delta c_0}{\Delta c_L}} = K_1 aAL \frac{\Delta c_0 - \Delta c_L}{\ln \dfrac{\Delta c_0}{\Delta c_L}} = K_1 aAL (\Delta c)_{\log} \qquad (IV.42)$$

which can be compared with equation (III.61).

In the foregoing, we have neglected the influence of liquid dispersion. If dispersion or longitudinal mixing occurs in the flows, the overall driving force in the apparatus will decrease (as we have already discussed in Section III.4.2). Equation (IV.39) does not adequately describe such a case. However, for practical purposes, cases in which dispersion has to be taken into account can still be written in the form of this equation:

$$\int_{c_1(x=0)}^{c_1(x=L)} \frac{dc_1}{c_1 - mc_2} = \frac{L}{HETU} \qquad (IV.43)$$

which is the definition of the 'height equivalent of a transfer unit' (HETU). This HETU is, in general, the sum of the HTU and of terms which account for the mixing in both phases.

As a second example, we treat again the situation of Figure IV.6, but now for the case that the two flows can be treated as plug flows in combination with a dispersion process with dispersion coefficients D_1 and D_2, respectively. For this case the following result is obtained:

$$\text{HETU} = \text{HTU} + \frac{D_1}{v_1} + \frac{D_2}{v_2} \qquad \text{(IV.44)}$$

The determination of these dispersion coefficients has already been discussed in Section II.7.6. There we saw (equation (II.112)) that the height of a mixing unit is given by:

$$\text{HMU} = \frac{L}{n}$$

where n is the number of perfect mixers in series yielding a corresponding residence time distribution. For small dispersion, the residence time distribution (see Section II.6) for n perfect mixers in series approximates to that of a pipe flow (velocity v, length L) with a dispersion coefficient D given by:

$$\frac{1}{n} \simeq 2\frac{D}{vL}$$

Consequently, in this case we find (analogous to equation (III.64)):

$$\text{HETU} = \text{HTU} + \tfrac{1}{2}(\text{HMU})_1 + \tfrac{1}{2}(\text{HMU})_2 \qquad \text{(IV.45)}$$

For the system water–air in a packed column, $(\text{HMU})_{\text{water}} \approx d_p$ and $d_p < (\text{HMU})_{\text{air}} < 5d_p$. Since the height of a transfer unit is $8d_p < \text{HTU} < 40d_p$, in practically all cases plug flow of the liquid phase can be assumed.

Of course, for the exceptional case that one of the phases is perfectly mixed, the foregoing analysis (which is restricted to small dispersion coefficients) is not of much help. In such a case, equation (IV.43) and the HETU can no longer be used. However, the approach here is again very simple. Considering that the bulk concentration of the well-mixed phase is uniform (and equal to its exit concentration) we can write down the mass balance for the other phase over a differential height dx. This equation can then be integrated in a straightforward way, because it contains only one concentration which is dependent on the location. If, as our third example, the other phase flows in plug flow, equation (IV.39) results again, with the index 2 now indicating the perfectly mixed phase and c_2 being constant, e.g. $c_2(x = L)$.

IV.4.4 The concept of theoretical plates

Another frequently used expression in the calculation of continuous mass exchangers is the 'height equivalent of a theoretical plate' (HETP). One HETP is that column height at which the outgoing flows concentration comes up to the equilibrium condition. This has been elucidated in Figure IV.7. which shows a material balance over a segment of a mass exchanger with the length of 1 HETP. So the equilibrium condition is:

$$c_1(x) = mc_2(x + \text{HETP})$$

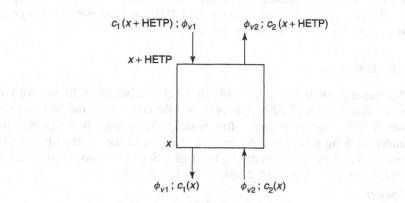

Figure IV.7 Material balance of a mass exchanger over a height of one HETP

In order to determine the correlation between HETP and HTU, use is made of the definition equation (IV.40) of the HTU:

$$-\frac{\text{HETP}}{\text{HTU}} = \int_x^{x+\text{HETP}} \frac{dc_1}{c_1 - mc_2} \tag{IV.46}$$

The integral on the right-hand side of equation (IV.46) can be calculated if it is borne in mind that a relationship between c_1 and c_2 can be found from a mass balance. For the case of plug flow, this mass balance reads:

$$-\phi_{v1}c_1 + \phi_{v2}c_2 = \text{constant}$$

Equation (IV.46) can then be elaborated to:

$$\frac{\text{HETP}}{\text{HTU}} = \frac{1}{1-E} \ln \left[\frac{(1-E)c_1(x + \text{HETP}) - mc_2(x + \text{HETP}) + Ec_1(x + \text{HETP})}{(1-E)c_1(x) - mc_2(x + \text{HETP}) + Ec_1(x + \text{HETP})} \right]$$

and using the definition of the HETP, $c_1(x) = mc_2(x + \text{HETP})$, this can be reduced to:

$$\frac{\text{HETP}}{\text{HTU}} = \frac{1}{1-E} \ln \left[\frac{1}{E} \frac{c_1(x + \text{HETP}) - mc_2(x + \text{HETP})}{c_1(x + \text{HETP}) - mc_2(x + \text{HETP})} \right] = \frac{\ln E}{E - 1} \tag{IV.47}$$

So only if the extraction factor approaches the value of 1, which will be chosen in many practical cases, will HETP $\approx$ HTU.

For packing materials such as Raschig rings and Berl saddles, the HTU is of the order of magnitude of 6–60 cm, with 15 cm being a good average. The reader may check for himself what this means for the HETP if E is respectively, 0.25, 0.50, 0.75 and 1.5.

Because the HETP is a stronger function of the extraction factor E than the HTU, the latter should be considered as a fundamentally more correct expression for describing the operation of an extraction column.

When heat and mass transfer are taking place simultaneously the concept of theoretical plates cannot be used because the HETP for mass transfer will generally differ from the HETP for heat transfer.

IV.4.5 Problems

1. A column is filled with a layer of silica gel to a height of 30 cm. Air with a moisture content of 1000 ppm is passed through. When the air flows very slowly the moisture content after passing the column is 0.1 ppm, while under working conditions the moisture content of the air after passing the column is 10 ppm. Calculate the layer of silica gel necessary to obtain air with a moisture content of 2 ppm.

   Answer: 41 cm

2. A H_2S-containing water stream (concentration of H_2S is 50 g/m^3, temperature 25 °C) is treated countercurrently with air in order to desorb the H_2S. At 25 °C the distribution coefficient of H_2S (= concentration in air: concentration in water) is 0.44.
   (a) Calculate the minimum amount of air necessary for fully degassing 1 m^3 water.
   (b) What is the lowest concentration of H_2S which can be achieved in the discharged water if 1 m^3 air is used per m^3 water?

   Answer: (a) 2.3 m^3/m^3
   (b) 28 g/m^3

3. Air flows through a column filled with benzoic acid spheres. The outgoing flow appears to be for 90% saturated with benzoic acid vapour. Now, instead of air, water is passed through the column in such a way that the Reynolds number is the same for both flows. What is the degree of saturation of the effluent water with benzoic acid?
   Data:

	Kinematic viscosity	Diffusion coefficient of benzoic acid
Water	1×10^{-6} m^2/s	1×10^{-9} m^2/s
Air	14×10^{-6} m^2/s	2×10^{-5} m^2/s

   Answer: 8.1%

*4. A water stream of 1000 l/h must be freed from Ca^{2+} ions. The initial concentration is 500 mg/l, while the required final concentration is 10 mg/l. This process is carried out in a column filled with ion exchanger consisting of almost spherical particles. The Ca^{2+} ion concentration, which is in equilibrium with the exchanger, is negligibly small. The rate-determining step during ion exchange is the mass transfer from liquid to the particles. Determine the product of length and diameter of the column required for this process.

Data: $d_p = 2 \times 10^{-3}$ m, volume fraction of voids, $\varepsilon = 0.4$
$$\mathbb{D}_{Ca^{2+}ions} = 10^{-9} \ m^2/s. \ \nu = 10^{-6} \ m^2/s$$

Answer: $DH = 0.12 \ m^2$

5. An extraction is carried out by causing droplets of water to descend through an oil layer. The resistance to mass transfer is entirely in the oil phase. The mass transfer can be described with the penetration theory by introducing as contact time: the time a droplet needs to cover a distance equal to twice its own diameter. The flow around the droplets can be considered as a laminar flow round a rigid sphere (Stokes' law applicable).
   (a) By what factor does the mass transfer coefficient increase or decrease when the droplets are made twice as small?
   (b) How much does the product ka (i.e. the product of transfer coefficient and total surface area of all droplets in 1 m^3 of the extraction column) increase or decrease if at this droplet size reduction the total water flow (m^3/s) remains unchanged?

Answer: (a) 0.71 times as great
   (b) 5.7 times as great

*6. An organic liquid has to be saturated to 90% with a certain substance A. However, only a saturated solution of A in water is available. The organic liquid which is insoluble in water is now caused to ascend in small rigid droplets (diameter: $d_p = 2$ mm) through the saturated solution of A in water (rate of rise: $v = 10^{-2}$ m/s). What should be the height of the water column to attain a 90% saturation of the organic liquid with A? The diffusion coefficient of A both in water and in the organic liquid is $10^{-9} \ m^2/s$ and the solubility of A in the organic liquid is 200 times as low as in water. The continuous water phase is present in excess and is not exhausted. The initial concentration of A in the droplets is zero.

Answer: 2 m

7. In order to saturate an air stream with water vapour, the air is passed through a wash bottle filled to a height of 10 cm with pure water at 25 °C. The air is divided through a sieve plate into bubbles with a diameter of 3 mm.

The bubbles behave as rigid spheres. The volume fraction of water in the dispersion is 0.8.

(a) Calculate the rate of rise of the swarm of bubbles by using the relationship of Richardson and Zaki (Section II.5.5).

(b) What liquid height is necessary to ensure a relative humidity of 99.9% in the outflowing air?

(c) If the water is originally air-free, how long does it take to saturate the liquid for 99% of air.

Data: $\mathbb{D}_{H_2O \text{ in air}} = 2.5 \times 10^{-5} \text{ m}^2/\text{s}$
$\mathbb{D}_{\text{air in } H_2O} = 5 \times 10^{-9} \text{ m}^2/\text{s}$

Answer: (a) 0.175 m/s
         (b) 1.1 cm
         (c) 121 s

8. A dry air stream must be saturated with 99% water vapour. This has to be attained by passing the air flow through a wetted wall column (a pipe in which a liquid film flows downwards along the inside wall). The thickness of the liquid film is very small compared with the column diameter.

Data: $\phi_{v \text{ air}} = 6.9 \times 10^{-3} \text{ m}^3/\text{s}$, $D_{\text{internal column}} = 5 \times 10^{-2} \text{ m}$
$\rho_{\text{air}} = 1 \text{ kg/m}^3$, $\eta_{\text{air}} = 17 \times 10^{-6} \text{ Ns/m}^2$
$\mathbb{D}_{H_2O \text{ in air}} = 2.5 \times 10^{-5} \text{ m}^2/\text{s}$

For the mass transfer coefficient in wetted wall columns, the following applies, according to Gilliand and Sherwood (*Ind. Eng. Chem.*, 1934, **26**, 516):

$$Sh = 0.023 \; Re^{0.83} \; Sc^{0.44} \left(\begin{matrix} 2 \times 10^3 < Re < 3.5 \times 10^4 \\ 0.60 < Sc < 2.5 \end{matrix} \right)$$

(a) What is the required length of the column?

(b) A physicist maintains that the column becomes undoubtedly much shorter if a recirculation of the gas flow over the column is applied. Show quantitatively whether his statement is correct for a recirculation ratio 1:1 (i.e. just as much gas is recirculated as leaves the apparatus).

(c) What column length would be calculated as the answer to question (a), if instead of the given correlation for Sh, use was made of the analogy between mass and heat transfers?

Answer: (a) 9.6 m; (b) 9.2 m; (c) 10.5 m

*9. An SO_3-air mixture containing 7 mol% SO_3 is introduced at the bottom of an absorption column which is filled with 1.5 in Raschig rings ($a = 140 \text{ m}^2/\text{m}^3$, $\varepsilon = 0.92$). The gas rate calculated on the empty column is $v_0 = 2$ m/s. Concentrated H_2SO_4 is passed downwards over the Raschig rings. The SO_3 equilibrium pressure of strong H_2SO_4 is almost zero.

(a) What length of column is required to absorb 98% of the SO_3 passed through if plug flow of the gas phase is assumed?

(b) What length of column would be required if measurements on a 2 m long column had shown that the residence time distribution resembled that of ten perfect mixers connected in series?

Answer: (a) 3.65 m (b) 4.05 m

10. An organic pollutant is removed from an effluent (water) by extracting it with an organic solvent which is immiscible with water. The extraction process takes place in an apparatus in which both liquids are separated by porous membranes. The membranes are shaped as hollow fibres. The pores of the membranes are wetted by the organic solvent which flows inside the fibres, while the effluent flows outside and perpendicular to the fibres. Consider the membrane to be a flat plate of which the diameter is the average of the inner and outer diameter of the membrane.

(a) Calculate the (average) mass transfer coefficients in the hollow fibres (k_i), in the membrane (k_m) and outside the membrane (k_e).

(b) Calculate the total mass transfer coefficient related to the water phase and to the outer diameter of the membrane.

(c) Show quantitatively why wetting of the membrane by the solvent is to be preferred.

Data: Inner diameter of hollow fiber: $d_i = 4 \times 10^{-4}$ m
Outer diameter of hollow fiber: $d_e = 5 \times 10^{-4}$ m
Length of hollow fibers: $L = 0.30$ m
Velocity of organic solvent: $v_i = 8 \times 10^{-2}$ m/s
Velocity of water: $v_e = 1.5$ m/s
Distribution coefficient of pollutant: $m = \dfrac{c(\text{solvent})}{c(\text{water})} = 200$
Diffusion coefficient of pollutant in water: $D_w = 1.2 \times 10^{-9}$ m^2/s
Diffusion coefficient of pollutant in solvent: $D_o = 1.2 \times 10^{-9}$ m^2/s
Effective diffusion coefficient of pollutant in membrane:
$D_m = 4 \times 10^{-10}$ m^2/s

Answer: (a) $\langle k_i \rangle = 1.6 \times 10^{-5}$ m/s; $k_m = 8 \times 10^{-6}$ m/s $\langle k_e \rangle = 3.5 \times 10^{-4}$ m/s
(b) $K_e = 2.5 \times 10^{-4}$ m/s
(c) $K_e = 7 \times 10^{-6}$ m/s, if the membrane is wetted by water instead of solvent

11. Air bubbles containing less than 10% CO_2 rise through a column filled with water. In the first case, the bubbles rise in clean water so that their surface is completely mobile. In the second case, the water contains a surface-active agent which makes the bubbles rigid. The mass transfer resistance inside the bubbles may be neglected.

(a) Make the best possible estimates of the ratio of the mass transfer coefficients for both cases.

(b) Calculate the fraction of CO_2 which is not absorbed after the bubbles leave the column by using the (estimated) mass transfer coefficients for both cases.

(c) Show that the mass transfer resistance inside the bubbles can be neglected.

Data: Diameter of the air bubbles for both cases: $d_b = 6$ mm
Velocity of rise of mobile bubbles: $v_1 = 0.25$ m/s
Velocity of rise of rigid bubbles: $v_2 = 0.18$ m/s
Water height in the column: $h = 2.0$ m
Distribution coefficient of CO_2 : $m = 1.0$

Answer: (a) $k_1/k_2 = 7$
(b) Case 1: ~0.10;   case 2: ~0.6

*12. Show that John reached the right conclusion during his deliberations about his cup of tea, discussed at the beginning of this section.

Comments on problems

Problem 4

A mass balance over a short height dx of the column yields after integration ($c = c_0$ at $H = 0$):

$$\frac{c}{c_0} = \exp\left(-\frac{ka\pi D^2 H}{4\phi_v}\right)$$

With $c/c_0 = 0.02$ and $a = 6(1 - \varepsilon)/d_p = 1800$ m^2/m^3 we find:

$$kD^2 H = 7.65 \times 10^{-7} \text{ m}^4/\text{s} \tag{1}$$

The mass transfer process is analogous to heat transfer during flow through a packed bed, so we can adapt equation (III.69a) for the estimation of the mass transfer coefficient:

$$\frac{\langle k \rangle d_p}{\mathbb{D}} = 1.8 \left(\frac{v_0 d_p}{\nu}\right)^{1/2} \left(\frac{\nu}{\mathbb{D}}\right)^{1/3}$$

and we find:

$$kD = 6.3 \times 10^{-6} \text{ m}^2/\text{s} \tag{2}$$

Combining this with equation (1), we obtain for the desired product:

$$DH = 0.12 \text{ m}^2$$

So a small column of, for example, $D = 0.2$ m and $H = 0.6$ m, would be suitable for this purpose.

Problem 6

The only difficulty in this problem is to decide which mass transfer coefficient is important. This problem has already been discussed in Section IV.1.2, so we can be very brief here. The following figure shows the concentrations that play a role and, since there is no accumulation of matter at the interface, we can write:

$$\phi''_m = k_1(mc_2^* - c_1) = k_2(c_2 - c_2^*)$$

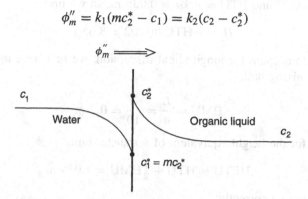

Eliminating the unknown c_2^*, we find for the mass flux:

$$\phi''_m = \frac{mc_2 - c_1}{\dfrac{m}{k_2} + \dfrac{1}{k_1}}$$

Since the diffusion coefficient of the transported compound is the same in both phases, $k_1 = k_2$, but because $m = c_1^*/c_2^* = 1/200$, the mass transport in the drop-let is rate-determining and equation (1) becomes:

$$\phi''_m = k_1(mc_2 - c_1)$$

The concentration equalization in the droplets is:

$$\frac{c_1 - \langle c \rangle}{c_1 - c_0} = \frac{c_2^* - 0.9c_2^*}{c_2^*} = 0.1$$

and we find from Figure III.8 that Fo $= \mathbb{D}t/d_p^2 = 0.05$ and $L = vt = 2$ m.

Problem 9

The rate of absorption is determined by the gas-phase mass transfer resistance because the SO_3 vapour pressure of concentrated H_2SO_4 is practically zero. This mass transfer coefficient can be calculated with the help of the boundary layer theory (equation (IV.32); see Section IV.3.2, case 5) by using instead of v_∞ the sum $v_i + v_\infty$. Here we can assume $v_i \ll v_\infty$ and we find ($v_\infty = v/\varepsilon = 2.18$ m/s):

$$k = 0.765 \times 10^{-2} \text{ m/s} \quad \text{and} \quad \langle k \rangle = 1.53 \times 10^{-2} \text{ m/s}$$

A mass balance over the differential height dx of the column yields after integration ($c = c_0$ at $H = 0$):

$$\frac{c}{c_0} = \exp\left(-\frac{ka\pi D^2}{4\phi_v}H\right) = \exp\left(-\frac{ka}{v}H\right) = \exp\left(-\frac{H}{\text{HTU}}\right)$$

Now $c/c_0 = 0.02$ and HTU $= v/ka = 0.935$ m, so we find:

$$H = -\text{HTU}\ln 0.02 = 3.65 \text{ m}$$

If we want to account for longitudinal dispersion, we first have to calculate the height of a mixing unit:

$$\text{HMU} = \frac{L}{n} = \frac{2}{10} = 0.2 \text{ m}$$

So we find for the height equivalent of a transfer unit:

$$\text{HETU} = \text{HTU} + \tfrac{1}{2}\text{HMU} = 1.035 \text{ m}$$

and, writing more correctly:

$$\frac{c}{c_0} = \exp\left(-\frac{H}{\text{HETU}}\right)$$

we find $H = 4.05$ m.

Problem 12 John and his cup of tea

We assume the mass transfer resistance to be situated outside the coal particles. A material balance over a differential height element dx of the coal column then leads (with $c = c_0$ at $H = 0$) to the expression:

$$\frac{c}{c_0} = \exp\left(-\frac{ka\pi D^2}{4\phi_v}\right) = \exp\left(-\frac{H}{\text{HTU}}\right)$$

The height of a transfer unit stays constant and from the information about the normal operation of the column ($c = 0.01c_0$) we find that $H_1 = 2.3$ HTU. With the impure water, for the same final concentration (now $c = 0.0033 \, c_0$) a column height of $H_2 = 3.4$ HTU is required, i.e. 48% more.

IV.5 Mass transfer with chemical reaction

John looked at the fiercely burning haystack. The police agent informed him that at 5 am, when he discovered the fire, the stack appeared to be 4 m high. Now, 3 hours later the stack was still 2.5 m high. John thought: the initial height of the stack was 6 m and therefore the fire must have started at 1 am or later. Lighting his pipe and carefully extinguishing the match, he inquired about what sum the hay was insured.

Mass transport can be increased if the substance transported into another phase reacts there chemically, but a chemical reaction can be slowed down by a mass transfer resistance (the transport rate is then lower than the conversion rate). We can distinguish here between homogeneous and heterogeneous reactions.

During homogeneous reactions, the chemical conversion occurs at all places where the reactants are present. The reaction rate term has to be included in the mass balance, equation (IV.1), which for the situations we shall discuss can be simplified to:

$$\mathbb{D}_A \left(\frac{d^2 c_A}{dx^2} \right) + R_A = 0 \qquad \text{(IV.48)}$$

(no concentration gradient in the y- and z-directions, no flow velocity, stationary state). Since the transported substance is converted at all places, the concentration gradient is increased and a sufficiently fast reaction will therefore increase mass transfer.

During heterogeneous reactions, diffusion and reaction are separated. The reaction occurs at a reaction interface only (e.g. at the catalyst surface) and the mass transfer coefficient is not influenced by the chemical reaction.

The volumetric production rate in equation (IV.48) is, for a first-order chemical reaction, given by:

$$R_A = -k_r c_A \qquad \text{(IV.49)}$$

For an nth-order reaction, this expression becomes:

$$R_A = -k_r c_A^n \qquad \text{(IV.50)}$$

whereas the reaction rate for a second-order bimolecular reaction between reactants A and B is given by:

$$R_A = -k_r c_A c_B \qquad \text{(IV.51)}$$

We will restrict ourselves here to irreversible homogeneous reactions of different order ($n = 1, 2, 3, \ldots$, etc., and an intermediate case between zero and first order) and to heterogeneous reactions. For analyzing these problems, we will make use of the film model (Section IV.1.4), i.e. we will assume that the bulk of the liquid is well mixed and that all mass transfer resistance is localized in a thin boundary layer of thickness δ_c. Thus, the mass transfer coefficient is given by $k = \mathbb{D}D/\delta_c$.[†]

[†] In the following, the subscript c will be omitted so as not to give rise to confusion.

For homogeneous reactions we can distinguish two extreme cases: the chemical reaction is so slow that it occurs mainly in the bulk of the liquid or the reaction is so fast that it occurs partly or completely in the boundary layer.

IV.5.1 Slow homogeneous first-order reactions

If the chemical reaction is slow, we can neglect the reaction occurring in the thin boundary layer and assume that the entire reaction takes place in the bulk of the liquid. In this case, we will find a concentration gradient over the laminar boundary layer, as shown in Figure IV.8. At the interface the concentration is c_{Ai} and on the other side of the film the concentration equals the bulk concentration. The assumption of negligible reaction in the film is justified, if the amount transported through the film ($\sim Ak_A c_{Ai}$) is much bigger than the amount of A converted in the film by chemical reaction ($\sim A\delta k_r c_{Ai}$). Thus, if:

$$A\delta k_r c_{Ai} \ll Ak_A c_{Ai}$$

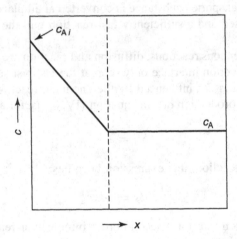

Figure IV.8 Concentration distribution during mass transfer with slow homogeneous first-order chemical reaction

or, by making use of the relationship $\delta = \mathbb{D}/k_A$, if:

$$\frac{\delta k_r}{k_A} \ll 1 \quad \text{or} \quad \frac{\mathbb{D}_A k_r}{k_A^2} \ll 1$$

This is generally written as:

$$\sqrt{\frac{\mathbb{D}_A k_r}{k_A^2}} = \text{Ha} \ll 1 \quad \text{(in practice} < 0.3)$$

The dimensionless group $\sqrt{\mathbb{D}_A k_r/k_A^2}$ is called the Hatta (Ha) number, which will be discussed in greater detail in the next section. The reader should furthermore realize that in most practical situations the film volume is very small when compared with the bulk volume. If we take, for example, a stirred gas–liquid reactor with $a = 1000$ m^2/m^3, we find $\delta = \mathbb{D}/k_A \approx 10^{-5}$ m and the total film volume is then 10^{-2} m^2/m^3, i.e. just 1% of the total volume.

Returning to the situation shown in Figure IV.8, we can now set up a macro-mass balance, realizing that in the stationary state all reactant A that diffuses through the boundary layer has to react in the bulk volume. Thus:

$$\phi_A = Ak_A(c_{Ai} - c_A) = k_r V c_A \tag{IV.52}$$

where A is the size of the interface and V the bulk volume. Eliminating the unknown bulk concentration, we find for the mass flux through the interface:

$$\phi_A'' = k_A c_{Ai}\left(\frac{Vk_r}{Vk_r + Ak_A}\right) = k_A c_{Ai}\left(\frac{1}{1 + \dfrac{Ak_A}{Vk_r}}\right) \tag{IV.53}$$

If the reaction is very slow, i.e. $Vk_r \ll Ak_A$ or $Ak_A/Vk_r \gg 1$, the bulk concentration will be almost equal to the concentration at the interface, $c_A \approx c_{Ai}$. From equation (IV.53) we find, then, that the mass flux is given by:

$$\phi_A'' = \frac{V}{A}k_r c_{Ai} \tag{IV.54}$$

This means that under these conditions the rate of the transport process is completely determined by the rate of the (slow) chemical reaction.

If, on the other hand, the chemical reaction is very fast, $Vk_r \gg Ak_A$ or $Ak_A/Vk_r \ll 1$, we find: if Ha < 0.3

$$\phi_A'' = k_A c_{Ai} \tag{IV.55}$$

because the bulk concentration c_A is now very small (due to the fast conversion of A). The rate of the transport process is in this case determined by the mass transport through the boundary layer.

IV.5.2 Fast homogeneous first-order reactions

If the reaction is so fast that all substance A is converted in the boundary layer, we can easily solve the problem of diffusion with chemical reaction by using the mass balance (equation (IV.48)):

$$\mathbb{D}\left(\frac{d^2 c_A}{dx^2}\right) + R_A = 0$$

where $R_A = -k_r c_A$. The solution of this differential equation with the boundary conditions $c_A = c_{Ai}$ at $x = 0$ and $c_A = 0$ for $x = \infty$ is:

$$c_A = c_{Ai} \exp\left(-x\sqrt{\frac{k_r}{\mathbb{D}_A}}\right) \qquad (IV.56)$$

This concentration distribution is shown in Figure IV.9. From equation (IV.56), we find for the mass flux through the boundary layer:

$$\phi_A'' = -\mathbb{D}_A \frac{dc_A}{dx}\bigg|_{x=0} = \sqrt{\mathbb{D}_A k_r}\, c_A \qquad (IV.57)$$

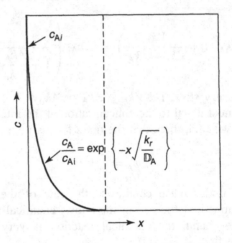

Figure IV.9 Concentration distribution during mass transfer with fast homogeneous first-order chemical reaction

In this case, the mass transfer coefficient is given by $\sqrt{\mathbb{D}_A k_r}$ and is solely dependent on the diffusion coefficient and the reaction rate constant k_r. The ratio $\sqrt{\mathbb{D}_A k_r / k_A^2}$, which represents the ratio between the mass transfer coefficients with and without chemical reaction, is the Hatta (Ha) number, which plays an important part in mass transfer with chemical reaction.

From the concentration distribution, equation (IV.56), we see that the penetration depth δ_r of the concentration distribution with chemical reaction is given approximately by:

$$\delta_r \simeq \sqrt{\mathbb{D}_A / k_r} \qquad (IV.58)$$

Consequently, the ratio of the thickness $\delta = \mathbb{D}/k_A$ of the boundary layer and the penetration depth δ_r is given by:

$$\frac{\delta}{\delta_r} = \sqrt{\frac{\mathbb{D}_A k_r}{k_A^2}} = \text{Ha}$$

which shows another physical meaning of the Ha number.

If the reaction occurs partly in the boundary layer and partly in the bulk of the liquid (Figure IV.10d) (where, because of the fast reaction, c_A is very low), we can show that the mass flux is given by:

$$\phi_A'' = \sqrt{k_r \mathbb{D}_A + k_A^2} \, c_{Ai} = k_A c_{Ai} \sqrt{1 + \text{Ha}^2} \qquad \text{(IV.59)}$$

If the reaction is very fast, i.e. $\text{Ha} = \sqrt{k_r \mathbb{D}_A / k_A^2} \gg 1$ (for all practical purpose $\text{Ha} > 3$), equation (IV.59) can be simplified to equation (IV.57), which we have already derived. In this case, the total reaction occurs in the boundary layer. If the reaction is relatively slow, $\text{Ha} = \sqrt{k_r \mathbb{D}_A / k_A^2} \ll 1 (\text{Ha} < 0.3)$, we find from equation (IV.59) the relationship for simple physical absorption:

$$\phi_A'' = k_A c_{Ai}$$

and only a negligible part of the reaction occurs in the boundary layer.

The Ha number is therefore the criterion for whether the reaction occurs completely in the bulk of the liquid ($\text{Ha} < 0.3$) or completely in the boundary layer ($\text{Ha} > 3$). If $0.3 < \text{Ha} < 3$, reaction in both the bulk and the film is important. Figure IV.10 shows the five possible cases of mass transport with homogeneous first-order chemical reaction which we have discussed. If we want to determine to which class a certain problem belongs we start by calculating the Ha number. If $\text{Ha} < 0.3$, as a second criterion the expression Ak_A/Vk_r, has to be determined (compare equation (IV.53)).

Equation (IV.59) and Figure IV.10 also show that the rate of physical absorption is increased by a factor:

$$F_c = \sqrt{1 + \text{Ha}^2} \qquad \text{(IV.60)}$$

by the occurrence of a homogeneous first-order chemical reaction. This increase in adsorption rate by a factor $\sqrt{1 + \text{Ha}^2}$ in the case of a chemical reaction is valid under all flow conditions. The thickness δ_r of the boundary layer where the chemical reaction occurs is so small that it is not influenced by the flow conditions. In addition, during turbulent flow, the eddies cannot penetrate this thin layer. The enhancement factor $F_c = \sqrt{1 + \text{Ha}^2}$ is therefore universally applicable.

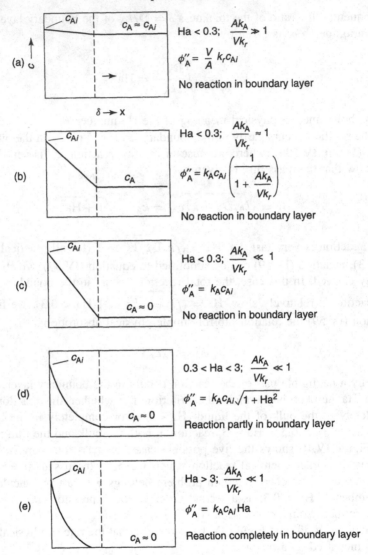

Figure IV.10 Five cases of mass transfer with homogeneous irreversible first-order reaction

IV.5.3 Non-linear homogeneous reactions

The analysis developed in Sections IV.5.1 and IV.5.2. can also be applied to nth-order chemical reactions, where the production term of equation (IV.48) is given by (equation (IV.50)):

$$R_A = -k_r c_A^n$$

The results of this analysis are completely analogous to that of first-order reactions and, again, the five situations summarized in Figure IV.10 can be distinguished.

For this type of reaction, the thickness of the reaction boundary layer (equation (IV.58)) becomes:

$$\delta_r = \sqrt{\frac{n+1}{2} \frac{\mathbb{D}_A}{k_r c_{Ai}^{n-1}}}$$

and the definition of the Hatta number is now:

$$\text{Ha} = \frac{1}{k_A} \sqrt{\frac{2}{n+1} k_r c_{Ai}^{n-1} \mathbb{D}_A} \qquad (\text{IV.61})$$

With the above Ha number, the relationship developed in Sections IV.5.1 and IV.5.2 for the first-order reactions also describe nth-order reactions. There is one important practical difference: for first-order reactions, δ_r and Ha are only (through k_r) dependent on temperature, but for nth-order reactions these parameters are also dependent on the concentration of A at the interface.

This may be illustrated further for biochemical processes, which are of increasing interest today. The production term for these reactions is quite often almost linear for small concentrations, but levels off with increased concentrations because of limiting factors in the metabolic rate. A "logistic" type relationship is often proposed (Michaelis–Menten kinetics):

$$R_A = -\frac{Bc_A}{K + c_A} \qquad (\text{IV.62})$$

From this it is obvious that the order of this reaction lies between $n = 0$ ($c_A/K \gg 1$) and $n = 1$($c_A/K \ll 1$), for which extremes equation (IV.61) offers the solution.

For intermediate concentrations, a new parameter, c_{A0}/K, enters the analysis in a rather complicated way. Equation (IV.48) can be integrated analytically (have a try), with the result showing the appropriate limits, but this will not be elaborated here. In line with equation (IV.61), the result can be written in the following form with reasonable approximation:

$$\text{Ha} = \frac{1}{k_A} \left(\frac{K}{B\mathbb{D}_A} + \frac{c_{A0}}{2B\mathbb{D}_A} \right)^{-1/2}$$

For the zero- and first-order limits, this equation gives exact solutions. In the intermediate region, the deviation from the exact solution is always smaller than 6%.

However, homogeneous biochemical reactions with enzymes as biocatalysts are scarce, because it is difficult to recover the expensive enzymes from the solution. Heterogeneous catalysis (see Section IV.5.5) is often preferred.

IV.5.4 Homogeneous second-order reactions

A component A is absorbed and reacts with a component B which is present in the bulk and which is not desorbed. The production term in equation (IV.48) is given by (equation (IV.51)):

$$R_A = -k_r c_A c_B$$

If the concentration of component B is uniform and equal to $c_{B\infty}$ (the bulk concentration) the solution to this problem is simple: the reaction can be regarded as being of a pseudo first-order type with a first-order reaction rate constant $k_{rp} = k_r c_{B\infty}$. Thus:

$$R_A = -k_{rp} c_A \tag{IV.63}$$

The solution to this problem has already been treated in Sections IV.5.1 and IV.5.2. There are two conditions under which the assumption of a pseudo first-order reaction is justified:

(a) If the reaction takes place in the bulk so that:

$$Ha = \sqrt{\frac{k_{rp} \mathbb{D}_A}{k_A^2}} = \sqrt{\frac{k_r c_{B\infty} \mathbb{D}_A}{k_A^2}} < 0.3$$

(b) If the reactant B can be transported quickly enough into the film where the reaction occurs so that the concentration of B in the film can be assumed to equal the bulk concentration (see Figure IV.11), i.e., if;

$$k_B c_{B\infty} \gg \sqrt{k_{rp} \mathbb{D}_A} c_{A0} = \sqrt{k_r c_{B\infty} \mathbb{D}_A} c_{A0} \tag{IV.64}$$

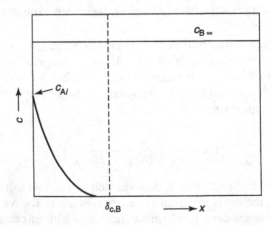

Figure IV.11 Concentration distribution during mass transfer and second-order reaction if $Ha > 3$. $k_B c_{B\infty} / K_A c_{Ai} > 10 Ha$

In practice, this condition is fulfilled if:

$$k_B c_{B\infty} > 10\sqrt{k_r c_{B\infty} \mathbb{D}_A c_{A0}}$$

or:

$$\frac{k_B c_{B\infty}}{k_A c_{Ai}} > 10\,\mathrm{Ha} \tag{IV.65}$$

If equation (IV.65) is fulfilled, the correction factor F_c for the acceleration of mass transport by chemical reaction is given by:

$$F_c = \sqrt{1 + \mathrm{Ha}^2} = \sqrt{1 + \frac{k_r c_{B\infty} \mathbb{D}_A}{k_A^2}} \tag{IV.66}$$

If the reaction is so fast that insufficient B is transported into the boundary layer, the transport of B will influence the reaction rate in the film. If:

$$\frac{k_B c_{B\infty}}{k_A c_{Ai}} \ll \mathrm{Ha}(< 0.1\mathrm{Ha})$$

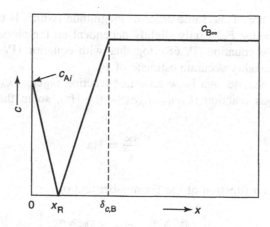

Figure IV.12 Concentration distribution during mass transfer with second-order reaction if $\mathrm{Ha} > 3$; $k_B c_{B\infty}/k_A c_{Ai} < 0.1\,\mathrm{Ha}$

the reaction is so fast that A and B cannot exist simultaneously at the same place. A reaction front is formed at $x = x_R$ (see Figure IV.12) where $c_A = c_B = 0$. The film model yields for the transport rates of both components to the reaction front:

$$\frac{\mathbb{D}_A}{x_R} c_{Ai} = \frac{\mathbb{D}_B}{\delta_{c,B} - x_B} c_{B\infty} \tag{IV.67}$$

Hence, the chemical acceleration factor is found to be, bearing in mind the definition of the Ha number given in Section IV.5.2 and realizing that for this case always Ha > 3:

$$F_c = \text{Ha} = \frac{\delta_A}{x_R} = \frac{\delta_A}{\delta_B}\left(1 + \frac{\mathbb{D}_B c_{B\infty}}{\mathbb{D}_A c_{Ai}}\right) \qquad \text{(IV.68)}$$

where δ_A and δ_B are the thicknesses of the boundary layers during physical absorption only. The penetration theory predicts that:

$$\frac{\delta_A}{\delta_B} = \frac{\text{Sh}_B}{\text{Sh}_A} = \left(\frac{\mathbb{D}_A}{\mathbb{D}_B}\right)^{1/2} \qquad \text{(IV.69)}$$

(see case 1 of Section IV.3.2) whereas the boundary layer and the Lévêque theory (cases 2 and 4) predict a relationship of the form:

$$\frac{\delta_A}{\delta_B} = \left(\frac{\mathbb{D}_A}{\mathbb{D}_B}\right)^{1/3} \qquad \text{(IV.70)}$$

If $\mathbb{D}_A$ and $\mathbb{D}_B$ are of the same order of magnitude (which is often the case) the enhancement factor F_c is only slightly dependent on the physical model chosen and we can use equation (IV.68), together with equation (IV.69) or (IV.70) to produce a reasonably accurate estimate of F_c.

In the intermediate area between a fast reaction ($k_B c_{B\infty}/k_A c_{Ai} > 10$ Ha) and an instantaneous reaction ($k_B c_{B\infty}/k_A c_{Ai} < 0.1$ Ha), so in the area where (see Figure IV.13):

$$\frac{k_B c_{B\infty}}{k_A c_A} = \text{Ha}$$

the F_c value is a function of the Ha number and of

$$\left[\left(\frac{\mathbb{D}_A}{\mathbb{D}_B}\right)^m + \frac{c_{B\infty}}{c_{Ai}}\left(\frac{\mathbb{D}_A}{\mathbb{D}_B}\right)^{m-1}\right]$$

where the exponent m again is a function of the physical model chosen ($m = \frac{1}{2}$ for penetration theory, $m = \frac{1}{3}$ for boundary layer theory). The dependency of F_c on these two expressions is shown in Figure IV.14; the latter can be used for estimating the chemical acceleration factor.

Table IV.3 gives a summary of the various solutions for mass transfer involving second-order reactions.

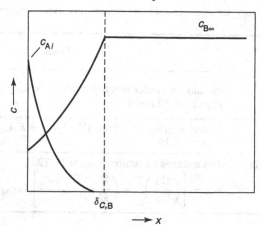

Figure IV.13 Concentration distribution during mass transfer with second-order reaction if Ha > 0.3; $k_B c_{B\infty}/k_A c_{Ai} \approx$ Ha

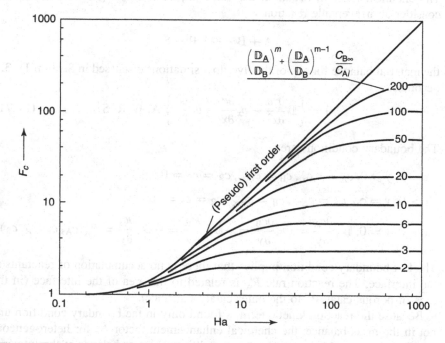

Figure IV.14 Values for the enhancement factor F_c for second-order reactions if Ha > 0.3, $k_B c_{B\infty}/k_A c_{Ai} \approx$ Ha

Table IV.3 Summary of mass transfer situations involving irreversible second-order reactions

Ha	$\dfrac{k_B c_{B\infty}}{k_A c_{Ai}}$	Solution
<0.3	—	Pseudo first-order reaction with $k_{rp} = k_r c_{B\infty}$: solutions of Figure IV.10 apply
—	>10 Ha	
>0.3	$\approx$ Ha	Situation given in Figure IV.13 $\phi''_A = F_c k_A c_{Ai}$ with F_c from Figure IV.14
>0.3	<0.1 Ha	Instantaneous reaction, Figure IV.12 $$\phi''_A = \left[\left(\frac{\mathbb{D}_A}{\mathbb{D}_B}\right)^m + \left(\frac{\mathbb{D}_A}{\mathbb{D}_B}\right)^{m-1}\frac{c_{B\infty}}{c_{Ai}}\right]k_A c_{A0} \text{ where } \tfrac{1}{3} < m < \tfrac{1}{2}$$

IV.5.5 Mass transfer with heterogeneous chemical reaction

The chemical reaction occurs at the solid–liquid or solid–gas interface. If we consider an irreversible reaction:

$$A + B \longrightarrow R + S$$

the material balance for any of the five flow situations discussed in Section IV.3.2 becomes:

$$0 = \left\{ \mathbb{D}\frac{\partial^2 c}{\partial x^2} - v_x \frac{\partial c}{\partial x} - v_y \frac{\partial c}{\partial y} \right\} \text{A, B, R, S} \qquad \text{(IV.71)}$$

The boundary conditions are:

$$x = 0,\ y \geqslant 0,\ c_A = c_{A\infty},\ c_B = c_{B\infty},\ c_R = c_S = 0$$

$$x > 0,\ y = \infty,\ c_A = c_{A\infty},\ c_B = c_{B\infty},\ c_R = c_S = 0$$

$$x > 0,\ y = 0,\ \mathbb{D}_A \frac{\partial c_A}{\partial y} = \mathbb{D}_B \frac{\partial c_B}{\partial y} = -\mathbb{D}_R \frac{\partial c_R}{\partial y} = -\mathbb{D}_S \frac{\partial c_s}{\partial y} = R'_A(c_A, c_B, c_R, c_S)$$

The last boundary condition implies that there is no accumulation of reactants at the interface. The reaction rate R'_A is related to the area of the interface (in the case of porous catalysts to the outer catalyst surface).

Because the reaction kinetic term is found only in the boundary condition and not in the mass balance, the chemical enhancement factor F_c for heterogeneous reactions is unity. Under steady-state conditions, a mass balance at the interface shows (no accumulation of reactants at interface):

$$k_A(c_{A\infty} - c_{Ai}) = k_B(c_{B\infty} - c_{Bi}) = k_R c_{Ri} = k_S c_{Si} = -R'_A(c_{Ai}, c_{Bi}, c_{Ri}, c_{Si})$$
$$\text{(IV.72)}$$

These equations are, in general, easily solved. The rate of conversion is, in all cases, given by:

$$\phi_A'' = k_A(c_{A\infty} - c_{Ai}) = k_A c_{A\infty}\left(1 - \frac{c_{Ai}}{c_{A\infty}}\right) \tag{IV.73}$$

We will now discuss the special cases of first- and second-order reactions.

(a) *First-order reaction.* Equation (IV.73) for this case reads:

$$\phi_A'' = k_A(c_{A\infty} - c_{Ai}) = -R_A' = k_r'c_{Ai} \tag{IV.74}$$

The reaction rate constant k_r' is here related to the outer size of the interface (macroscopic reaction rate constant) and has the dimension (m/s). It is related to the well-known chemical reaction rate constant $k_r(s^{-1})$ via the specific interface area a (m^2/m^3) as follows:

$$k_r = ak_r'^{\dagger} \tag{IV.74a}$$

From equation (IV.74) the ratio $c_{Ai}/c_{A\infty}$ can be calculated to be:

$$\frac{c_{Ai}}{c_{A\infty}} = \frac{k_A}{k_r' + k_A} = \frac{1}{1 + \dfrac{k_r'}{k_A}}$$

and the overall conversion rate is found from equation (IV.74) as:

$$\phi_A'' = k_A c_{A\infty}\left(\frac{1}{1 + \dfrac{k_A}{k_r'}}\right) \tag{IV.75}$$

Apparently, here the quotient k_A/k_r' fulfils the role the expression Ak_A/Vk_r played during homogeneous reactions (Section IV.5.1). We can distinguish three principal situations:

A. $\qquad\qquad k_r' \ll k_A, c_{Ai} \approx c_{A\infty}, \phi_A'' = k_r'c_{A\infty}$

The reaction is so slow compared with the mass transfer process that mass transfer does not decrease the reaction rate and the concentration of A is uniform and equals $c_{A\infty}$.

B. $\qquad\qquad k_r' \gg k_A, c_{Ai} \approx 0, \phi_A'' = k_A c_{A\infty}$

---

† In the case of heterogeneous catalysis, often a microkinetic reaction rate constant k_r'' related to the total pore surface of the catalyst is used; k_r'' is related to k_r' and k_r by:

$$k_r = a_p k_r'' = ak_r'$$

where a = outer surface of catalyst particles and a_p = specific pore area of catalyst (m^2/m^3).

The reaction is so fast that the concentration of A at the interface is practically zero and physical mass transport determines the overall reaction rate.

C.
$$k_r' \approx k_A, \quad 0 < c_{Ai} < c_{A\infty}, \quad \phi_A'' = k_A c_{A\infty} \left(\dfrac{1}{1 + \dfrac{k_A}{k_r'}} \right)$$

Both chemical reaction and mass transfer determine the overall conversion rate.

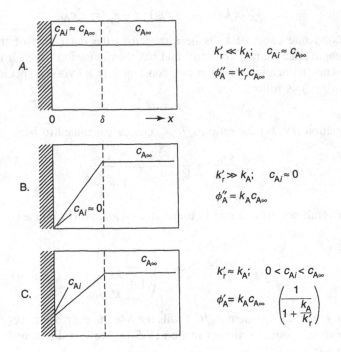

A.
$$k_r' \ll k_A; \quad c_{Ai} \approx c_{A\infty}$$
$$\phi_A'' = k_r' c_{A\infty}$$

B.
$$k_r' \gg k_A; \quad c_{Ai} \approx 0$$
$$\phi_A'' = k_A c_{A\infty}$$

C.
$$k_r' \approx k_A; \quad 0 < c_{Ai} < c_{A\infty}$$
$$\phi_A'' = k_A c_{A\infty} \left(\dfrac{1}{1 + \dfrac{k_A}{k_r'}} \right)$$

Figure IV.15   Three cases of irreversible heterogeneous first-order reactions

Figure IV.15 gives a summary of these three situations and shows the concentration distributions which occur. The reader should compare this figure with the first three cases of Figure IV.10, which refer to homogeneous first-order reactions.

(b) *Second-order heterogeneous reactions.* Equation (IV.73) for this case reads:

$$\phi_A'' = k_A(c_{A\infty} - c_{Ai}) = k_B(c_{B\infty} - c_{Bi}) = k_r' c_{Ai} c_{Bi}$$

Elimination of c_{Bi} leads to the following expression for $c_{Ai}/c_{A\infty}$:

$$x\phi \left(\frac{c_{Ai}}{c_{A\infty}} \right)^2 + \{x(1 - \phi) + 1\} \frac{c_{Ai}}{c_{A\infty}} - 1 = 0 \qquad \text{(IV.76)}$$

where:

$$x = \frac{k_r' c_{B\infty}}{k_A} \quad \text{and} \quad \phi = \frac{k_A c_{A\infty}}{k_B c_{B\infty}}$$

Calculation of $c_{Ai}/c_{A\infty}$ from equation (IV.76) enables us to find the rate of conversion as:

$$\phi_A'' = k_A c_{A0} \left(1 - \frac{c_{Ai}}{c_{A\infty}}\right) \tag{IV.77}$$

The reader may check that for $\phi \longrightarrow 0$, second-order heterogeneous reactions can be regarded as pseudo first-order reactions.

(c) *Michaelis–Menten heterogeneous catalysis.* Starting with equation (IV.62) we now get:

$$\phi_A'' = k_A (c_{A0} - c_{Ai}) = \frac{B' c_{Ai}}{K + c_{Ai}}^\dagger$$

Eliminating c_{Ai} from these two equations and solving the resulting second-order algebraic equation for ϕ_A'' gives the desired result, which will not be reported here, because the results for the regimes A, B and C of Figure IV.15 follow from first principles or by analogy:

A. $c_{Ai} \cong c_{A\infty} : \phi_A'' = \dfrac{B' c_{A\infty}}{K + c_{A\infty}}$

with the two obvious extremes:

$$\phi_A'' = B'(K/c_{A\infty} \longrightarrow 0, \text{ zero order, regime A1}) \text{ and}$$

$$\phi_A'' = \frac{B' c_{A\infty}}{k}(c_{A\infty}/K \longrightarrow 0, \text{ first order, regime A2}).$$

B. $\dfrac{c_{Ai}}{c_{A\infty}} \cong 0 \left(\dfrac{K}{c_{A\infty}} \longrightarrow 0\right) : \phi_A'' = k_A c_{A\infty} (\text{regime B})$

C. In analogy with the previous case C:

$$0 < c_{Ai} < c_{A\infty}, \ k_r' = \frac{B'}{K + c_{A\infty}} \cong k_A$$

$$\phi_A'' = \frac{k_A c_{A\infty}}{1 + \dfrac{k_A K}{B'}\left(1 + \dfrac{c_{A\infty}}{K}\right)}$$

$^\dagger aB' = B$ (see equation (IV.74a) and accompanying notes).

which holds when compared with the exact solution, if:

$$\frac{3k_A c_{A\infty}}{B'} \ll 1 + \frac{k_A K^\dagger}{B'}$$

This leads to the following two regimes:

$$B'/K \longrightarrow 0, \quad \phi_A'' = \frac{B' c_{A\infty}}{K\left(1 + \dfrac{c_{A\infty}}{K}\right)} \qquad \text{(regime } C_1 \text{ with } A_2 \text{ as a limit)}$$

and:

$$K/c_{A\infty} \longrightarrow 0, \quad \phi_A'' = \frac{k_A c_{A\infty}}{1 + \dfrac{k_A c_{A\infty}}{B'}} \qquad \text{(regime } C_2 \text{ with } A_1 \text{ as a limit)}$$

In all these cases, knowledge of the physical mass transfer coefficient, k_A, is essential (see Section IV.3). The proviso in the treatment given is that the external surface of the heterogeneous matrix is active and not its bulk (e.g. via pores).

With the knowledge now acquired on mass transfer with chemical reaction, simple isothermal reactors can be designed. However, because of the heat production during chemical reactions, reactors can seldom be considered as isothermal. This demands combined calculations on (mass transfer with) chemical reaction and heat transfer. The interplay between these two processes is rather complex in most cases. For such examples, the reader is referred to advanced handbooks on chemical reaction engineering, e.g. K. R. Westerterp, W. P. M. van Swaaij and A. A. C. M. Beenackers, *Chemical Reactor Design and Operation*, John Wiley & Sons, 1987.

IV.5.6 Problems

1. On the bottom of a cylindrical vessel (diameter $D = 1$ m, height $H = 1$ m) with a stirrer is a 1 cm layer of solid substance. To remove this layer the vessel is filled with a liquid in which the solid dissolves. In solution the solid decomposes according to a first-order chemical reaction.

    Data: mass transfer coefficient (without chemical reaction) $k = 10^{-4}$ m/s
    solubility of the solid substance in the liquid $= 50$ kg/m^3
    density of the solid substance $\rho = 2000$ kg/m^3
    diffusivity $\mathbb{D} = 10^{-9}$ m^2/s
    reaction rate constant $k_r = 10^{-2}$ s^{-1}.

    (a) Show that the mean concentration of the dissolved solid in the liquid is much lower than the solubility of the solid substance.

---

† If this inequality does not hold, the full solution has to be used, regime C_3 (see problem 13 below).

(b) Show that the conversion by chemical reaction in the boundary layer is negligible.

(c) How long does it take for the layer of solid substance to be dissolved completely?

(d) A chemist states that the process can no doubt be accelerated by adding a catalyst which makes the reaction rate constant 1000 times as great. Demonstrate quantitatively whether his statement is correct.

Answer: (a) $c = 0.01 \, c_i$

       (b) $\delta_c = 10^{-5}$ m; $<0.3\%$ of the total conversion in film

       (c) 4000 s

       (d) $\sqrt{2}$ times as fast

2. If under otherwise equal conditions pure oxygen or oxygen from the surrounding air is absorbed in a strong sodium sulphite (Na_2SO_3) solution to which a large amount of cobalt salts have been added, the difference in rate of adsorption appears to be a factor of 10. Determine the order of the reaction between O_2 and Na_2SO_3 with respect to the oxygen (the reaction is very fast, $Ha > 3$).

Answer: 2

3. In a laminar liquid film (contact time 0.1 s) CO_2 ($m = 1$) is absorbed from the pure gas (20 °C, 1 atm). The liquid film contains 0.04 kmol/m^3 sodium hydroxide (NaOH). Under the prevailing conditions the second-order reaction rate constant between CO_2 and NaOH is 6×10^3 m^3/kmol s. Calculate the chemical enhancement factor and the mole flux in this situation and draw the concentration distribution in the film. Assume $D_{CO_2} = D_{OH^-} = 2 \times 10^{-9}$ m^2/s; because 1 mole of CO_2 uses two moles of NaOH, for $c_{B\infty}$ in this case $\frac{1}{2}c_{NaOH}$ must be used.

Answer: $F_c = 1.5$, $\phi''_{mol} = 9.85 \times 10^{-6}$ kmol/m^2 s

4. What will be the chemical enhancement factor and the mole flux if in the situation of problem 3 a temperature of 70 °C is taken ($k_r \longrightarrow 2 \times 10^5$ m^3/kmol s) and the concentration of NaOH is chosen equal to 0.4 kmol/m^3 ($m = 0.42$)?

Answer: $F_c = 13.5$; $\phi_{mol} = 5.6 \times 10^{-5}$ kmol/m^2 s.

*5. In a packed column of catalyst particles a component from the liquid stream is converted by a chemical reaction. The reaction is an irreversible heterogeneous reaction of the first order and proceeds at the surface of the particles. How high should be the catalyst bed to ensure a degree of conversion of 0.63 ($= 1 - e^{-1}$)?

Data: reaction rate constant $k_r = 4 \times 10^{-5}$ m/s (macrokinetic)

       diffusion coefficient $\mathbb{D} = 10^{-9}$ m^2/s

superficial flow rate in the bed $v_0 = 4$ cm/s
kinematic viscosity of the liquid $\nu = 10^{-6}$ m^2/s
bed porosity $\varepsilon = 0.40$
particle diameter $d_p = 1.0$ mm

Answer: 0.375 m

*6. During fat hydrogenation pure H$_2$ is absorbed in liquid fat containing cata-
lyst particles at which the reaction between H$_2$ and unsaturated fatty acid
takes place. The rate of conversion per unit liquid volume related to H$_2$
is independent of the degree of conversion of the oil and appears to be
linearly proportional to the solubility of H$_2$. In a certain experiment the
ratio between these two constants appears to be 0.03 min^{-1}.
If the interface between gas and liquid is increased by a factor of 2, the
specific rate of conversion will be 20% higher. If the catalyst concentration
is increased by a factor of 2, the specific rate of conversion will be 50%
higher. Calculate for this experiment the gas–liquid interfacial area per unit
liquid volume and the (macro)reaction rate constant of the reaction between
H$_2$ and unsaturated fat.

Data: the partial mass transfer coefficients in the liquid at the interface with
the gas, $k_1 = 10^{-4}$ m/s, and at the interface with the catalyst, $k_2 =$
10^{-5} m/s; the interface of the catalyst per unit volume of liquid $=$
150 m^2/m^3

Answer: 15 m^2/m^3; 10^{-5} m/s

7. One of the components (A) of a gas mixture is absorbed in a solution in
which it reacts very fast with a component B which is present in excess,
according to a second-order reaction of the type $R = kc_A c_B$ (i.e. the reaction
takes place entirely in the boundary layer). At a certain concentration of
B, the overall mass transfer coefficient is found to be $K = 10^{-4}$ m/s. In
order to find for this case the partial mass transfer coefficient in the gas
phase, the overall mass transfer coefficient is measured again at a four times
higher concentration of B. The result is $K = 1.5 \times 10^{-4}$ m/s. Calculate the
partial mass transfer coefficient in the gas phase if the given mass transfer
coefficients are related to the gas phase.

Answer: $k_g = 3 \times 10^{-4}$ m/s

*8. Reith carried out the following tests:

Two equal streams of O$_2$ and N$_2$ were mixed and passed through a stirred
gas–liquid reactor containing a solution with an excess of sulphite and
sufficient cobalt ions to realize chemical absorption in the range of the
fast reaction (Ha > 3).
Subsequently, he passed under otherwise the same conditions the two
streams O$_2$ and N$_2$ separately into the gas–liquid reactor.

In the second experiment he found absorption rates which were 40% higher than in the first test. How do you explain this?

Answer: There is complete segregation in the bubble phase under the conditions of this experiment.

9. The chemical absorption of CO_2 (A) in aqueous NaOH solutions can to a first approximation be described by a second-order reaction:

$$R_{CO_2} = k_r(CO_2)(OH^-) \quad \text{or} \quad R_A = -k_r c_A c_B$$

In a laminar jet (diameter 0.66 mm, $v_i = 5$ m/s, $T = 20\,°C$) the following absorption rates ϕ_m are measured at various jet lengths L:

L (m)	ϕ_m (kg/s)
2.0×10^{-2}	2.04×10^{-8}
4.5×10^{-2}	4.60×10^{-8}

where $2c_{B\infty} = (OH^-)_\infty = 1.05$ kmol/m³, $c_{A0} = CO_2 = 0.238$ kg/m³, $\mathbb{D}_A = 1.9 m^2/s$ and $\mathbb{D}_B = 1.6 \times 10^{-9}$ m²/s. Calculate the reaction rate constant k_r (at 20 °C) from these data.

Explain why in this case $2c_{B\infty} = (OH^-)_\infty$.

Answer: $k_r = 4.3 \times 10^3$ m³/kmol s

*10. CO_2 has to be washed out of a continuous stream of hydrogen at 30 °C and 7 atm absolute pressure in a packed column by means of a countercurrently flowing monoethanolamine (MEA = $HO-C_2H_4-NH_2$); one mole CO_2 uses two moles MEA) solution, which is reactivated and recycled. If the CO_2 concentration in the hydrogen has to be decreased from 20 to 0.01 vol% and if the initial concentration of free MEA in the wash solution is 3 kmol/m³, how high should the packing in the column be?

Data: column diameter $D = 1$ m
     packing: Raschig rings, $d_p = 7.5$ cm ($a = 60$ m²/m³, $\varepsilon = 0.95$)
     gas flow rate $\phi_g = 4000$ Nm³/h
     liquid flow rate $\phi_L = 30$ m³/h
     solubility of CO_2 in MEA solution, $m = 1$
     second-order reaction rate constant between CO_2 and MEA,
     $k_r = 7500$ m³/kmol s
     $\rho_g = 0.56$ kg/m³ (at 7 atm), $\rho_L = 1000$ kg/m³
     $\eta_g = 9.3 \times 10^{-6}$ Ns/m²
     $\mathbb{D}_{CO_2 \text{ in } H_2} = 5.5 \times 10^{-5}$ m²/s
     $\mathbb{D}_{CO_2 \text{ in } H_2O} = 2 \times 10^{-9}$ m²/s

$$\mathbb{D}_{\text{MEA in } H_2O} = 10^{-9} \text{ m}^2/\text{s}$$
$$\nu_L = 10^{-6} \text{ m}^2/\text{s}$$

Answer: 15.4 m

11. CO_2 reacts with monoethanolamine (MEA) in an aqueous solution by a second-order reaction (first order in CO_2 and first order in MEA). In order to determine the reaction rate constant the following experiment is performed: An MEA solution in a cylindrical beaker is being stirred. Above the free liquid surface a gas mixture consisting of an inert gas with 10 mol% CO_2 is introduced at atmospheric pressure. The absorption rate of CO_2 in the liquid is measured at a temperature of 20 °C. The result is:

$$\phi_{\text{mol A}} = 1.61 \times 10^{-7} \text{ kmol/s}.$$

(a) Why does the following relationship hold for the ratio between de mass transfer coefficients of CO_2 (A) and MEA (B) in the case of a mobile liquid surface:

$$\frac{k_A}{k_B} = \left(\frac{\mathbb{D}_A}{\mathbb{D}_B}\right)^{1/2} ?$$

(b) Calculate the second-order reaction rate constant of the reaction between CO_2 and MEA if the gas phase resistance for CO_2 is negligible.
(c) Why is a gas mixture with 10 mol% CO_2 to be prefered to pure CO_2?

Data: Inner diameter of the beaker: $D = 0.10$ m
Bulk concentration of MEA: $c_{B\infty} = 3.5 \text{ kmol/m}^3$
Diffusion coefficient of CO_2 in the solution: $\mathbb{D}_A = 1.9 \times 10^{-9} \text{ m}^2/\text{s}$
Diffusion coefficient of MEA in the solution: $\mathbb{D}_B = 3.8 \times 10^{-9} \text{ m}^2/\text{s}$
Partition coefficient of CO_2: $m = \dfrac{c_A(\text{liquid})}{c_A(\text{gas})} = 0.92$
Mass transfer coefficient of CO_2 without chemical reaction:
$k_A = 5 \times 10^{-4}$ m/s
Answer: (b) $k_r = 4.3 \times 10^3 \text{ m}^3/(\text{kmol s})$

12. Small spherical particles of benzoic acid are dissolved in a solution of sodium hydroxide (NaOH). The chemical reaction between benzoic acid (A) and the hydroxylic ion (B) can be considered to be infinitely fast.
(a) Show that the following relationship holds for the enhancement factor F_c in the case of spherical geometry:

$$F_c = 1 + \frac{\mathbb{D}_B c_{B\infty}}{\mathbb{D}_A c_{A0}}$$

if the relative velocity between particles and solution is neglected (i.e. diffusion only).
(b) Calculate the time in which complete dissolution of the particles takes place.

(c) Is neglect of the relative velocity between particles and solution permisseble?

Data: $\mathbb{D}_A = 8 \times 10^{-10}$ m^2/s; $\mathbb{D}_B = 2 \times 10^{-9}$ m^2/s

Solubility of benzoic acid: $c_{A0} = 1.6 \times 10^{-2}$ kmol/m^3

Bulk concentration of sodium hydroxide: $c_{B\infty} = 1.7$ kmol/m^3

Original diameter of the particles: $d_p = 4 \times 10^{-4}$ m

Density of the particles: $\rho_A = 1075$ kg/m^3

Molecular mass of benzoic acid: $M_A = 122$ kg/kmol

Density of the solution: $\rho_L = 1072$ kg/m^3

Viscosity of the solution: $\eta = 1.5 \times 10^{-3}$ Pa s

Answer: (b) 17 s

(c) no; Sh ≈ 4 instead of 2 for the original particle diameter and without chemical reaction.

13. Prove that the molar flux per unit area of A for a heterogeneous reaction of the Michaelis–Menten type is given by:

$$\phi_A'' = \frac{B'}{2}X\left[1 - \left(1 - 4\frac{k_A c_{A\infty}}{B'X^2}\right)^{1/2}\right]$$

with

$$X = 1 + \frac{k_A K}{B'}\left(1 + \frac{c_{A\infty}}{K}\right)$$

with the symbols having the meaning as in the text. Derive from this expression for ϕ_A'' the six regimes discussed in Section IV.5.5.

14. A batch hydrolysis of lactose by homogeneous catalysis (enzyme lactase) shows the following concentration change with time, t ($t = 0$, $c_A = c_{A0}$):

$$\frac{\ln(c_{A0}/c_A)}{c_{A0} - c_A} = \frac{1}{6}\frac{t}{c_{A0} - c_A} - 1.32 \quad \text{(l/mol), } t\text{(h)}$$

Determine the values of B and K for this reaction, assuming Michaelis–Menten kinetics.

Answer: $K = 0.76$ mol/l, $B = 0.13$ mol/l h

15. Polysaccharides (pectines) in fruit juices hinder their filtration; enzymes (pectinases) split up these polysaccharides. In a juice with a relatively high concentration of pectine, its concentration is reduced by 90% in 70 min. How long will it take to reduce the pectine content by 90% in a juice with a 40% higher initial concentration of pectine, assuming zero-order Michaelis–Menten kinetics?

Answer: $t = 98$ min

16. For the design of a production plant for baker's yeast from glycose it has been established that the cells (volume fraction 0.16, diameter 4 μm) need

3×10^{-3} kg/m^3 s glycose and 10^{-3} kg/m^{3}s oxygen at the maximal growth rate. The glycose concentration in the bulk is 0.8 kg/m^3 (its diffusion coefficient is 2×10^{-10} m^2/s), and the oxygen concentration 10^{-3} kg/m^3 (its diffusion coefficient is 2×10^{-9} m^2/s). The yeast cells float in the medium and can only grow uninhibited as long as the oxygen concentration at their surface exeeds 10^{-4} kg/m^3. Prove that there are no transfer limitation from the medium to the cells.

Answer: Sh ≥ 2, k(glycose) $\geq 10^{-4}$ m/s, k(O$_2$) $\geq 10^{-3}$ m/s; $a = 2.4 \times 10^5$ m^2/m^3

The surface concentration of glycose and oxygen are nearly equal to their bulk concentrations.

17. If in the case of problem 16 (above), the oxygen concentration in the bulk has to be maintained by bubbling air through the medium (bubble diameter, 1.5 mm), what is the minimum volume fraction of air that is then needed? How realistic is this design concept?

Answer: Re $= 3 \times 10^2$; Sh $= 92$; k(O$_2$) $= 1.23 \times 10^{-4}$ m/s; $a = 4 \times 10^{-3}(1 - \varepsilon)$; $1 - \varepsilon > 0.25$.

At such high volume fractions of air these small bubbles will coalesce, unless surface-ative material present prevents this effectively.

18. A heterogeneous biocatalytic reaction is executed in a stirred tank ($c_{A\infty} = K$, Michaelis–Menten kinetics). It is found that the conversion rate is 91% of the maximal transport rate ($c_{A\infty} \gg K$). What is the ratio of the maximal reaction rate and the maximal physical transport rate (known as the Damköhler number, Da)? In which region does the process take place?

Answer: $X = 1.1 + \dfrac{2}{Da}$ or Da $= 20$; case B of Figure IV.15

*19. Show that John drew the right conclusion in the case of the burning haystack related at the beginning of this Section.

Comments on problems

Problem 5

A material balance over a short height dh of the packed column reads (A = cross-sectional area of column):

$$-\phi_L \, dc_L = ka(c_L - c_i)A \, dh = k_r a c_i A \, dh$$

and isolating and substituting the unknown interface concentration c_i we obtain:

$$-\phi_L \, dc_L = \frac{ac_L A}{\dfrac{1}{k} + \dfrac{1}{k_r}} \, dh$$

or, after integration between $h = 0, c_{L0}$ and h, c_L:

$$\frac{c_L}{c_{L0}} = \exp\left[\frac{-ah}{v_0\left(\dfrac{1}{k} + \dfrac{1}{k_r}\right)}\right]$$

Now the degree of conversion was given as $(c_{L0} - c_L)/c_{L0} = 1 - e^{-1}$, so $c_I/c_{L0} = e^{-1}$ and therefore:

$$h = \frac{v_0}{a}\left(\frac{1}{k} + \frac{1}{k_r}\right)$$

We find the mass transfer coefficient with the aid of equation (III.69a) (making use of the analogy between heat and mass transfer) to be $k = 1.14 \times 10^{-4}$ m/s and with $a = 6(1 - \varepsilon)/d_p = 3600$ m^2/m^3 we finally obtain $h = 0.376$ m.

Problem 6

The reaction rate between hydrogen and fat is given by a relationship of the form ($c_L = H_2$ concentration in liquid phase):

$$-R_A = k_r'' c_L^\alpha c_{fat}^\beta$$

Now it is stated that the reaction rate is linearly dependent on the hydrogen concentration, so $\alpha = 1$, and independent of the degree of conversion of the fat, hence $\beta = 0$, i.e. the reaction is of the pseudo first-order type. We can write:

$$-R_A = k_r' c_L \quad \text{(kmol/m}^3 \text{ s)}$$

Hydrogen has to be transported from the gas phase (concentration c_g, mass transfer coefficient k_g, specific interface a_g) to the liquid phase (concentration c_L) and from there to the surface of the catalyst particles (concentration c_i, mass transfer coefficient k_L, specific interface a_i) where it reacts chemically. Under steady-state conditions, all of these mass transfer steps proceed at the same rate (no accumulation of H$_2$ in any phase) and we can write ($k_r' = a_i k_r$, where k_r = mascroscopic rate constant in m/s and k_r' = chemical rate constant in s^{-1}) that:

$$\frac{\phi_{mol}}{V} = a_g k_g(mc_g - c_L) = a_i k_L(c_L - c_i) = a_i k_r c_i \tag{1}$$

From this equation we isolate and eliminate the unknown concentrations c_L and c_i and obtain:

$$\frac{\phi_{mol}}{V} = \frac{mc_g}{\dfrac{1}{k_g a_g} + \dfrac{1}{k_L a_i} + \dfrac{1}{k_r a_i}} \tag{2}$$

This equation completely describes the overall hydrogenation process. With the aid of the information given we can evaluate the various terms of this equation as follows:

given: $\dfrac{\phi_{mol}}{Vmc_g} = 0.03\ \text{min}^{-1} = 5 \times 10^{-4}\ \text{s}^{-1} = \dfrac{1}{\dfrac{1}{k_g a_g} + \dfrac{1}{k_L a_i} + \dfrac{1}{k_r a_i}}$ (3)

given: if a_g is doubled, ϕ_{mol} increases by 20%, so:

$$1.2 = \dfrac{\dfrac{1}{k_g a_g} + \dfrac{1}{k_L a_i} + \dfrac{1}{k_r a_i}}{\dfrac{1}{2k_g a_g} + \dfrac{1}{k_L a_i} + \dfrac{1}{k_r a_i}}$$ (4)

given: if a_i is doubled (via catalyst concentration), ϕ_{mol} increases by 50%, so:

$$1.5 = \dfrac{\dfrac{1}{k_g a_g} + \dfrac{1}{k_L a_i} + \dfrac{1}{k_r a_i}}{\dfrac{1}{k_g a_g} + \dfrac{1}{2k_L a_i} + \dfrac{1}{2k_r a_i}}$$ (5)

When this equation is worked out it appears to be identical with equation (4), so the information given is consistent.

From equations (3) and (4), $k_g a_g$ can be calculated and is found to be $k_g a_g = 1.5 \times 10^{-3}\ \text{s}^{-1}$, so that with the given k_g value we find $a_g = 15\ \text{m}^2/\text{m}^3$ (which is very low for a gas–liquid reactor). Using the values given for k_L and a_i we find from equation (4) that the macroscopic rate constant $k_r = 10^{-5}$ m/s, so the chemical rate constant is given by $k_r' = 1.5 \times 10^{-3}\ \text{s}^{-1}$.

Problem 8

Sodium sulphite is oxidized in solution to sodium sulphate by oxygen if cobalt ions (which act as catalysts) are present:

$$O_2 + 2SO_3{}^{2-} \xrightarrow{\ Co^{2+}\ } 2SO_4{}^{2-}$$

At a pH of 7–8 and 15–35 °C the reaction is:

  first order in Co^{2+} if $3 \times 10^{-6} < c_{Co^{2+}} < 10^{-3}\ \text{kmol/m}^3$,
  second order in O_2 (see problem 2),
  first order in SO_3^{2-} if $c_{SO_3^{2-}} < 0.4\ \text{kmol/m}^3$,
  zero order in SO_3^{2-} if $c_{SO_3^{2-}} > 0.4\ \text{kmol/m}^3$.

For the above range of Co^{2+} concentrations and $c_{SO_3^{2-}} > 0.4\ \text{kmol/m}^3$, the pseudo second-order reaction rate constant is:

$$2 \times 10^5 < k_r < 10^8\ \text{m}^3/\text{kmol s}$$

The only explanation of the phenomenon that Reith measured is segregation of the gas bubbles which influences the driving force of the mass transfer process. If we neglect the concentration decrease in the gas bubbles we can write for Ha > 3:

$$\phi_{mol} = Ac_i\sqrt{\tfrac{2}{3}k_rc_i\mathbb{D}} = Am\sqrt{\tfrac{2}{3}k_rm\mathbb{D}c_g^{3/2}}$$

In Reith's first experiment, $c_g = 50$ vol% in all bubbles, i.e. all A. In the second experiment, $c_g = 100$ vol% in half of the bubbles, so in 50% of A if complete segregation occurs. The ratio of the two mass flux rates would be:

$$\frac{\phi_{mol\ separate\ streams}}{\phi_{mol\ mixture}} = \frac{\tfrac{1}{2} \times 1^{3/2}}{(\tfrac{1}{2})^{3/2} \times 1} = \sqrt{2} = 1.41$$

Theoretically, for complete segregation of the bubbles we would expect a 41% higher absorption rate during the second experiment, which indeed was measured. This strong segregation (no coalescence of bubbles) is caused by the high ion concentration of the sulphite system. In distilled water, and in most organic solvents, considerable coalescence and redispersion of bubbles would occur.

Problem 10

The drawing of the absorption column shows the gas and liquid flows and the corresponding concentrations in all streams. To begin with we assume that the reaction between CO_2 and MEA proceeds in the boundary layer, i.e. that Ha > 3. We will check this assumption later on, when we calculate the Ha number.

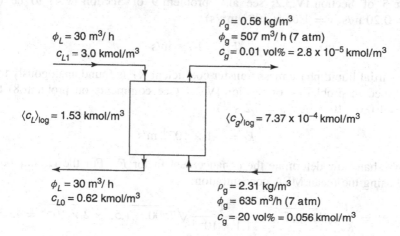

$\phi_L = 30$ m³/ h
$c_{L1} = 3.0$ kmol/m³

$\rho_g = 0.56$ kg/m³
$\phi_g = 507$ m³/ h (7 atm)
$c_g = 0.01$ vol% = 2.8 x 10⁻⁵ kmol/m³

$\langle c_L\rangle_{log} = 1.53$ kmol/m³

$\langle c_g\rangle_{log} = 7.37$ x 10⁻⁴ kmol/m³

$\phi_L = 30$ m³/ h
$c_{L0} = 0.62$ kmol/m³

$\rho_g = 2.31$ kg/m³
$\phi_g = 635$ m³/h (7 atm)
$c_g = 20$ vol% = 0.056 kmol/m³

Since the height of the column will be greater than 10 packing particles, we can assume countercurrent plug flow of both fluids, and the mass flow rate of

CO_2 into the MEA solution is given by equation (IV.53) as:

$$\phi_m = K_L aAL \frac{\Delta c_0 - \Delta c_1}{\ln \dfrac{\Delta c_0}{\Delta c_1}}$$

where K_L is the overall mass transfer coefficient related to the liquid phase and Δc_0, and Δc_1 the concentration differences at the column entrance and exit. Thus:

$$\Delta c_0 = mc_{g0} - c_{L0} = 0.056 - 0 = 0.056 \text{ kmol/m}^3$$

$$\Delta c_1 = mc_{g1} - c_{L1} = 2.8 \times 10^{-5} - 0 = 2.8 \times 10^{-5} \text{ kmol/m}^3$$

since the CO_2 concentration in the bulk liquid is zero.

Consequently, the logarithmic mean driving force is given by:

$$(\Delta c)_{\log} = \frac{\Delta c_0 - \Delta c_1}{\ln \dfrac{\Delta c_0}{\Delta c_1}} = 7.37 \times 10^{-3} \text{ kmol/m}^3$$

We shall now calculate the overall mass transfer coefficient K_L which is given by (see equation (IV.15b)):

$$K_L = \left(\frac{m}{k_g} + \frac{1}{k_L F_c} \right)^{-1}$$

since the mass transport is enhanced by chemical reaction. The partial gas–phase mass transfer coefficient k_g is found with the aid of the boundary layer theory (case 5 of Section IV.3.2; see also problem 9 of Section IV.4) to be (with $v_0 = 0.20$ m/s, $\nu = 1.66 \times 10^{-5}$ m^2/s):

$$\langle k_g \rangle = 10^{-2} \text{ m/s}$$

The partial liquid phase mass transfer coefficient K_L is found analogously to that described in problem 8 of Section IV.2.1 (see comments on problem 8) to be ($v_0 = 1.05 \times 10^{-2}$ m/s, $s = 6.2$ s^{-1})

$$k_L = 1.1 \times 10^{-4} \text{ m/s}$$

We shall now determine the enhancement factor F_c. For the Ha number, we find using the mean MEA concentration:

$$\text{Ha} = \frac{1}{K_L}\sqrt{k_r c_{B\infty} \mathbb{D}_A} = \frac{1}{1.1 \times 10^{-4}}\sqrt{7500 \times 1.53 \times 2 \times 10^{-9}} = 43.5$$

This confirms our initial assumption that Ha > 3 was correct and the reaction proceeds in the boundary layer.

At the column entrance (bottom) (see Section IV.5.4):

$$\frac{k_B c_{B\infty}}{k_A c_{Ai}} = \frac{D_B}{D_A} \frac{c_{B\infty}}{c_{Ai}} = \frac{1}{2}\left(\frac{0.62}{0.056}\right) = 5.53 \approx Ha$$

and we have to use Figure IV.14 in order to find F_c. With:

$$\left(\frac{D_A}{D_B}\right)^{1/2} + \frac{c_{B\infty}}{c_{Ai}}\left(\frac{D_A}{D_B}\right)^{-(1/2)} = 1.4 + \frac{0.62}{0.056 \times 1.4} = 9.3$$

we find from this figure that $F_c \approx 9.0$. At the column exit (top):

$$\frac{k_B c_{B\infty}}{k_A c_{Ai}} = \frac{1}{2}\left(\frac{3}{2.8 \times 10^{-5}}\right) = 53.500$$

which is much larger than 10 Ha, so here we have a pseudo first-order reaction with $F_c = Ha = 43.5$. For further calculations we will use as a rough approximation the logarithmic mean F_c value of $\langle F_c \rangle_{\log} = 20.8$.

With the foregoing results we find for the overall mass transfer coefficient:

$$K_L = \left\{\frac{m}{k_g} + \frac{1}{k_L F_c}\right\}^{-1} = \left\{\frac{1}{10^{-2}} + \frac{1}{1.1 \times 10^{-4} \times 20.8}\right\}^{-1} = 1.86 \times 10^{-3} \text{ m/s}$$

Since the mass flow of CO_2 from the gas to the liquid phase is $\phi_m = \phi_g \Delta c_g = 635 \times 0.056/3600 = 9.9 \times 10^{-3}$ kmol/s, the necessary height of the column is found as:

$$L = \frac{\phi_m}{K_L a A(\Delta c)_{\log}} = \frac{9.9 \times 10^{-3}}{1.86 \times 10^{-3} \times 60\frac{\pi}{4}7.37 \times 10^{-3}} = 15.4 \text{ m}$$

Problem 19

Let us assume that the haystack has the form of a half-sphere of radius r, that all burning occurs at the surface of the stack and that the rate of burning is completely controlled by transfer of oxygen from the air. With these assumptions, the rate of volume decrease of the haystack is dependent on the oxygen transport from the air:

$$\frac{dV}{dt} = 4r^2\frac{dr}{dt} \sim k_g 2\pi r^2(c_g - c_i)$$

or, assuming $c_g - c_i = $ constant:

$$\frac{dr}{dt} = \text{constant}, \quad r = C_1 t + C_2$$

Evaluating the constants with the information given, i.e. $r = 6$ m at $t = 0$, $r = 4$ m at $t = x$, and $r = 2.5$ m at $t = x + 3$, we find $x = 4$ h, so the fire must have started at 1 am.

If the haystack has a form other than that of a hemisphere, it will have a higher surface/volume ratio and will therefore burn down more rapidly.

IV.6 Combined heat and mass transport

The transport of matter from one phase to another is, in principle, attended by a heat effect. When, for example, on absorption a certain component from the gas phase dissolves in the absorbing liquid, heat of solution is released; in addition, an extra heat effect may occur if the dissolved component reacts with the liquid or with the reagents in it.

The temperature of the interface adjusts itself at a value which depends on the possibility of heat removal (or supply in the case of a negative effect) through the two phases. The absorption rate, however, is influenced by the interfacial temperature (e.g. via the solubility), so that in such a case the mass transfer and the heat transport are coupled.

Examples in which the influence of the heat transport on the mass transfer rate could be neglected were given in Section IV.2 (problem 4), and in Section IV.3 (problem 8).

IV.6.1 Drying

A special case of combined heat and mass transfer is obtained if, for example, a drop of liquid (or an inert object moistened with liquid) is put in a gas flow which has not been saturated with the vapour of the liquid. Assume that the liquid originally has the same temperature as the gas. Because the gas is not saturated at that temperature, evaporation occurs from the liquid surface (mass transfer). As a result, heat of evaporation is withdrawn from the liquid, so that its temperature begins to fall. The lower the temperature of the liquid, the greater the temperature difference between gas and liquid, and the more heat will be transported, from the gas to the liquid phase. After some time, a stationary state is reached in which the temperature, and hence also the heat content of the liquid no longer changes with time; all heat required for evaporation is then supplied by the passing gas. In this situation of dynamic equilibrium, the following energy balance applies (see Figure IV.16):

$$\phi_H'' = \phi_{mol\ A}'' \Delta H_{e,A} \qquad (IV.78)$$

where $\Delta H_{e,A}$ = the molar heat of evaporation of A at T_w.

For the local heat flux, ϕ_H'' to the surface can be written:

$$\phi_H'' = h(T_g - T_w)$$

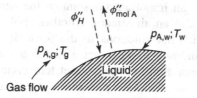

Figure IV.16 Evaporation from a wet surface

For the local rate of evaporation per unit of area, if $p_{A,w}$ is sufficiently small with respect to the total pressure, the following applies:

$$\phi''_{mol\ A} = \frac{k}{RT}(p_{A,w} - p_{A,g}) \tag{IV.79}$$

where for T the mean between T_g and T_w can be taken. By substitution of these equations for ϕ''_H and $\phi''_{mol\ A}$ in the energy balance (equation (IV.78)), we find:

$$\frac{p_{A,w} - p_{A,g}}{T_g - T_w} = \frac{RT}{\Delta H_{e,A}}\frac{h}{k} \tag{IV.80}$$

In this relationship, $p_{A,w}$ and T_w are coupled via the vapour pressure curve of liquid A. If the evaporating drop is so small that the values of $p_{A,g}$ and T_g do not change appreciably, h/k in this relationship can be replaced by the quotient of the mean transfer coefficients over the entire area $\langle h \rangle / \langle k \rangle$. The value of this quotient can now be calculated with the help of the analogy between heat and mass transports. Both $\langle h \rangle$ and $\langle k \rangle$ refer to the same geometric situation and to the same flow condition (the same value of Re). If Re is not too small, the analogous relationship (Section IV.3.1) can be applied, which leads to:

$$\frac{\langle h \rangle}{\langle k \rangle} = c_p \rho (\mathrm{Le})^{2/3} \tag{IV.81}$$

where the product $c_p \rho$ must be taken at the mean temperature T, and Le $= a/\mathbb{D} = \mathrm{Sc/Pr}$ is the dimensionless Lewis number. For gas mixtures the value of Le lies between 0.5 and ca. 2, and for liquids, appreciably higher than 1.

With this relationship, the following is found as the ultimate result:

$$\frac{p_{A,w} - p_{A,g}}{T_g - T_w} = \frac{RT c_p \rho}{\Delta H_{e,A}}(\mathrm{Le})^{2/3} \tag{IV.82}$$

The temperature T_w is called the 'wet-bulb temperature', because this temperature is reached by a moist sphere if it is placed in a gas flow.[†]

[†] In principle, the same effect occurs in mass transfer in a liquid flow, e.g. on dissolving crystals. Ascertain that in this case usually very small temperature differences occur.

This temperature is an important constant in the description and calculation of drying of materials in an air stream. Another application is found in spray-drying towers in which hot solutions in the form of droplets or liquid layers are brought countercurrently into contact with the outside air; the droplets cool down because of evaporation but their outlet temperature cannot be lower than the wet-bulb temperature, belonging to the temperature and the partial water vapour pressure of the ambient air.

The wet-bulb temperature takes its name from psychrometry, according to which T_g and T_w are measured with a 'dry' and a 'wet' thermometer, respectively, in a gas stream. Using equation (IV.82), $p_{A,g}$ can then be calculated. The psychrometer is usually used for the air/water system, but application to other liquids is in principle also possible. The method is particularly useful because T_w is independent of the gas flow rate. However, in such a measurement we should make sure that the heat flow to the 'wet' sphere by conduction or radiation is small with respect to the heat flow from the gas.

The wet-bulb temperature must not be confused with the so-called adiabatic saturation temperature, T_s. This is the temperature a gas (e.g. air with a temperature T_g and water vapour pressure $p_{A,g}$) attains if it is saturated with vapour, the gas itself supplying the required heat of evaporation by cooling from T_g to T_s. The calculation of T_s is based on an energy balance over a unit of mass of the gas, in which the transport of heat and mass does not play a role. Derive that for T_s (and the relevant equilibrium vapour pressure $p_{A,s}$), the following applies:

$$\frac{p_{A,s} - p_{A,g}}{T_g - T_s} = \frac{RT c_p \rho}{\Delta H_{e,A}} \tag{IV.83}$$

Except for the factor $(Le)^{2/3}$, equation (IV.83) corresponds with equation (IV.82). For all gas mixtures Le lies near 1, and for the system air/water vapour $(Le)^{2/3} = 0.95$. Therefore, sometimes (wrongly) no distinction is made between T_w and T_s, although with regard to mechanism, they are of completely different origin.

In many drying problems, humid air plays an important role. A humidity or psychrometric chart can be quite helpful in calculating T_s, T_w and absolute or relative humidity (see Figure IV.17). The adiabatic saturation temperature of air, given its original temperature and humidity, can be found as follows:

Determine the point in the chart which corresponds with the original temperature and humidity. Follow the adiabatic saturation line on which this point lies until the line of 100% relative humidity is reached; the intersection gives T_s. As T_s and T_w are almost equal, T_w can also be found in this way with good approximation. Check that air at 60 °C and 10% relative humidity has a T_s of 30 °C, and therefore that T_w of a wet object in this air, when flowing, has a value of approximately 30 °C.

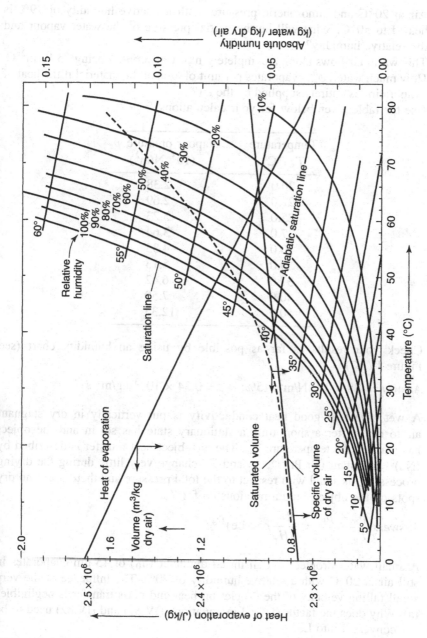

Figure IV.17 A humidity chart

IV.6.2 Problems

*1. Air at 20 °C and atmospheric pressure, with a relative humidity of 79% is heated to 50 °C. What will be the partial pressure of the water vapour and the relative humidity?

This warm air flows along a completely moist material, h being 35 W/m^2°C. How much water, ϕ_m'', evaporates per unit of area of the material if the heat of evaporation is entirely supplied by the air?

Use the table given below for your calculations.

Temperature T_s (°C)	Vapour pressure p_s (10^3 N/m^2)
20.0	2.38
22.0	2.69
25.0	3.23
27.0	3.64
30.0	4.33
34.0	5.13
37.0	6.42
40.0	7.50
50.0	12.33

Check your results as far as possible by using an humidity chart (see Figure IV.17).

Answer: 1.88×10^3 N/m^2; 15%; $\phi_m'' = 0.34 \times 10^{-3}$ kg/m^2 s

2. A wet object of good heat conductivity is put vertically in dry stagnant air to dry. After a short time a stationary state has set in and the object has reached the temperature T_w. The air-object heat transfer is described by $\langle$Nu$\rangle$ = constant (Gr Pr)$^{1/4}$, p_g and T_g change very little during the drying process, p_w is small with respect to the total pressure and there occur no dry spots on the object. Find a relationship for T_w.

Answer: $\dfrac{p_w - p_g}{T_g - T_w} = \dfrac{RT}{\Delta H_e} \rho c_p (\text{Le})^{3/4}$

3. A small water droplet with an initial diameter (d_0) of 15 μm evaporates in still air at 20 °C with a relative humidity of 70%. The influence of the very small falling velocity of the droplet on heat and mass transfer is negligible.
   (a) Why does the factor (Le)$^{2/3}$ in equations (IV.81) and (IV.82) need to be converted into Le?
   (b) What is the driving force (in kg/m^3) for the evaporation of water from the droplet?

(c) Derive an equation for the time of complete evaporation, and calculate this time.

Answer: (b) $\Delta\rho_A = 1.51 \times 10^{-3}$ kg/m³

(c) $\dfrac{\rho(\text{water})d_0^2}{8\mathbb{D}(\rho_{A,w} - \rho_{A,g})}$; 0.73 s

Comments on problems

Problem 1

The partial water vapour pressure in air at 20 °C and 79% saturation is $p_{H_2O} = 0.79 \times 2.38 \times 10^3 = 1.88 \times 10^3$ N/m². If this air is heated to 50 °C, p_{H_2O} stays constant, but the saturation becomes $1.88 \times 10^3 / 12.33 \times 10^3 = 15\%$. The wall temperature is found via:

$$\frac{p_w - p_g}{T_g - T_w} = \frac{p_w - 1.86 \times 10^3}{50 - T_w} = \frac{RT c_p \rho}{\Delta H_e} \mathrm{Le}^{2/3} = 69$$

(by trial and error) to be $T_w = 26$ °C and $p_w = 3.37 \times 10^3$ N/m². The mass flux of the evaporating water is given by equation (IV.79) and with the aid of equation (IV.81) we find:

$$\phi''_{\text{mol}} = \frac{k}{RT}(p_w - p_g) = \frac{h(p_w - p_g)}{c_p \rho (\mathrm{Le})^{2/3} RT} = 18.6 \times 10^{-6} \text{ kmol/m}^2 \text{ s}$$

$$= 3.35 \times 10^{-4} \text{ kg/m}^2 \text{ s}$$

Notation

Symbol	Term	Units
A	area	m^2
a	interfacial area per unit volume	m^2/m^3
a	thermal diffusivity	m^2/s
a	absorption coefficient for radiation	—
a	radius of inner cylinder	m
B	parameter in homogeneous Michaelis–Menten equation	$kmol/m^3 s$
B'	parameter in heterogeneous Michaelis–Menten equation	$kmol/m^2 s$
b	radius of outer cylinder	m
C	flow coefficient	—
C_c	coefficient of contraction	—
C_f	friction factor	—
C_w	drag coefficient	—
c	concentration	$kmol/m^3$
c_A	concentration of component A	$kmol/m^3$
c_A^*	equilibrium concentration of component A	$kmol/m^3$
c_p	heat capacity at constant pressure	$J/kg\ K$
c_v	heat capacity at constant volume	$J/kg\ K$
D	diameter	m
D	dispersion coefficient	m^2/s
D_h	hydraulic diameter	m
D_t	tank (vessel) diameter	m
$\mathbb{D}$	diffusion coefficient	m^2/s
d	diameter	m
d_b	bubble diameter	m
d_p	particle diameter	m
E	eddy diffusivity	m^2/s
E	extraction factor	—
E_{fr}	frictional specific energy	J/kg
E_t	total energy density	J/m^3
$E(t)$	exit age distribution[†]	$1/s$
e	emission coefficient for radiation	—

[†] See Section II.6.3.

Symbol	Term	Units
F	force	N
F	Helmholtz free energy	J/kg
F_c	chemical enhancement factor for mass transfer	—
F_{jk}	view factor for radiation	—
F_{f-w}	force from fluid on wall	N
F_{w-f}	force from wall on fluid	N
F''	momentum flux	N/m^2
F''_{xy}	y-momentum flux in x-direction	N/m^2
$F(t)$	volumetric part of the effluent with a residence time smaller than t[†]	—
Fo	Fourier number	—
f	Fanning friction factor	—
f_D	Stefan correction factor for diffusion	—
G	Gibbs free energy	J/kg
Gr	Grashof number	—
Gz	Graetz number	—
g	(gravitational) acceleration, field force density	m/s^2
H	height	m
H	enthalpy	J/kg
HMU	height (length) of a mixing unit	m
HTU	height (length) of a transfer unit	m
HETU	height equivalent of a transfer unit	m
HETP	height equivalent of a theoretical plate	m
Ha	Hatta number	—
ΔH_e	molar heat of evaporization	J/kmol
ΔH_m	heat of melting	J/kg
ΔH_v	heat of evaporization	J/kg
h	height	m
h	heat transfer coefficient	W/m^{2}k
h_r	heat transfer coefficient for radiation	W/m^{2}k
j_D	Chilton — Colburn factor for mass transfer	—
j_H	Chilton — Colburn factor for heat transfer	—
K	overall mass transfer coefficient	m/s
K	parameter in Michaelis–Menten equation	kmol/m^3
k	mass transfer coefficient	m/s
k_g	mass transfer coefficient based on partial pressure difference	kmol/m^{2}Pa s

---

[†] See Section II.6.3.

Symbol	Term	Units
k_r	reaction rate constant (dimension dependent on reaction order n)	$\dfrac{m^{3n-1}}{kmol^{3n-1}s}$
k_{rp}	pseudo first-order reaction rate constant	1/s
k_r'	reaction rate constant for heterogeneous reactions	m/s
L	length	m
L	length dimension	—
Le	Lewis number	—
M	molar mass	kg/kmol
M	mass dimension	—
M_0	flow coefficient for laminar flow	—
m	distribution coefficient	—
Nu	Nusselt number	—
n	coordinate perpendicular to plane	m
n	Stirrer speed	1/s
n^*	minimal stirrer speed for suspension	1/s
P	power	W
Po	power number	—
Pr	Prandtl number	—
p	pressure	Pa
p_A	partial pressure of component A	Pa
Δp	pressure drop	Pa
Q	heat per unit mass	J/kg
q	heat production per unit volume	W/m^3
R	gas constant	J/kmol k
R	radius	m
R_A	volumetric production rate of component A	$kmol/m^3 s$
R_A'	production rate of component A per unit area	$kmol/m^2 s$
R_c	flow resistance of filter cake	$1/m^2$
R_h	hydraulic radius	m
R_m	flow resistance of filter medium	1/m
r	radial coordinate	m
r_X	volumetric production rate of X	—[†]
S	entropy	J/kg k
S	circumference	m
S_v	surface area per unit volume of cavities	m^2/m^3
Sc	Schmidt number	—
Sh	Sherwood number	—
s	frequency of surface renewal	1/s

[†] Dimension: quantity per m^3 and per s.

Symbol	Term	Units
T	temperature	K
T	time dimension	—
T_r	reference temperature	K
T_s	adiabatic saturation temperature	K
T_w	wet-bulb temperature	K
t	time	s
t_m	mixing time	s
U	internal energy	J/kg
U	overall heat transfer coefficient	W/m^{2}k
V	volume	m^3
V_{cr}	critical volume of droplet	m^3
v	velocity	m/s
v_r	relative velocity	m/s
v_s	terminal velocity	m/s
$(v_s)_s$	sedimentation rate of a swarm of particles	m/s
v_0	superficial velocity	m/s
W	work per unit mass	J/kg
W	width	m
We	Weber number	—
w	weight fraction	—
X	volumetric concentration	—[†]
X	quantity	—
x	coordinate	m
$\bar{x}$	average wall roughness	m
y	coordinate	m
z	coordinate	m

Greek Symbols

α	flow coefficient for orifices and flow meters	—
α	volume fraction of dispersed phase	—
β	eigenvalue	—
β	coefficient of expansion	—
δ	boundary layer thickness; film thickness	m
δ_c	concentration boundary layer thickness	m
δ_h	hydrodynamic boundary layer thickness	m
δ_r	concentration boundary layer thickness with chemical reaction	m
δ_T	thermal boundary layer thickness	m

[†] Dimension: quantity per m^3.

Symbol	Term	Units
ε	energy dissipation per unit mass	W/kg
ε	porosity; volume fraction of continuous phase	—
η	dynamic viscosity	Pa s
Θ	temperature dimension	
Θ	cylindrical coordinate; angle	radians
Θ	dimensionless time	—
Θ	dimensionless temperature	—
κ	C_p/C_v	—
λ	thermal conductivity	W/m K
λ	wavelength	m
λ_k	Kolmogoroff-scale	m
ν	kinematic viscosity	m^2/s
ξ	dimensionless x-coordinate	—
ρ	density	kg/m^3
ρ_A	mass concentration of component A	kg/m^3
σ	surface tension	N/m
σ	standard deviation	—
σ	Stefan–Boltzmann constant	W/m^2 K^4
τ	mean residence time	s
τ	shear stress	N/m^2
τ_w	shear stress at wall	N/m^2
τ_{xy}	shear stress in y-direction caused by gradient of v_y in x-direction	N/m^2
τ_0	yield stress	N/m^2
ϕ	flow rate	—[†]
ϕ	total heat flow rate by radiation	W
ϕ'	flow rate per unit length	—[†]
ϕ''	flux	—[§]
ϕ_A	flow rate of mechanical energy	W
ϕ_H	heat flow rate	W
ϕ_H''	heat flux	W/m^2
ϕ_i	heat flow rate by foreign irradiation	W
ϕ_n	heat flow rate by net radiation	W
ϕ_m	mass flow rate	kg/s
ϕ_m''	mass flux	kg/m^{2}s

[†] Dimension: quantity per S.
[†] Dimension: quantity per m and per S.
[§] Dimension: quantity per m^2 and per S.

Symbol	Term	Units
ϕ_{mol}	molar flow rate	kmol/s
ϕ''_{mol}	molar flux	kmol/m^{2}s
ϕ_v	volume flow rate	m^3/s
ϕ_X	flow rate of quantity X	—[†]
ϕ_z	heat flow rate by radiation from a black body	W
$\psi(t)$	age distribution function	—

Subscripts

$A, B, \ldots$	component $A, B, \ldots,$ etc.
e	final; external
f	fluid
eff	effective
g	gas
i	internal; at interface
in	ingoing
l	liquid
log	logarithmic mean
max	maximum
out	outgoing
p	particle
p	Pressure
r	reference
v	volume
w	wall
X	quantity X
x	x-direction
y	y-direction
z	z-direction
0	initial value
∞	bulk value
$\perp$	perpendicular to

Miscellaneous Symbols

$\bar{x}$	average value of, e.g. x
$\langle\ \rangle$	flow averaged value over a cross-section
Δ	difference

[†] Dimension: quantity per s.

Index